T0212196

# NEUROTRANSMITTER RECEPTORS

## Mechanisms of Action and Regulation

# ADVANCES IN EXPERIMENTAL MEDICINE AND BIOLOGY



---

Recent Volumes in this Series

Volume 169
OXYGEN TRANSPORT TO TISSUE – V
Edited by D. W. Lübbers, H. Acker, E. Leniger-Follert, and T. K. Goldstick

Volume 170
CONTRACTILE MECHANISMS IN MUSCLE
Edited by Gerald H. Pollack and Haruo Sugi

Volume 171
GLUCOCORTICOID EFFECTS AND THEIR BIOLOGICAL CONSEQUENCES
Edited by Louis V. Avioli, Carlo Gennari, and Bruno Imbimbo

Volume 172
EUKARYOTIC CELL CULTURES: Basics and Applications
Edited by Ronald T. Acton and J. Daniel Lynn

Volume 173
MOLECULAR BIOLOGY AND PATHOGENESIS OF CORONAVIRUSES
Edited by P. J. M. Rottier, B. A. M. van der Zeijst, W. J. M. Spaan, and M. C. Horzinek

Volume 174
GANGLIOSIDE STRUCTURE, FUNCTION AND BIOMEDICAL POTENTIAL
Edited by Robert W. Ledeen, Robert K. Yu, Maurice M. Rapport, and Kunihiko Suzuki

Volume 175
NEUROTRANSMITTER RECEPTORS: Mechanisms of Action and Regulation
Edited by Shozo Kito, Tomio Segawa, Kinya Kuriyama, Henry I. Yamamura, and Richard W. Olsen

Volume 176
HUMAN TROPHOBLAST NEOPLASMS
Edited by Roland A. Pattillo and Robert O. Hussa

---

A Continuation Order Plan is available for this series. A continuation order will bring delivery of each new volume immediately upon publication. Volumes are billed only upon actual shipment. For further information please contact the publisher.

# NEUROTRANSMITTER RECEPTORS

## Mechanisms of Action and Regulation

Edited by

Shozo Kito
Tomio Segawa

Hiroshima University School of Medicine
Hiroshima, Japan

Kinya Kuriyama

Kyoto Prefectural University of Medicine
Kyoto, Japan

Henry I. Yamamura

University of Arizona Health Sciences Center
Tucson, Arizona

and

Richard W. Olsen

University of California
Los Angeles, California

Springer Science+Business Media, LLC

Library of Congress Cataloging in Publication Data

Main entry under title:

Neurotransmitter receptors.

(Advances in experimental medicine and biology; v. 175)
"Proceedings of a meeting commemorating Dr. Kito's 10th anniversary as a professor of the Third Department of Internal Medicine, held October 6-8, 1983, in Hiroshima, Japan"—T.p. verso.
Includes bibliographies and index.
1. Neurotransmitter receptors—Congresses. I. Kitō, Shōzō, date- . II. Series.
[DNLM: 1. Synaptic Receptors—congresses. W1 AD559 v.175/WL 102.8 N4944 1983]
QP364.7.N474 1984 599′.0188 84-11507
ISBN 978-1-4684-4807-8 ISBN 978-1-4684-4805-4 (eBook)
DOI 10.1007/978-1-4684-4805-4

Proceedings of a meeting commemorating Dr. Kito's 10th Anniversary as a Professor of the Third Department of Internal Medicine at Hiroshima University School of Medicine, held October 6-8, 1983, in Hiroshima, Japan


Originally published by Plenum Press in 1984
Softcover reprint of the hardcover 1st edition 1984

## PREFACE

This meeting was held commemorating Dr. Kito's 10th Anniversary as Professor of the Third Department of Internal Medicine, Hiroshima University School of Medicine. Dr. Kito was born in 1927 in Nagoya, graduated from Tokyo University School of Medicine and received his M.D. in 1951. He spent his first academic years as a research associate (1952 - 1968) at the Third Department of Internal Medicine, Tokyo University School of Medicine. During this period he studied for one year (1952 - 1953) at Illinois University School of Medicine, and acquired his Ph.D. in 1959. In 1968 he became Instructor and in 1971 he was appointed as Assistant Professor of Tokyo Women's Medical College. In 1973, he became Professor of the Third Department of Internal Medicine, Hiroshima University School of Medicine. Dr. Kito is a clinician but he is always enthusiastic about basic medicine. His major research field concerns neurotransmitters and their receptors in the central nervous system. He prefers a combination of neurotransmitter immunohistochemistry and receptor autoradiography as research techniques. He is also engaged in biochemical studies on amyloid proteins. When the Eighth International Congress of Pharmacology was held in Tokyo in 1981, Dr. Segawa, Dr. Yamamura, and Dr. Kuriyama organized a Satellite Symposium on Neurotransmitter Receptors in Hiroshima. Dr. Kito attended this meeting and was deeply impressed by the active presentations and discussions.

In order to make some contribution to the progress of neurosciences, Dr. Kito chose to have a scientific meeting commemorating his 10th Anniversary as Professor, instead of just having a routine party. For this purpose he asked Dr. Segawa to join in organizing this symposium. The symposium was held in Hiroshima on October 6-8, 1983, and entitled "Neurotransmitter Receptor Regulation, Interactions, and Coupling to the Effectors." Twenty-three invited speakers from five countries and a poster session covered the latest advances in research on virtually all of the neurotransmitter receptors. Emphasis was on effector coupling, especially the receptors which negatively regulate adenylate cyclase, but other areas included interactions of receptors with each other and ion channels, and new techniques such as autoradiography and molecular biology of receptors. In all likelihood, these subjects will remain at the cutting edge of neuroscience for the next several years.

Shozo Kito
Tomio Segawa
Kinya Kuriyama
Henry I. Yamamura
Richard W. Olsen

# ACKNOWLEDGMENTS

Acknowledgements are due to many friends and colleagues for their assistance.

It was Dr. Eiko Itoga, former assistant professor of the Third Department of Internal Medicine, Hiroshima University, who first had the idea of having this symposium during her stay at Dr. Yamamura's laboratory in Tucson, and Dr. Yamamura agreed with her. She had devoted herself to establishment of the Department and morphological studies on neurotransmitters and their receptors until she left the laboratory recently. We tender our warmest acknowledgements to her.

Thanks are also expressed to Miss Masae Inokawa, instructor of Hiroshima University for her excellent secretarial work. Nancy Morrison, at UCLA, had the monumental task of assembling this volume and did a superb job in very rapid fashion.

Finally, this symposium was financially supported by the Japanese Educational Ministry and the Commemorative Association for the Japan World Exposition (1970).

CONTENTS

## COUPLING OF NEUROTRANSMITTER RECEPTORS TO ADENYLATE CYCLASE

## RECEPTOR STRUCTURE, LOCALIZATION, AND ION CHANNELS

# SELECTIVE BLOCKAGE BY ISLET-ACTIVATING PROTEIN, PERTUSSIS TOXIN, OF NEGATIVE SIGNAL TRANSDUCTION FROM RECEPTORS TO ADENYLATE CYCLASE

Michio Ui, Toshiaki Katada, Toshihiko Murayama and Hitoshi Kurose

Department of Physiological Chemistry
Faculty of Pharmaceutical Sciences
Hokkaido University, Sapporo 060, Japan

## OVERVIEW

The adenylate cyclase-cyclic AMP system is one of the signal transduction processes that has been the subject of extensive studies in recent years. Stimulation of some membrane receptors ("activatory" receptors) by particular agonists produces, via activation of adenylate cyclase, cyclic AMP, which acts as an intracellular second messenger of the receptors by activating one of the several kinds of cytosolic protein kinases. The receptor-linked adenylate cyclase system is now known to consist of three components; the guanine nucleotide-binding protein usually referred to as N, G, or G/F, has been identified as the functional communicator between receptors (R) and the catalytic unit of the cyclase (C). A growing body of evidence has accumulated in support of the concept that N plays the pivotal role in receptor-mediated activation of adenylate cyclase, as reviewed by Ross and Gilman (1), Limbird (2) and Spiegel and Downs (3). When no extracellular signal is available to the cell, the species of the guanine nucleotide bound to N is GDP, which allows functional dissociation of the three components, R, N, and C. If information is transmitted by an agonist (A) which occupies R, the ternary complex, A-R-N is formed as the "turn-on" reaction which leads to replacement of GDP by intracellular GTP at the specific binding sites on N. The adenylate cyclase enzyme is then activated as a result of association of the GTP-bound N with C. Synthesis of cyclic AMP continues until GTP is hydrolyzed by the GTPase "turn-off" reaction, leaving GDP on N.

Numerous kinds of approaches have been employed successfully to afford experimental evidence for the above-outlined sequence of receptor-linked activation of adenylate cyclase. One of these approaches is the use of cholera toxin (see reference 4 for review). Exposure of a variety of mammalian cells to cholera toxin resulted in enormous increases in adenylate cyclase activity of membranes prepared therefrom, even in the absence of receptor agonists or guanine nucleotides. It has been established that the active fragment of the toxin oligomeric protein is an enzyme catalyzing ADP-ribosylation of a subunit of the N protein. N loses its GTPase activity upon ADP-ribosylation. Consequently, GTP continues to occupy the guanine nucleotide-binding sites on N even after agonists have detached from R. Thus, the adenylate cyclase enzyme retains its high activity irreversibly in membranes from cholera toxin-treated cells.

Apart from the above mentioned activation of adenylate cyclase, stimulation of several other receptors leads to inhibition, rather than activation, of adenylate cyclase (5-7). These "inhibitory" receptors include muscarinic cholinergic, alpha(2)-adrenergic, adenosine (A1) and opiate receptors. GTP is also an essential factor for the inhibition to occur in membrane preparations, suggesting that guanine nucleotide binding proteins are involved in receptor-mediated inhibition as well as activation of the cyclase. It is likely, for reasons described later, that the guanine nucleotide regulatory protein mediating the inhibition is an entity distinct from the N protein involved in the activation. The former is now referred to as $N_i$ and the latter as $N_s$, where subscripts "s" and "i" stand for "stimulatory" and "inhibitory", respectively. As to the mechanism whereby the signal given to receptors is transferred to the cyclase enzyme via N, much more remains to be solved for $N_i$ than for $N_s$. An agent that could modify the function of $N_i$ specifically, if any, would contribute to solving the problem. Islet-activating protein (IAP) (8,9), the exotoxin produced by *Bordetella pertussis*, is the most promising probe in this regard, since our recent studies have shown that receptor-mediated inhibition of adenylate cyclase is reversed by exposure of a variety of cells to IAP (10-13). The present review will summarize the experimental data that IAP directly interacts with $N_i$ in such a fashion as to block the negative signal transduction. It contrasts sharply with cholera toxin that is a specific modifier of $N_s$.

## A BRIEF HISTORY LEADING TO DISCOVERY OF ISLET-ACTIVATING PROTEIN

The heat-labile exotoxin present in the culture medium of multiplying *Bordetella pertussis*, the pathogenic bacteria for whooping cough, has long been known for its potent biological activities (14). Among these are increases in the number of circulating lymphocytes and increases in the histamine-induced death (induction of histamine hypersensitivity) that are observed upon the injection of pertussis vaccine into mice; a putative factor involved often has been referred to as a lymphocytosis-promoting factor (LPF) or a histamine-sensitizing factor (HSF). In addition, the injection of pertussis vaccine into rats was very effective in suppressing epinephrine-induced hyperglycemia; the catecholamine was no longer hyperglycemic in rats that had been injected with the vaccine 2 to 7 days before. The earlier attempts to unify these diverse actions of pertussis vaccine resulted in an hypothesis that a toxin present in the vaccine blocks *beta*-adrenergic receptors; it was based on both the notion that stimulation of *beta*-receptors by injected epinephrine is responsible for hyperglycemia and the experimental data that the histamine-sensitizing activity of the vaccine was somehow mimicked by *beta*-adrenergic antagonists.

The first stimulus that urged us to start our own study of pertussis toxin in 1977 was our earlier interest in epinephrine hyperglycemia. Up to that time, we had noticed that both *alpha*- and *beta*-adrenergic receptors are responsible for hyperglycemia following epinephrine injection (15,16) and had found that shift of the balance between the two classes of adrenergic receptors in favor of either one was very effective in attenuating hyperglycemia. The balance was effectively shifted by changes in pH of the body fluids (17,18), by injection of adrenocortical hormones (19), by induction of the hypo- or hyperthyroid state (20), or by forced exercise (21); epinephrine-induced hyperglycemia was markedly blunted under these particular conditions. The injection of pertussis vaccine into rats was an additional means effective in suppressing the epinephrine action to cause hyperglycemia (22,23). This action of pertussis vaccine was characterized by its long duration; the rat was unresponsive to epinephrine over a week after a single vaccination, indicating irreversible modification of certain bodily functions. In any case, a shift in balance between *alpha*- and *beta*-adrenergic receptor functions was postulated to occur in the animals treated with pertussis vaccine as our working hypothesis. We used insulin secretion as a much better index of adrenergic functions than glycemic changes.

Extraordinary modification of adrenergic functions by pertussis treatment was revealed by perfusion experiments in which the perfusate of rat pancreas was supplemented with epinephrine with or without further addition of an *alpha*- or *beta*-adrenergic antagonist to stimulate either class of receptors selectively (24). It was found that *alpha*-receptors are predominant over *beta*-receptors in the normal (not treated with vaccine) rat pancreas in such a fashion as to render the infused epinephrine inhibitory to insulin secretion, whereas *beta*-adrenergic stimulation of insulin secretion is the only event observed in pancreas isolated from pertussis-treated rats. *Alpha*-adrenergic receptors did not function any longer in pancreas after pertussis treatment of the organ-donor rats. In response to other stimuli such as infusion of glucose, amino acids, or glucagon, much more insulin was also released from the pancreas of pertussis-treated rats than from pancreas of non-treated rats (24). A search for the factor responsible for this unique action in pertussis vaccine resulted in successful purification of a protein with a Mr value of 117,000 from the 2-day culture supernatant of B. pertussis (Phase I, Tohama strain) (8). This protein, referred to as islet-activating protein (IAP) after its extraordinary action to enhance insulin secretory responses of pancreatic islets, possesses LPF and HSF activities and suppresses epinephrine-induced hyperglycemia due to accompanying hyperinsulinemia, when injected into rodents (9,25). Thus, IAP should be the same entity as the exotoxin termed pertussis toxin (26) or pertussigen (27).

## IAP AS A SPECIFIC MODIFIER OF THE RECEPTOR-ADENYLATE CYCLASE LINKAGE

Receptor mediated changes in insulin release from isolated rat islets are usually associated with changes in the cellular cyclic AMP content in the same direction, reflecting the fact that cyclic AMP is the second messenger of these receptors in islet cells (28-30). Modification by IAP of receptor-linked insulin release was also accompanied by the same-directional modification of cyclic AMP changes in islets even when the degradation of cyclic AMP was prevented by methylxanthines or Ro-20-1724. This finding, together with the failure of IAP to affect dibutyryl cyclic AMP-induced insulin release (29), indicates that the process(es) responsible for receptor-linked regulation of adenylate cyclase activity should be the sole site of IAP action. Actually, membrane adenylate cyclase was markedly inhibited by the addition of epinephrine via alpha(2)-adrenergic receptors (31) when the membranes had been prepared from control islet cells, while this inhibition was of much less magnitude when membrane donor cells had been exposed to IAP (12). Moreover, glucagon- or adenosine (A2)-induced activation of adenylate cyclase in islet cell membranes was enhanced, though only slightly, by prior treatment of the cells with IAP (11,32).

The IAP-sensitive cell is not restricted to islets. Exposure of rat heart (13), fat (33) or C6 glioma cells (34), mouse 3T3 fibroblasts (35) or rat glioma x mouse neuroblastoma hybrid NG108-15 cells (36) to IAP resulted either in enhancement of increases in the cellular cyclic AMP content (or activation of membrane adenylate cyclase) via beta-adrenergic, glucagon or prostaglandin E receptors or in abolition (or marked attenuation) of decreases in cellular cyclic AMP (or inhibition of membrane adenylate cyclase) via alpha(2)-adrenergic, muscarinic cholinergic, adenosine (A1), or opiate receptors. It is noteworthy that, in all the IAP-sensitive cell types studied so far, neither the baseline content of cyclic AMP in the cell nor the basal activity of membrane adenylate cyclase that is assessed without stimulation of coupled receptors was altered by IAP treatment; it was their responses to receptor stimulation that was modified by IAP. Hence, IAP is unlikely to interact directly with the adenylate cyclase catalytic protein (C). Furthermore, IAP-induced modification (enhancement or attenuation) of the cellular cyclic AMP (or membrane adenylate cyclase) response to receptor agonists is invariably observed in such a manner that the agonist concentration-response curve is shifted "upwards" rather than "to the left"; i.e., the degree of response to agonists is modified, without changes in the affinity of agonists for receptors, by IAP treatment (11-13,32-36). It is thus unlikely that the receptor protein (R) is a direct target of IAP action. As direct evidence for this notion, the affinity for antagonist binding to membrane receptors was not affected by the treatment of membrane-donor cells with IAP (36,37). In certain cases, GTP-induced activation of membrane adenylate cyclase even in the absence of receptor agonists was enhanced by IAP (34,35), suggesting that IAP could interact with the guanine nucleotide regulatory protein (N).

## ADP-RIBOSYLATION OF A MEMBRANE PROTEIN BY IAP AS A MECHANISM FOR MODIFICATION OF THE RECEPTOR-ADENYLATE CYCLASE SYSTEM

All the data mentioned above were obtained with intact cells exposed to IAP for times as long as 2-24 h. Direct addition of IAP by itself to the cell-free membrane preparations failed to exert any influence on the membrane adenylate cyclase system even after a long incubation, suggesting that a cytosolic factor would be essential for membranes to be affected by IAP. Such a factor was later identified as NAD. GTP-dependent (and isoproterenol-stimulated) adenylate cyclase activity of membranes from C6 cells was markedly enhanced after the membranes were incubated for a short while with IAP only if the incubation medium was further fortified with NAD and ATP (38). NAD served as the substrate of an IAP-catalyzed novel reaction; labeled NAD was incorporated into a membrane protein with an Mr value of 41,000 upon incubation of membranes with IAP.

This IAP-catalyzed labeling of the membrane Mr = 41,000 protein exhibited the following characteristics (39). 1. The radioactivity was detected in this protein when the ADP-ribose moiety of NAD was labeled with $^{32}P$, $^{14}C$ or $^{3}H$, whereas no radioactivity was incorporated from [carbonyl-$^{14}C$]NAD. Moreover, [α-$^{32}P$]NAD once incorporated into the Mr = 41,000 protein in the presence of IAP was liberated upon subsequent incubation of membranes with snake venom phosphodiesterase as 5'-AMP. Thus, IAP displays ADP-ribosyltransferase activity with the Mr = 41,000 protein as substrate. 2. Tryptic digestion of the IAP-labeled Mr = 41,000 protein was markedly interfered with by the prior incubation of membranes with Gpp(NH)p or NaF, specific ligands of the N protein, probably due to a change in the peptide conformation. Thus, the IAP specific substrate, the Mr = 41,000 protein, must be one of the subunits of the N protein. 3. Incubation of membranes with labeled NAD in the presence of the

$A_1$ (active) component of cholera toxin resulted in ADP-ribosylation of membrane proteins with Mr values of 45,000 and 48,000/49,000 (doublet), proteins distinct from the substrate protein of the IAP-catalyzed reaction. 4. The degree of IAP-catalyzed ADP-ribosylation of the Mr = 41,000 protein was strictly correlated with IAP-induced enhancement of receptor-dependent adenylate cyclase activity under various conditions. Such was also the case when intact cells were exposed to IAP.

The higher the concentration of IAP in the medium for incubation of cells, the smaller the amount of ADP-ribose incorporated into the Mr = 41,000 protein during the subsequent incubation with IAP of membranes isolated therefrom. No ADP-ribose was incorporated any longer into the membrane Mr = 41,000 protein, if membrane-donor cells had been exposed to a saturating concentration of IAP. Enhancement of receptor-mediated adenylate cyclase activity in membranes as induced by the prior treatment of the membrane-donor cells with IAP was inversely correlated in magnitude with [$^{32}$P]ADP-ribosylation of the Mr = 41,000 protein caused by IAP during the incubation of another batch of the same membranes. This is because the ADP-ribosyl moiety of the intracellular NAD is transferred to the membrane Mr = 41,000 protein during exposure of intact cells to IAP; as a result, no further incorporation of [$^{32}$P]ADP-ribose occurs to the same protein molecule during the subsequent incubation of the isolated membranes. Thus, ADP-ribosylation of the Mr = 41,000 protein is invariably associated with the occurrence of the unique action of IAP on the membrane receptor-adenylate cyclase system either in intact cells or in isolated membranes.

These characteristics afford convincing evidence for the idea that IAP exerts its unique influence on the membrane signal transmission processes as a result of ADP-ribosylation of a subunit of the guanine nucleotide regulatory protein (N) which differs from the peptide serving as the substrate of the cholera toxin-catalyzed similar reaction.

## BLOCKADE BY IAP OF $N_i$-MEDIATED NEGATIVE SIGNAL TRANSDUCTION

The function of the N protein is bidirectional; in one direction, it associates with the receptor protein (R) when R is occupied by an agonist and N is bound with GDP, while, in another direction, it activates (in the case of $N_s$ as a positive signal) or inhibits (in the case of $N_i$ as a negative signal) the adenylate cyclase enzyme (C) when bound GDP is replaced by GTP subsequent to agonist binding to R. We have studied how these functions are affected by IAP, and concluded that IAP blocks the negative signal transduction via $N_i$. The experimental basis for this conclusion is as follows.

1. Adenylate cyclase of adipocyte membranes exhibits biphasic

responses to GTP in the presence of isoproterenol. The plot of the enzymic activities against GTP concentrations gave a bell-shaped pattern with a progressive rise below, and a progressive fall beyond, 10 μM. The rise is due to activation via $N_s$ while the fall is due to inhibition via $N_i$ (6). IAP treatment of membranes resulted in complete abolition of the $N_i$-mediated fall without a change in the activatory phase via $N_s$ (33). 2. The affinity of agonist binding to alpha(2)-adrenergic or muscarinic receptors of membranes from NG108-15 cells was lowered by Gpp(NH)p, reflecting dissociation of the high-affinity ternary complex. If the same experiment was repeated with IAP-treated membranes, the affinity of agonists for these "inhibitory" receptors was lower than the control membranes in the absence of Gpp(NH)p, the further addition of which was without effect, suggesting that $N_i$ is uncoupled from these receptors in IAP-treated membranes (36). 3. The adenylate cyclase activity of membranes of mouse 3T3 fibroblasts was maintained at a very high level in the presence of Gpp(NH)p or NaF or after treatment of membranes with cholera toxin. Such a high enzymic activity is undoubtedly due to activation of the cyclase C protein via $N_s$. The addition of GTP to the membranes under these conditions led to a decrease in the cyclase activity as a result of its selective binding to $N_i$. This GTP-induced inhibition of membrane adenylate cyclase was abolished by prior treatment of membrane-donor cells with IAP, indicating that the communication between $N_i$ and C was also blocked by IAP (35). 4. Stimulation by enkephalin of opiate receptors in NG108-15 cell membranes gave rise to increased GTP hydrolysis via $N_i$. This agonist-induced GTPase was abolished by treatment of membranes with the A protomer (see below for the toxin's subunit assembly) of IAP (40), in accordance with the report by Burns et al. (41). This is another line of evidence for IAP-induced blockade of the receptor-$N_i$ linkage. 5. In rat adipocyte membranes, there was a lag period prior to onset of the steady-state Gpp(NH)p activation of adenylate cyclase. This lag time was shortened by isoproterenol or cholera toxin as a result of facilitation of Gpp(NH)p binding to $N_s$ in exchange for the previously bound GDP (33). THe lag time was not shortened by IAP, however. Nor was the action of a beta-stimulant to shorten the lag time affected by IAP. Thus, the possibility was excluded that IAP interacts with $N_s$ (33).

These findings afford convincing evidence for the blockade by IAP of the communication between receptors and $N_i$.

## DIFFERENTIAL EFFECTS OF IAP AND CHOLERA TOXIN ON THE N-PROTEINS

The best way to assess the coupling of receptors with N is to estimate the receptor agonist-induced GTP-GDP exchange reaction. The technique to estimate the exchange reaction consists of three successive procedures. First, membranes are incubated with a receptor agonist and [$^3$H]GTP to replace the endogenously bound GDP by the radioactive GTP at the binding sites on N. Second, the labeled membranes are washed with non-radioactive GTP and the receptor antagonist to minimize the nonspecific binding of radioactive GTP. During this washing procedure, the bound [$^3$H]GTP is converted to [$^3$H]GDP due to the membrane GTPase activity. Third, the specifically labeled membranes are again incubated with an agonist of the N-coupled receptors and GTP to estimate the release of [$^3$H]GDP in exchange for added GTP. With the use of this technique, we have studied the differential effects of IAP and cholera toxin on the receptor coupling with N in membranes of rat or hamster adipocytes (42).

Rat adipocyte membranes were first labeled with [$^3$H]GTP via stimulation of beta-adrenergic receptors. $N_s$ should have been labeled with radioactive GTP under these conditions. Radioactive GDP release from these membranes during the second incubation was: (i) stimulated by agonists of activatory receptors such as glucagon and secretin as well as isoproterenol; (ii) not stimulated by any agonist of inhibitory receptors such as prostaglandin E or adenosine; (iii) stimulated by the active subunit of cholera toxin plus NAD even in the absence of stimulatory agonists; (iv) not stimulated by the A protomer of IAP plus NAD in either the presence or the absence of stimulatory agonists. Thus, a common single pool of $N_s$ is linked to multiple activatory receptors, but is not linked to inhibitory receptors. This $N_s$ pool is insusceptible to IAP, but its ADP-ribosylation by cholera toxin results in spontaneous release of GDP in exchange for added GTP without stimulation of coupled receptors.

There are both alpha(2)- and beta-adrenergic receptors in hamster adipocyte membranes. When the membranes were labeled with radioactive GTP via beta-receptors, the subsequent release of radioactive GDP was stimulated by beta-agonists much more strongly than alpha-agonists. No GDP release was stimulated at all by agonists of other inhibitory receptors: e.g., by prostaglandin E or nicotinic acid. In contrast, GDP release from the membranes that had been labeled with the radioactive nucleotide by stimulation of alpha(2)-receptors was promoted by not only an alpha-agonist itself but also by prostaglandin E or nicotinic acid. Much lower amounts of GDP were released by a beta-agonist. Thus, the $N_s$ and $N_i$ proteins are distinctly different entities from each other. Cholera toxin was without effect on GDP release from $N_i$ in either the presence or the absence of an agonist of coupled receptors. In the case of IAP-catalyzed ADP-ribosylation, however, neither spontaneous nor agonist-stimulated release of GDP was observed from the chemically modified $N_i$. Thus, IAP blocks the communication between inhibitory receptors and $N_i$.

Now it is established that cholera toxin is a modifier of $N_s$ while IAP is a modifier of $N_i$. Both catalyze ADP-ribosylation of one of the subunits of the respective N proteins. When $N_s$ is ADP-ribosylated by cholera toxin, it is readily occupied by GTP in exchange for GDP even in the absence of coupled receptor stimulation. Thus, cholera toxin-catalyzed ADP-ribosylation stabilizes $N_s$ in its active form. The ADP-ribosylation of $N_s$ appears to interfere somehow with its coupling with receptors (37). On the contrary, $N_i$ is retained in its inactive state after IAP-catalyzed ADP-ribosylation, in the sense that no GTP binds to $N_i$ even upon stimulation of inhibitory receptors by agonists. The important question as to how the same ADP-ribosylation reaction exerts differential influences on $N_s$ and $N_i$ is the subject for further investigations.

Stimulation of beta-adrenergic (13,34) or glucagon (13) receptors of intact cells resulted in much more accumulation of cyclic AMP in IAP-treated cells than in control cells. Stimulation of these receptors in membrane preparations, however, caused the same degree of activation of adenylate cyclase between IAP-treated and control membranes (34,37). This apparent disparity would suggest that, in intact cells, both $N_s$ and $N_i$ are somehow coupled with the activatory receptors, and that receptor-mediated cyclic AMP generation reflects a balance between $N_s$-linked activation and $N_i$-linked inhibition of adeylate cyclase. Thus, the abolition of the $N_i$-linked inhibition of the cyclase by IAP must lead to enhanced accumulation of cyclic AMP following receptor stimulation in intact cells (33). Probably, such functional association of $N_i$ and $N_s$ in membranes of intact cells would be disturbed during isolation of membranes from the cells (37). IAP-specific exaggeration of insulin secretory responses to a variety of secretagogues is accounted for by the accumulation of cyclic AMP in islet cells by this mechanism.

## PHOSPHOLIPASE ACTIVATION AS A PROCESS OF MEMBRANE SIGNAL TRANSDUCTION AND ITS MODIFICATION BY IAP

Histamine is released from mast cells in response to stimulation of their IgE receptors by antigens or to an addition of compound 48/80, a specific histamine releaser, concanavalin A or somatostatin. IAP was a modifier of this histamine secretory response of mast cells as well; much less histamine was released from IAP-treated rat peritoneal mast cells than from control non-treated cells under these conditions (43). This IAP-induced inhibition of receptor-mediated histamine release was strictly paralleled by inhibition of the [$^{14}$C]arachidonic acid release which reflects receptor-linked phospholipase A2 activation (44). Unexpectedly, changes in the cellular content of cyclic AMP, if any occurred, were not correlated with the degrees of IAP inhibition of histamine release under a variety of conditions (43). Thus, the adenylate cyclase-cyclic AMP system does not appear to be involved in IAP-induced inhibition of histamine release. Instead, receptor-mediated phospholipid metabolism and accompanying calcium ion flux is in all likelihood the IAP-sensitive process of signal transduction in mast cells (44).

Stimulation of mastocyte receptors which are responsible for histamine release causes increased phosphatidyl inositol breakdown, influx of extracellular calcium ions, and phospholipase $A_2$ activation leading to the liberation of arachidonic acid. Probably, calcium ions and certain arachidonate metabolites will serve as an intracellular trigger for histamine release from mast cells. A23187, a potent calcium ionophore, provokes histamine release, without stimulation of receptors, as a result of calcium ion entry and phospholipase $A_2$ activation. The A23187-induced histamine release, in contrast to the receptor-mediated one, was not inhibited by the IAP treatment of mast cells (44). It is likely, therefore, that the site of IAP interference with receptor-mediated histamine release from mast cells is prior to the entry of calcium ions.

Of particular interest is the finding that this IAP-induced inhibition of receptor-mediated phospholipase $A_2$ activation and of the subsequent histamine release was associated with ADP-ribosylation of the Mr = 41,000 protein present in the mast cell cytosolic fraction (44). It would be likely, therefore, that a protein similar in characteristics to a subunit of the $N_i$ protein in the membrane receptor-adenylate cyclase system may play an important role as well in the phospholipid-calcium system in blood cells such as mast cells. Hence, both signal transduction systems were blocked by ADP-ribosylation of this protein. Receptor-induced stimulation of arachidonic acid release was inhibited by IAP in 3T3 cells (Murayama and Ui, unpublished) and in neutrophils (Okajima and Ui, unpublished) as well. In 3T3 cells, stimulation of the membrane receptors for thrombin and bradykinin resulted in decreases in cyclic AMP and in increases in arachidonic acid release in the cell. Both were suppressed by the treatment of the cells with IAP. A possible relationship between adenylate cyclase inhibition and phospholipase activation both elicited by stimulation of a single class of receptors such as muscarinic ones would be clarified by further studies with the use of IAP as the most promising probe.

## IAP AS ONE OF THE A-B TOXINS

As the final section of this review article, a brief mention will be made of the structure-activity relationship of IAP, which would be closely related to the development of the unique action of this bacterial toxin on many cell types. Analysis of the subunit structure of the IAP molecule revealed that it is a hexamer composed of five dissimilar peptides (45). These subunits are associated together in a non-covalent manner. The subunit assembly of IAP is such that a pentamer comprised of two dimers (D-1 and D-2), connected by means of the smallest C (Connecting) subunit is further associated with the biggest subunit. The biggest subunit (Mr = 28,000) displayed NAD-glycohydrolase activity in the absence of any cellular component (46) and catalyzed the transfer of the ADP-ribose moiety of NAD to the membrane Mr = 41,000 protein upon direct addition to the cell-free membrane preparation (45). Thus, it is referred to as an A (Active) protomer. The A protomer was effective on intact cells only when it was associated with the enzymatically inactive pentamer as the native IAP. The action of the native IAP on intact cells was competitively antagonized by the pentamer, suggesting that IAP binds to the cell surface via its pentamer moiety (47). The pentamer should be then a B (Binding) oligomer. Evidence has been provided for ready dissociation of IAP to the A protomer and the B oligomer moieties in a concentrated urea solution (45).

Thus, IAP is one of the A-B toxins (48); the B component binds to a particular receptor protein on the cell surface, thereby permitting the A component to traverse the plasma membrane and to reach the intracellular face of the membrane. The liberated A protomer then exhibits its biological activity by catalyzing ADP-ribosylation of one of the membrane proteins with cytosolic NAD as substrate. The lag period invariably preceding the onset of the IAP action on intact cells (10) may reflect the time required for its A protomer to traverse membranes, since there was no lag upon addition of the A protomer to isolated membranes (38).

The native IAP was as effective as its A protomer when they were directly added to membranes prepared from C6, NG108-15, 3T3 or fat cells, whereas it was the A protomer only that was effective on membranes from rat heart cells, islet cells, platelets or reticulocytes (unpublished data). Conceivably, IAP-sensitive cells are equipped with certain "processing" enzyme(s) that are responsible for liberation of the active A protomer from IAP; this enzymic activity was probably lost during procedures of membrane isolation from the latter cell types.

The B oligomer binds to the cell via its two dimers, D-1 and D-2. Such a "divalent" attachment to the cell surface causes crosslinkage of membrane glycoproteins, thereby exerting a mitogenic action on lymphocytes and an insulin-like action on adipocytes, just as has been frequently reported for some lectins such as concanavalin A (47). Thus, IAP is distinct from other A-B toxins in that its B oligomer moiety exhibits a particular biological activity in addition to its action common to all the B components of the A-B toxins to transport the A component across the membranes. This extraordinary action of the B oligomer of IAP appears to be of particular importance to the mechanisms involved in such diverse biological actions of a single protein, pertussis toxin, as to be termed as LPF, HSF, and IAP.

## CONCLUSION

There are now several lines of convincing evidence for our original proposal that islet-activating protein (IAP), pertussis toxin, is a specific modifier of the guanine nucleotide regulatory protein ($N_i$) that is involved in receptor-mediated inhibition of adenylate cyclase. IAP-catalyzed ADP-ribosylation of the Mr = 41,000 protein, one of the subunits of $N_i$, results in a total loss of the $N_i$ function to communicate between R and C in an inhibitory fashion. Thus, IAP is comparable to, and contrasts with, cholera toxin, that is a specific modifier of $N_s$. Widespread use of IAP will contribute to elucidation of the mechanism whereby $N_i$ mediates receptor-linked adenylate cyclase inhibition, just as has been the case with cholera toxin as a probe for $N_s$ functions.

## ACKNOWLEDGEMENTS

This work has been supported in part by the research grants from the Scientific Research Fund of the Ministry of Education, Science and Culture, Japan.

## REFERENCES

1. E. M. Ross and A. G. Gilman, Biochemical properties of hormone-sensitive adenylate cyclase, Ann. Rev. Biochem. 49:533 (1980).
2. L. E. Limbird, Activation and attenuation of adenylate cyclase. The role of GTP-binding protein as macromolecular messengers in receptor-cyclase coupling, Biochem. J. 195:1 (1981).
3. A. M. Spiegel and R. W. Downs, Guanine nucleotides: Key regulators of hormone receptor-adenylate cyclase interaction, Endocrine Rev. 2:275 (1981).
4. J. Moss and M. Vaughan, Activation of adenylate cyclase by choleragen, Ann. Rev. Biochem. 48:581 (1979).
5. K. H. Jacobs, Inhibition of adenylate cyclase by hormones and neurotransmitters, Mol. Cell. Endocrinol. 61:147 (1979).
6. M. Rodbell, The role of hormone receptors and GTP-regulatory proteins in membrane transduction, Nature 284:17 (1980).
7. D. M. F. Cooper, Bimodal regulation of adenylate cyclase, FEBS Lett. 138:157 (1982).
8. M. Yajima, K. Hosoda, Y. Kanbayashi, T. Nakamura, K. Nogimori, Y. Nakase, and M. Ui, Islet-activating protein (IAP) in Bordetella pertussis that potentiates insulin secretory responses of rats. Purification and characterization, J. Biochem. 83:295 (1978).

9. M. Yajima, K. Hosoda, Y. Kanbayashi, T. Nakamura, I. Takahashi, and M. Ui, Biological properties of islet-activating protein (IAP) purified from the culture medium of Bordetella pertussis, J. Biochem. 83:305 (1978).
10. T. Katada and M. Ui, Slow interaction of islet-activating protein with pancreatic islets during primary culture to cause reversal of alpha-adrenergic inhibition of insulin secretion, J. Biol. Chem. 255:9580 (1980).
11. T. Katada and M. Ui, In vitro effects of islet-activating protein on cultured rat pancreatic islets. Enhancement of insulin secretion, cyclic AMP accumulation and Ca flux, J. Biochem. 89:979 (1981).
12. T. Katada and M. Ui, Islet-activating protein. A modifier of receptor-mediated regulation of rat islet adenylate cyclase, J. Biol. Chem. 256:8310 (1981).
13. O. Hazeki and M. Ui, Modification by islet-activating protein of receptor-mediated regulation of cyclic AMP accumulation in isolated rat heart cells, J. Biol. Chem. 256:2856 (1981).
14. A. C. Wardlaw and R. Parton, Bordetella pertussis toxins, Pharmacol. Ther. 19:1 (1983).
15. H. Shikama and M. Ui, Metabolic background for glucose tolerance: mechanism for epinephrine-induced impairment, Am. J. Physiol. 229:955 (1975).
16. H. Shikama and M. Ui, Adrenergic receptor and epinephrine-induced hyperglycemia and glucose tolerance, Am. J. Physiol. 229:962 (1975).
17. M. Yajima and M. Ui, Carbohydrate metabolism and its response to catecholamines as modified in alkalotic rat, Am. J. Physiol. 228:1046 (1975).
18. M. Yajima and M. Ui, Hypoglycemia induced by alpha-adrenergic stimulation during alkalosis, Europ. J. Pharmacol. 41:93 (1977).
19. M. Yajima and M. Ui, Hydrocortisone restoration of the pH-dependent metabolic responses to catecholamines, Am. J. Physiol. 228:1053 (1975).
20. F. Okajima and M. Ui, Adrenergic modulation of insulin secretion in vivo dependent on thyroid states, Am. J. Physiol. 234:E106 (1978).
21. M. Yajima, T. Hosokawa, and M. Ui, An involvement of alpha-adrenergic stimulation in exercise-induced hypoglycemia, Europ. J. Pharmacol. 42:1 (1977).
22. T. Sumi and M. Ui, Potentiation of the adrenergic beta-receptor-mediated insulin secretion in pertussis-sensitized rats, Endocrinology 97:352 (1975).
23. T. Katada and M. Ui, Accelerated turnover of blood glucose in pertussis-sensitized rats due to combined actions of endogenous insulin and adrenergic beta-stimulation, Biochim. Biophys. Acta 421:57 (1976).

24. T. Katada and M. Ui, Perfusion of the pancreas isolated from pertussis-sensitized rats: potentiation of insulin secretory responses due to beta-adrenergic stimulation, Endocrinology 101:1247 (1977).

25. M. Ui, T. Katada, and M. Yajima, Islet-activating protein in Bordetella pertussis: Purification and mechanism of action, in: "International Symposium on Pertussis," C. R. Manclark and J. C. Hills, eds., p. 166, DHEW Publication No. (NIH)79-1830, Bethesda (1979).

26. M. Pittman, Pertussis toxin: The cause of the harmful effects and prolonged immunity of whooping cough. A hypothesis, Rev. Infect. Dis. 1:401 (1979).

27. J. J. Munoz and R. K. Bergman, Biological activity of Bordetella pertussis, in: "International Symposium on Pertussis," C. R. Manclark and J. C. Hills, eds., p. 143, DHEW Publication No. (NIH) 79-1830, Bethesda (1979).

28. M. Ui and T. Katada, A novel action of the islet-activating protein (IAP) to modify adrenergic regulation of insulin secretion, in: ""Proinsulin, Insulin and C-Peptide," S. Baba, T. Kaneko, and N. Yanaibara, eds., p. 124, Excerpta Medica, Amsterdam-Oxford (1979).

29. T. Katada and M. Ui, Islet-activating protein. Enhanced insulin secretion and cyclic AMP accumulation in pancreatic islets due to activation of native calcium ionophores, J. Biol. Chem. 254:469 (1979).

30. T. Katada and M. Ui, Effect of in vivo pretreatment of rats with a new protein purified from Bordetella pertussis on in vitro secretion of insulin: role of calcium, Endocrinology 104:1822 (1979).

31. S. Yamazaki, T. Katada, and M. Ui, Alpha(2)-adrenergic inhibition of insulin secretion via interference with cyclic AMP generation in rat pancreatic islets, Mol. Pharmacol. 21:648 (1982).

32. O. Hazeki, T. Katada, H. Kurose, and M. Ui, Effect of adenosine on cyclic AMP accumulation in cardiac and other cells and its modification by islet-activating protein, in: "Physiology and Pharmacology of Adenosine Derivatives," J. W. Daly, Y. Kuroda, J. W. Phillis, H. Shimizu, and M. Ui, eds., p. 41, Raven Press, New York (1983).

33. T. Murayama and M. Ui, Loss of the inhibitory function of the guanine nucleotide regulatory component of adenylate cyclase due to its ADP-ribosylation by islet-activating protein, pertussis toxin, in adipocyte membranes, J. Biol. Chem. 258:3319 (1983).

34. T. Katada, T. Amano, and M. Ui, Modulation by islet-activating protein of adenylate cyclase activity in C6 glioma cells, J. Biol. Chem. 257:3739 (1982).

35. T. Murayama, T. Katada, and M. Ui, Guanine nucleotide activation and inhibition of adenylate cyclase as modified by islet-activating protein, pertussis toxin, in mouse 3T3 fibroblasts, Arch. Biochem. Biophys. 221:381 (1983).
36. H. Kurose, T. Katada, T. Amano, and M. Ui, Specific uncoupling by islet-activating protein, pertussis toxin, of negative signal transduction via alpha-adrenergic, cholinergic, and opiate receptors in neuroblastoma x glioma hybrid cells, J. Biol. Chem. 258:4870 (1983).
37. H. Kurose and M. Ui, Functional uncoupling of muscarinic receptors from adenylate cyclase in rat cardiac membranes by the active component of islet-activating protein, pertussis toxin, submitted.
38. T. Katada and M. Ui, Direct modification of the membrane adenylate cyclase system by islet-activating protein due to ADP-ribosylation of a membrane protein, Proc. Natl. Acad. Sci. USA 79:3129 (1982).
39. T. Katada and M. Ui, ADP-ribosylation of the specific membrane protein of C6 cells by islet-activating protein associated with modification of adenylate cyclase activity, J. Biol. Chem. 257:7210 (1982).
40. M. Ui, T. Katada, T. Murayama, H. Kurose, M. Yajima, M. Tamura, T. Nakamura, and K. Nogimori, Islet-activating protein, pertussis toxin: a specific uncoupler of receptor-mediated inhibition of adenylate cyclase, in: "Advances in Cyclic Nucleotide Research," vol. 16, P. Greengard, G. A. Robison, and R. Paoletti, eds., Raven Press, New York, (1984).
41. D. L. Burns, E. L. Hewlett, J. Moss, and M. Vaughan, Pertussis toxin inhibits enkephalin stimulation of GTPase of NG108-15 cells, J. Biol. Chem. 258:1435 (1983).
42. T. Murayama and M. Ui, GDP release from rat and hamster adipocyte membranes independently linked to receptors involved in activation or inhibition of adenylate cyclase. Differential susceptibility to two bacterial toxins, J. Biol. Chem. 259: in press.
43. T. Nakamura and M. Ui, Suppression of passive cutaneous anaphylaxis by pertussis toxin, an islet-activating protein, as a result of inhibition of histamine release from mast cells, Biochem. Pharmacol. 32: in press.
44. T. Nakamura and M. Ui, Mast cell phospholipase inhibition by islet-activating protein, pertussis toxin, as associated with suppressed histamine release, submitted.
45. M. Tamura, K. Nogimori, S. Murai, M. Yajima, K. Ito, T. Katada, M. Ui, and S. Ishii, Subunit structure of islet-activating protein, pertussis toxin, in conformity with the A-B model, Biochemistry 21:5516 (1982).
46. T. Katada, M. Tamura, and M. Ui, The A protomer of islet-activating protein, pertussis toxin, as an active peptide catalyzing ADP-ribosylation of a membrane protein, Arch. Biochem. Biophys. 224:290 (1983).

47. M. Tamura, K. Nogimori, Y. Yajima, K. Ase, and M. Ui, A role of the B-oligomer moiety of islet-activating protein, pertussis toxin, in development of the biological effects on intact cells, J. Biol. Chem. 258:6756 (1983).
48. D. M. Gill, Seven toxin peptides that cross cell membranes, in: "Bacterial Toxins and Cell Membranes," J. Jeljaszewicz, and T. Wadstrom, eds., p. 291, Academic Press, New York (1978).

# A CENTRAL ROLE FOR MAGNESIUM IN THE REGULATION OF INHIBITORY ADENOSINE RECEPTORS

Dermot M. F. Cooper, Siu-Mei Helena Yeung,
Edward Perez-Reyes and Linda H. Fossom

University of Colorado School of Medicine
Department of Pharmacology
Denver, Colorado

Many adenylate cyclase systems are being recognized to be subject to both stimulation and inhibition of activity, which is mediated by distinct sets of receptors (1,2). Evidence is accumulating that distinct GTP regulatory proteins are associated with both classes of receptor, although it is not clear whether the GTP regulatory proteins are discrete or part of a large complex. The GTP regulatory proteins serve two major functions: firstly, they transmit the signal of receptor occupancy into either an increase or a decrease in catalytic activity; secondly, they regulate binding of effectors to their receptors (3). The ability to measure these two functions separately uncovers anomalies in our understanding of the communication between receptor, GTP regulatory protein and catalytic unit. Discrepancies are generally encountered between the properties of receptors when judged from the regulation of cyclase activity, compared with their properties as judged by guanine nucleotide regulation of receptor binding. This is particularly apparent in the case of receptors which mediate inhibition of adenylate cyclase. The non-hydrolyzable guanine nucleotide analogues are as effective as GTP at regulating receptor binding; these analogues also promote a characteristic transient inhibition of activity, although they are ineffective in the promotion of receptor-mediated inhibition of activity (1,2). Distinct concentration ranges are encountered for the regulation of receptor binding compared with the inhibition of activity. Monovalent cations generally promote inhibition of activity, although in many cases they will not affect binding (1). In order to accommodate some of these discontinuities, a number of years ago, Rodbell and his colleagues felt that it was necessary to propose that there were two GTP binding sites on each GTP regulatory protein (or two distinct

GTP binding proteins), referred to as $N_1$ and $N_2$, one of which regulated binding, while the other was involved in the interaction with the catalytic unit (3). Studies with the purified "G/F" indicates that it contains only one GTP binding site per molecule. Therefore, if there is a simple stoichiometric relationship between one receptor molecule and one GTP regulatory protein, and if this relationship pertains in both the regulation of binding and of catalytic activity, then the $N_1$-$N_2$ hypothesis seems untenable. However, if higher order associations, such as dimerization, are involved in a complete regulatory assembly, then of course non-identical properties could be displayed between equivalent GTP binding sites, through allosteric interactions.

The present study of inhibitory adenosine receptors was undertaken against this background. Particular attention was paid to the interaction of magnesium ion and GTP in the regulation of binding, since divalent cations have a selective effect on eliminating inhibition of activity by adenosine analogues and guanine nucleotides (1,2). Our studies indicate that a conformational option is available to the adenosine $R_i$-receptor, which is promoted by magnesium in the presence of GTP, which is absolutely required for receptor-mediated inhibition of activity, but which is not necessary for guanine-nucleotide mediated inhibition of activity. The magnesium-promoted state of the receptor is labile to treatment with low concentrations of N-ethylmaleimide (NEM) and will not withstand solubilization with sodium cholate, although GTP regulation of binding is retained in both cases.

## METHODS

[$^3$H]$N^6$-cyclohexyladenosine (CHA: 13.5 Ci/mmol), [$^{32}$P]ATP and [$^3$H]cAMP were from New England Nuclear. $N^6$-phenylisopropyladenosine was from Boehringer Mannheim. The sources of other materials and the preparation of fat cell and brain cortex membranes have been previously described (4,5).

### Adenosine Receptor Binding Assay

a. Binding to membranes. Adenosine receptor binding is monitored by using the high affinity agonist, $N^6$-cyclohexyladenosine (CHA) in incubations with cortical and fat cell membranes by the method of Trost and Schwabe (6). Membranes are incubated with 4 nM [$^3$H]CHA, 2 mM $MgCl_2$, 30 mM Tris HCl, pH 7.4, 5 mM creatine phosphate, 25 U/ml creatine phosphokinase (nucleotide regenerating system), 1 U/ml adenosine deaminase (to metabolize any endogenous adenosine present in the preparation) for 120 min at 24°. Incubations are terminated by addition of cold Tris HCl (50 mM), pH 7.4, containing 2 mM $MgCl_2$ and rapid filtration through Whatman glass fiber disks (GF/C) followed by two cold washes. Nonspecific binding is determined in the presence of 100 μM $N^6$-phenylisopropyladenosine. In experiments when the $MgCl_2$ concentration in the incubation was varied, the same concentration was used in the washing buffer.

b. Binding to solubilized preparations. The solubilized preparations (80-100 μg/assay) are incubated (2 hr, 24° C) in 30 mM Tris HCl, pH 7.4, containing 2 mM $MgCl_2$ in a final volume of 100 μl (final cholate concentration is approximately 5 mM). At the end of 2 hr, the samples are placed on ice, 20 μl of BSA (10%) are added. Eighty μl of polyethylene glycol (PEG, 25%) are added and incubated for 5 min on ice. Then 4 ml of 50 mM Tris HCl, pH 7.4, 2 mM $MgCl_2$, containing 10% PEG (buffer A) are added to each sample, vortexed and filtered through GF/C filters with 2 washes of 4 ml ice-cold buffer A. The filter paper is transferred to a 20 ml scintillation vial and 1 ml of NCS (0.6 N) is added to solubilize the radioactivity. It is incubated at 60° C for 1.5 hr. Then 0.8 ml of 1N GCl are added to neutralize the NCS, followed by 10 ml of scintillation fluid and the sample is counted in a liquid scintillation counter. Nonspecific binding is determined in the presence of 100 μM PIA.

## Solubilization of Adenosine Receptors

Sodium cholate (15 mM) is added to cortical membrane preparations at approximately 5 mg/ml and stirred on ice for 30 min. The buffer used is 50 mM Tris HCl, pH 7.4, 0.1 mM diethylenetetramine-penta-acetic acid (DTPA), 1 mM dithiothreitol (DTT), 0.1 mM phenyl methanesulfonyl fluoride (PMSF). The preparation is centrifuged (100,000 X $g_{av}$, 60 min) in a SW-40 rotor. The supernatant (approximately 2 mg/ml) is frozen directly in liquid $N_2$. Generally, up to 60% of the receptors of the native membrane are solubilized by this procedure.

## Adenylate Cyclase Assay

Adenylate cyclase is assayed at 24° to facilitate expression of both stimulation and inhibition (1) with 1-5 μg membrane preparation, using α-[$^{32}$P]ATP (0.1 mM; 1 μCi); 0.1 mM cyclic AMP, 30 mM Tris-HCl, pH 7.4, 1 mM $MgCl_2$, 2 mM creatine phosphate, 25 U/ml creatine phosphokinase,, 0.1 U/ml adenosine deaminase in an incubation volume of 100 μl for 10 min as previously described (4).

## Membrane Pretreatment with NEM

Fat cell membranes (mg/ml) were incubated with an equal volume of various concentrations of NEM (as indicated) in 20 mM Tris-HCl buffer, pH 7.4, containing 1 mM EDTA (Buffer B). The incubation was performed for 15 min at 25° C. The reaction was stopped by the addition of 2 volumes of a concentration of mercaptoethanol equal to that of NEM in Buffer B and kept on ice for 10 min. Under the conditions utilized, mercaptoethanol up to the highest concentration studied did not affect [$^3$H]CHA binding. The mixture was centrifuged for 10 min at 10,000 rpm (12,000 X g, 4°). After removal of the supernatant, the membranes were resuspended in 20 mM Tris-HCl buffer, pH 7.4. Control preparations were done in the same way except the NEM (and mercaptoethanol) were replaced with Buffer B.

## RESULTS

In initial binding experiments with fat cell membranes, magnesium ion was included, in order to mirror as closely as possible adenylate cyclase assay conditions. When the effect of GTP was examined on the CHA binding isotherm, a somewhat unusual effect of the nucleotide was observed (Fig. 1). GTP increased the amount of CHA bound over the entire isotherm. Transformation of this data into a Scatchard plot (Fig. 2) revealed that the increased binding was due to an increase in the number of receptors, without an appreciable alteration in receptor affinity. The only other instance that was in the literature at the early phase of these studies, which also indicated a GTP-mediated increase in inhibitory receptor binding, was that of a guanine-nucleotide-mediated increase in $\alpha_2$-adrenergic receptor binding in rat cerebellum. This report from U'Prichard and Snyder (7) indicated that the effect was dependent upon the ambient divalent cation concentration. Consequently, the effect on CHA binding of varying divalent cation concentrations was investigated. When magnesium ion concentrations were varied from 0 - 10 mM, two remarkable features were observed. Firstly, in the absence of guanine nucleotide, overall binding was decreased by magnesium ion; secondly, in the presence of GTP, binding was increased by magnesium ion, such that two phases of guanine nucleotide regulation were observed. At low magnesium concentrations, GTP decreased binding to the receptor, while at higher concentrations, GTP increased binding (Fig. 3).

Scatchard analysis of the effect of GTP in the absence of magnesium ion demonstrates that in this situation, GTP decreases receptor affinity about two-fold without affecting receptor number (Fig. 4). Thus, these results suggest an intimate interdependence between magnesium and GTP in the form of the GTP regulation of binding.

The functional significance of this interrelationship is addressed later, however, a primary issue to address was to determine whether the GTP- and Mg-mediated events could be separated. The alkyating agent, N-ethylmaleimide, was investigated because of its widely described effect on hormonal coupling mechanisms (8). Treatment of fat cell membranes with progressively increasing NEM concentrations revealed a perturbation in the Mg-GTP relationship in subsequent binding experiments. In the absence of GTP, NEM treatment did not appear to have perturbed seriously adenosine receptor

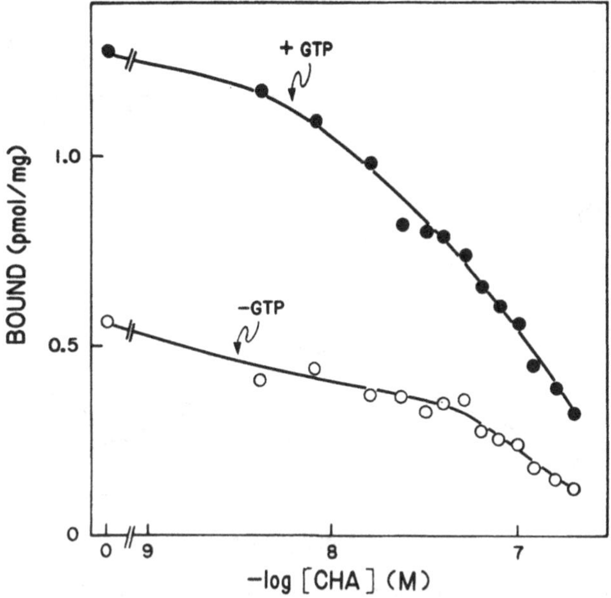


Figure 1: CHA binding to fat cell membranes in the presence (●) and absence (○) of 0.1 mM GTP in the presence of 2 mM $MgCl_2$.

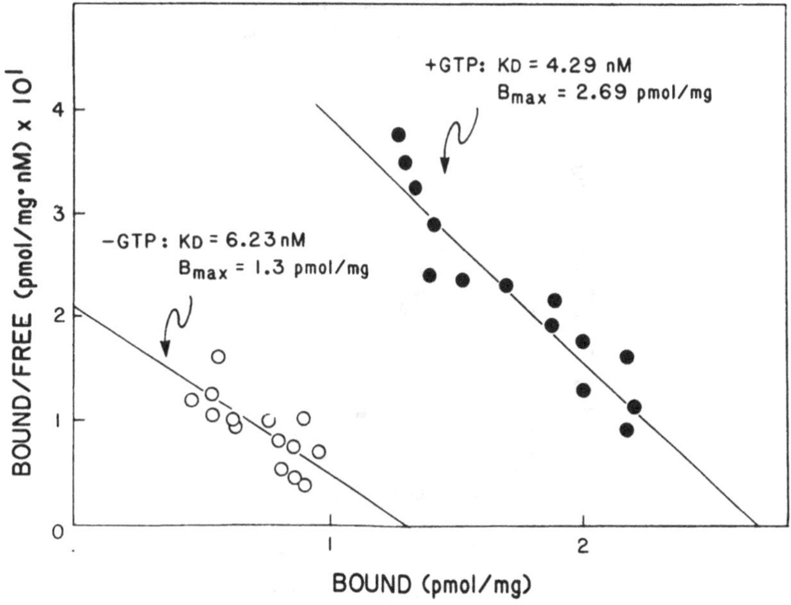


Figure 2: Scatchard analysis of the $^3$H-CHA binding data to fat cell membranes from Figure 1. ○, control; ●, 0.1 mM GTP.

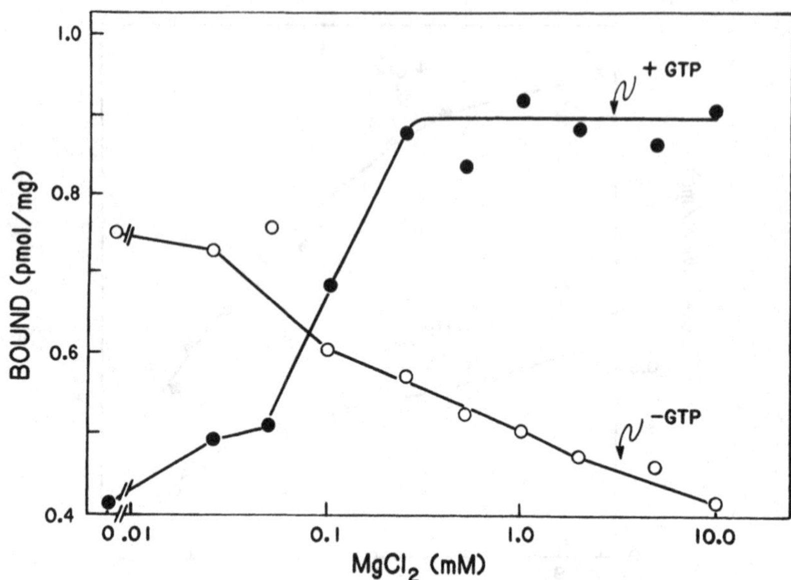


Figure 3: Concentration-response curves of the effects of $Mg^{++}$ on $^3$H-CHA binding in the presence (●) and absence (○) of 0.1 mM GTP to fat cell membranes.

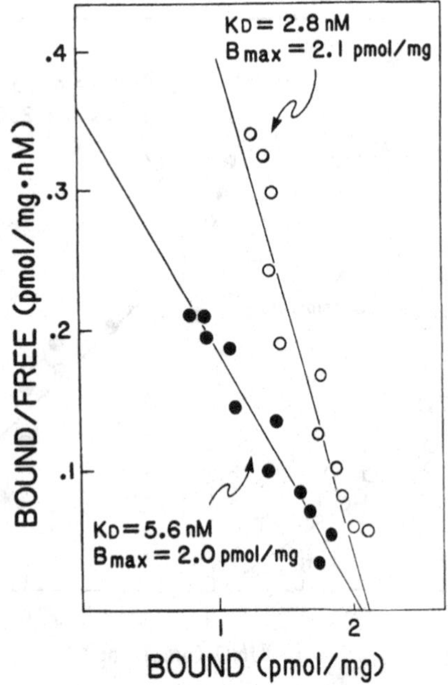


Figure 4: Scatchard analysis of the effect of GTP in the absence of $MgCl_2$. Fat cell membranes were incubated with (●) or without (○) 0.1 mM GTP.

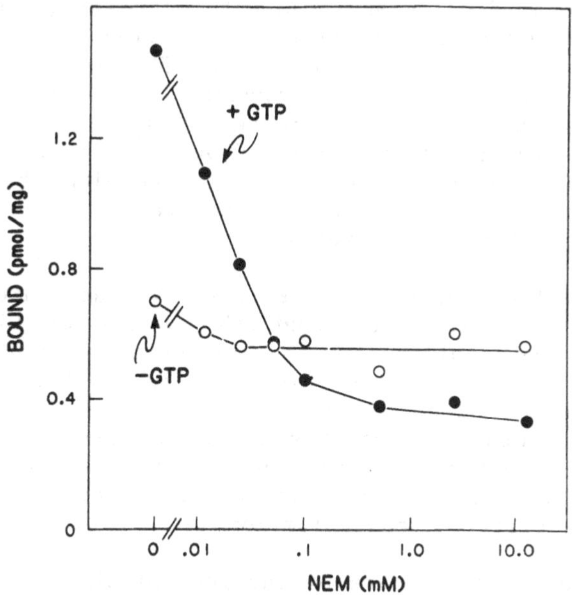


Figure 5: Effect of NEM pretreatment on $^3$H-CHA binding to fat cell membranes in the presence (●) or absence (O) of 0.1 mM GTP. 2 mM $MgCl_2$ was included in the assay.

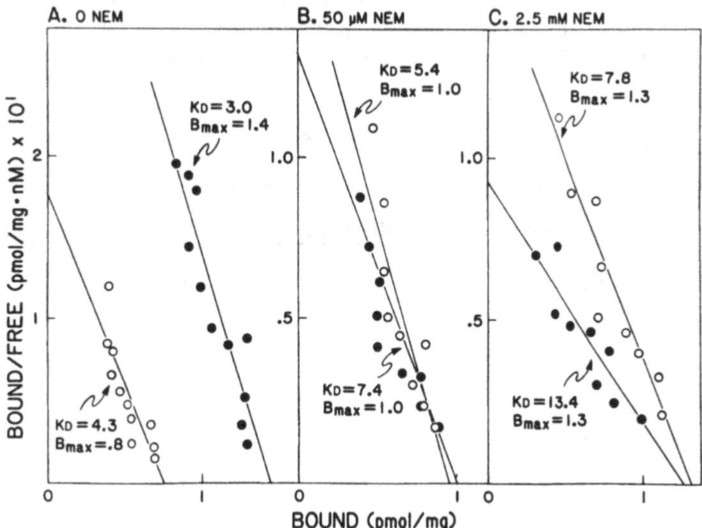


Figure 6: Scatchard analysis of $^3$H-CHA binding to fat cell membranes in the presence of 2 mM $MgCl_2$. A, control; B, 50 μM NEM-treated; C, 2.5 mM NEM-treated preparations as described in the Methods. In the absence (O) or in the presence (●) of 0.1 mM GTP.

binding. A gradual decline was seen with increasing concentrations of the agent. However, the effect on the binding in the presence of GTP was profound. Low concentrations of NEM in the pre-treatment (up to 50 μM) resulted in almost a complete loss in the GTP-promoted increase in binding (Fig. 5). That this was not a complete loss, but only a crossover point in passing from a GTP-dependent increase to a GTP-dependent decrease, is seen by examining the effects of higher treatments of NEM. It can be seen that at concentrations as high as 10 mM, a GTP effect is retained (Fig. 5).

The NEM treatment appears to selectively inactivate the magnesium-dependent GTP effect, although the effect of GTP alone is retained. This is confirmed in a Scatchard examination of the consequences of low (50 μM) and high (2.5 mM) NEM treatment (Fig. 6). Here it can be seen that following 2.5 mM NEM treatment, an effect of GTP is retained, but this is to decrease the affinity for adenosine, even though this experiment is performed in the presence of magnesium. Thus, it appears that the effect of NEM treatment is to eliminate the effect of magnesium ion.

The functional consequence of the loss of the magnesium effect is examined in Table 1. The adenylate cyclase activity of the fat cell membranes which were treated with a range of NEM concentrations were assessed under two conditions: activity was measured in the presence of forskolin with or without the addition of Gpp(NH)p; in dually regulated systems, the nucleotide inhibits the forskolin-stimulated activity, and provides a measure of $N_i$-C interaction (9). Activity was also measured in the presence of GTP with or without the addition of $N^6$-phenylisopropyladenosine, which provides a

Table 1. Effect of NEM pretreatment of fat cell adenylate cyclase.

| | Control | % Inhibition | 50 μM NEM-Treated | % Inhibition |
|---|---|---|---|---|
| Forskolin | 2128 | | 1490 | |
| Forskolin + Gpp(NH)p | 679 | 68 | 725 | 52 |
| GTP | 121 | | 167 | |
| GTP + PIA | 59 | 52 | 168 | 0 |

Fat cell membranes were treated with or without 50 μM NEM as indicated in the Methods. Activities (pmol/mg/10 min) were determined in the presence of 100 μM forskolin with or without 1 μM Gpp(NH)p or in the presence of 10 μM GTP with or without 1 μM PIA, as indicated.

measure of $R_i$-$N_i$-C interactions (1). It can be seen that the $N_i$-C interactions are retained up to moderately high concentrations of NEM, although catalytic activity is being destroyed. However, at low treatments of NEM the casualty is the PIA inhibition of enzyme activity. Thus, we can conclude that the functional consequence of the alteration in the regulatory properties of the receptor, which is induced by treatment with low concentrations of NEM, is the loss of receptor-mediated ihibition of activity. A corollary of this conclusion is that the regulatory property, which was lost, was essential to the inhibitory receptor regulation of activity.

Another inhibitory adenosine receptor system was examined, in order to determine whether this phenomenon was common to all adenosine receptors and also as a prelude to attempting to solubilize the receptors and studying molecular size transitions which might be mediated by divalent cations or GTP.

When the effect of GTP and magnesium are examined on the binding of CHA to cortical membranes, GTP is seen to decrease binding in the absence of $Mg^{++}$ (Fig. 7). Examination of this behavior by Scatchard analysis reveals that GTP decreases the Kd for CHA without affecting receptor number, whereas magnesium increases receptor number without affecting Kd, whether in the absence or presence of GTP (Fig. 8). When the magnesium concentration dependency for this effect is examined (Fig. 9), it is again evident that magnesium increases the total binding in the absence or presence of GTP. These results with the cortex show one major difference and two important similarities with those described for the fat cell above. The difference is that in the fat cell, magnesium alone decreased binding, whereas in the cortex, the effect was the opposite. The similarities are that GTP alone decreases receptor affinity in both tissues and that magnesium ion, when added in the presence of GTP, increased receptor number in both tissues. Since the regulatory consequence for adenylate cyclase of receptor occupancy (in the presence of magnesium) is the same in both sources, i.e., the inhibition of activity, it is perhaps comforting to observe these similarities. The discrepant behavior of magnesium alone may be informative in that it may suggest that different steps of a regulatory cycle involving magnesium and GTP may be rate-limiting in the two tissues.

Our recent studies with solubilized adenosine receptors from brain cortex reinforce these conclusions. The receptor retains its pharmacological properties and high affinity binding of adenosine analogues. Somewhat to our surprise, the receptor retained its ability to be regulated by guanine nucleotides (Fig. 10). However, the ability of magnesium ion to regulate receptor binding was diminished (Fig. 10). Thus, whereas in the native membrane, the cation could increase receptor number in the absence or presence of GTP; following solubilization, the cation could no longer overcome

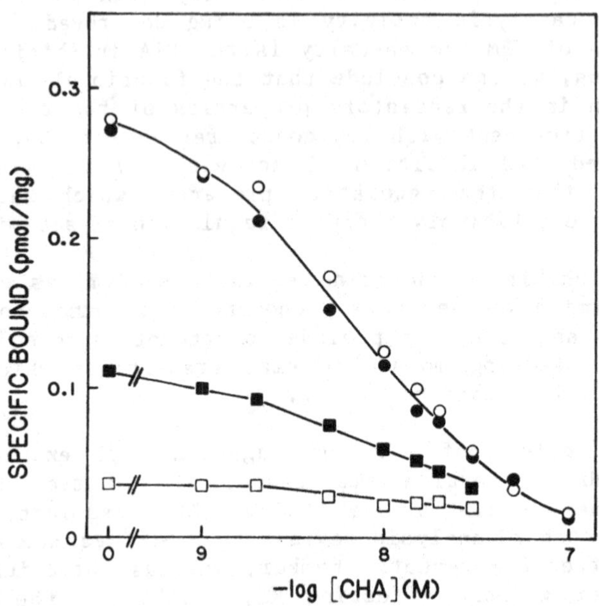


Figure 7: Effect of GTP and $Mg^{++}$ on $^3$H-CHA binding to cortical membranes. Membranes were incubated without (■) or with 0.1 mM GTP (□) in the absence (○) or presence (●) of 2 mM $MgCl_2$.

Table 2. Effect of solubilization on cortical adenylate cyclase.

| | Membrane | % Inhibition (pmol/min mg) | Solubilized | % Inhibition |
|---|---|---|---|---|
| Basal | 55 | | 21 | |
| Basal + Gpp(NH)p | 21 | 62 | 15 | 29 |
| GTP | 32 | | 20 | |
| GTP + PIA | 20 | 38 | 19 | 0 |

Cortical membranes before or following solubilization, as described under Methods, were incubated in the absence or presence of 1 μM Gpp(NH)p, or in the presence of 10 μM GTP with or without 1 μM PIA, as indicated.

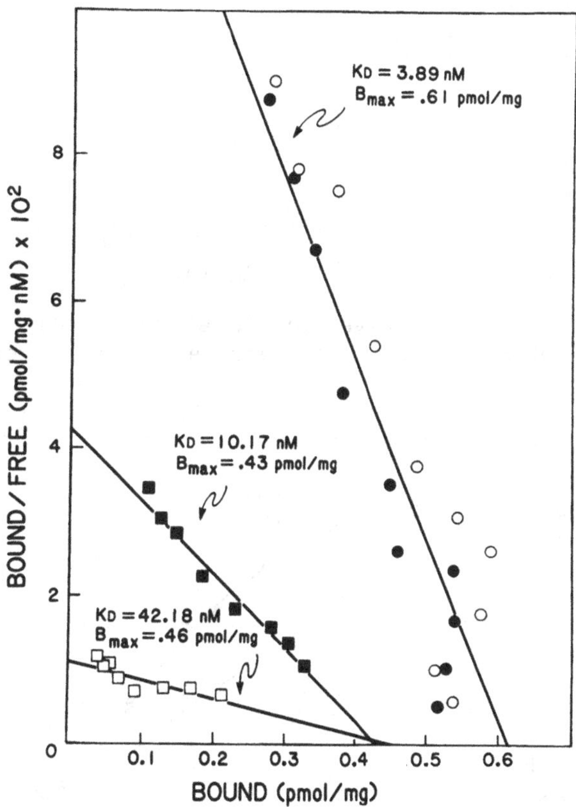


Figure 8: Scatchard analysis of the data in Figure 7. Control (■) in the presence of GTP (□), with (●) or without (○) 2 mM $MgCl_2$.

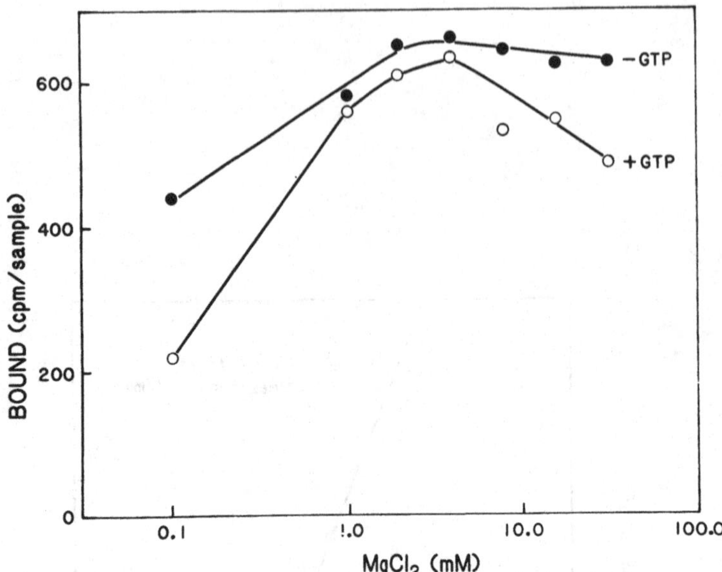


Figure 9: Dose-response curves of the effect of $Mg^{++}$ on $^3$H-CHA binding to cortical membranes in the presence (○) or absence (●) of 0.1 mM GTP.

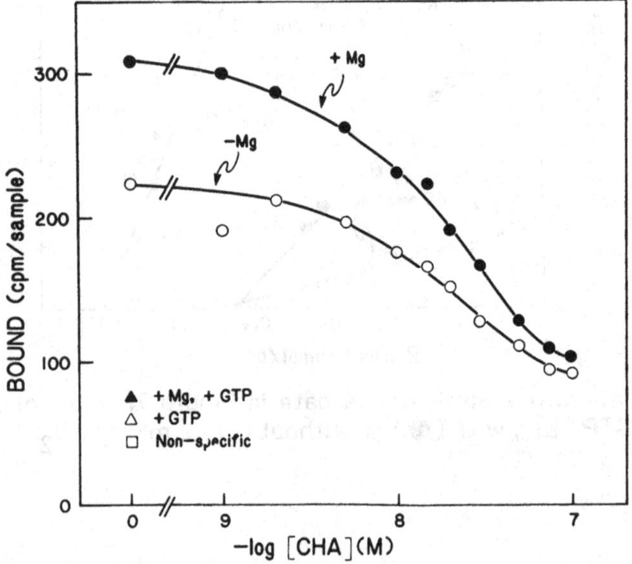


Figure 10: Effect of $Mg^{++}$ and GTP on $^3$H-CHA binding to solubilized cortical preparations. ○, control; ●, 2mM $Mg^{++}$; △, 0.1 mM GTP; ▲, 0.1 mM GTP + 2 mM $Mg^{++}$; , non-specific binding defined in the presence of 0.1 mM L-PIA was the same in the presence or absence of 2 mM $MgCl_2$.

the ability of GTP to decrease affinity. When the cyclase activity is examined in this preparation, similar results were encountered as had been found upon NEM-treatment of fat cell membranes described above (Table 1), i.e., although Gpp(NH)p inhibition of activity was retained, indicating competent $N_i$-C interaction, inhibition by $N^6$-PIA of GTP-promoted activity was lost (Table 2), indicating a loss of $R_i$-$N_i$-C communication.

CONCLUSIONS

The present study indicates a central role for a magnesium site or component in the interaction of adenosine, acting through inhibitory $R_i$-type receptors, with the putative $N_i$-GTP regulatory element and the catalytic unit of adenylate cyclase. Two adenosine-$R_i$ systems were chosen initially to seek uniformity in the results, and to permit generalizations to be made. The observation of some differences in the properties of the two systems focussed attention on the common properties, since these would be expected to be most pertinent to the common regulatory events.

In both fat cell and cortex membranes, GTP alone decreased adenosine receptor affinity, whereas magnesium increased receptor number in the presence of GTP. In the former tissue, magnesium alone decreased receptor binding whereas in the latter, it increased receptor number. In our studies of the fat cell system, the magnesium effect was eliminated upon treatment with low concentrations of NEM, whereas in the cortex, the effect was diminished upon solubilization. In both cases, however, the effect of GTP was retained. Similarly, in both cases, $N_i$-C interactions were unaffected by the loss of the magnesium regulation. The major component affected was the receptor-mediated inhibition of activity; i.e., $R_i$-$N_i$-C communication was lost.

The similarities in the behavior of the two systems force speculation on possible molecular mechanisms underlying this behavior with our current limited experimental information. A theory which accommodates certain of these observations would be that monomer-dimer transitions, which were governed by magnesium, were part of the normal regulatory cycles of these $R_i$-$N_i$ interactions. This suggestion is prompted by the two-fold differences in receptor number brought about by GTP in the case of the fat cell or by magnesium in the case of the cortex. It is not unlikely that NEM exerts its effect by alkylating sulfhydryl groups which could be involved in such a Mg-promoted dimerization. Similarly, it is easy to visualize how solubilization might preserve R-N interactions, but that weaker R-R associations might be disrupted. The foregoing are obviously somewhat fanciful speculations, which need to be evaluated by future experimentation. The solubilization studies are currently being extended and sucrose gradient centrifugation analysis is in progress to determine the apparent sizes of the solubilized components from the cortex; identification of the sedimentation behavior of the components and the role of regulators thereon may allow more intelligent speculation to be made on the regulatory cycles involved. However, it is conceivable that regulatory properties which are lost upon solubilization may be observed only in membranes. In this case, target size analysis is one of the few methods that would seem likely to yield insights into the various molecular options.

In summary, this study indicates that transitions in receptor number, which are governed by magnesium, are part of the regulatory cycle which is involved in receptor-mediated inhibition of adenylate cyclase activity.

## REFERENCES


1. D. M. F. Cooper, Bimodal regulation of adenylate cyclase, FEBS Lett. 138:157 (1982).
2. K. H. Jakobs, Inhibition of adenylate cyclase by hormones and neurotransmitters, Mol. Cell Endocrinol. 16:147 (1979).
3. M. Rodbell, The role of hormone receptors and GTP-regulatory proteins in membrane transduction, Nature 284:17 (1980).
4. D. M. F. Cooper, W. Schlegel, M. C. Lin, and M. Rodbell, The fat cell adenylate cylase system. Characterization and manipulation of its bimodal regulation by GTP, J. Biol. Chem. 254:8927 (1979).
5. D. M. F. Cooper, C. Londos, and M. Rodbell, Adenosine receptor-mediated inhibition of rat cerebral cortical adenylate cyclase by a GTP-dependent process, Mol. Pharmacol. 18:598 (1980).
6. T. Trost and U. Schwabe, Adenosine receptors in fat cells: Identification by (-)-$N^6$-[$^3$H]phenylisopropyladenosine binding, Mol. Pharmacol. 19:228 (1981).
7. D. C. U'Prichard and S. H. Snyder, Interactions of divalent cations and guanine nucleotides at $\alpha_2$-noradrenergic receptor binding sites in bovine brain mechanisms, J. Neurochem. 34:385 (1980).
8. E. M. Ross and A. G. Gilman, Biochemical properties of hormone-sensitive adenylate cyclase, Ann. Review Biochem. 49:533 (1980).
9. K. B. Seamon and J. W. Daly, Guanosine 5'-(β,γ-imido) triphosphate inhibition of forskolin-activated adenylate cyclase is mediated by the putative inhibitory guanine nucleotide regulatory protein, J. Biol. Chem. 257:11591 (1981).

# ROLE OF HORMONE-SENSITIVE GTPases IN ADENYLATE CYLASE REGULATION

Klaus Aktories, Günter Schultz and Karl H.Jakobs

Pharmakologisches Institut der Universität Heidelberg
Im Neuenheimer Feld 366
D-6900 Heidelberg, Federal Republic of Germany

## INTRODUCTION

A large variety of hormones and neurotransmitters transmit their regulatory signals into target cells via stimulation or inhibition of adenylate cyclase and the consequent increase or decrease of intracellular cyclic AMP levels. Because of the importance of adenylate cyclase as a signal transduction system, interest in adenylate cyclase has centered on the mechanisms and components involved in its hormonal control. The complete transduction system is composed of at least the specific stimulatory and inhibitory hormone receptors, the adenylate cyclase itself and two guanine nucleotide-binding, regulatory components, $N_s$ and $N_i$, which act as coupler between stimulatory and inhibitory hormone receptors, respectively, and adenylate cyclase (1,2). Both types of hormones, adenylate cyclase stimulatory and inhibitory, have been shown to stimulate GTPase(s) in membrane preparations. In this review, the role of this hormone-stimulated GTP hydrolysis in the bidirectional regulation of adenylate cyclase will be considered.

## STIMULATION OF GTP HYDROLYSIS BY ADENYLATE CYCLASE STIMULATORY HORMONES

The existence of a membrane-bound GTPase, which is stimulated by hormones, was first demonstrated by Cassel and Selinger (3) in turkey erythrocyte membranes. These authors reported that β-adrenergic agonists at similar concentrations as required for stimulation of adenylate cyclase increase the activity of a GTPase, which exhibits a high affinity for GTP ($K_m \simeq 0.1$ μM) and which requires the presence of $Mg^{2+}$ or $Mn^{2+}$ for activity.

Later on, hormonal stimulations of a low $K_m$ GTPase were also described in membranes of frog erythrocytes (4), rat pancreas (5), rat liver (6) and human platelets (7) by β-adrenergic agonists, hormones of the pancreozymin family, glucagon and prostaglandin $E_1$ ($PGE_1$), respectively. In all these studies, hormone-induced stimulation of GTP hydrolysis correlated well with respective stimulations of the adenylate cyclase.

Based on the data obtained in turkey erythrocyte membranes, Selinger and associates (3,8,9) proposed a model of adenylate cyclase stimulation by hormones which includes a regulatory GTPase cycle. According to this model, hormonal factors stimulate adenylate cylase by introducing GTP into the guanine nucleotide-binding, regulatory component, $N_s$, which in its GTP-bound form activates the enzyme (turn-on reaction). The activation is terminated by hydrolysis of the $N_s$-bound GTP to GDP and $P_i$, catalyzed by an $N_s$-associated high-affinity GTPase (turn-off reaction). Further activation of adenylate cyclase requires the release of the $N_s$-bound GDP and the renewed binding of GTP to $N_s$, which process is catalyzed by the hormone-receptor complex. Thus, the activity state of the adenylate cyclase is in a constant turn-over and depends on the amount of GTP-bound $N_s$. Under basal conditions, the fraction of the GTP-bound $N_s$, which activates adenylate cyclase, is low. Hormonal factors enhance the turn-on reaction and, thereby, increase the fraction of GTP-bound $N_s$. The hormone-induced increase in GTP hydrolysis is, according to this model, due to an increase in the substrate availability of the GTPase and is an expression of an increased turn-over of $N_s$ between its active, GTP-bound state and its inactive, GDP-bound state.

GTP analogs such as GTPγS and Gpp(NH)p also bind to $N_s$ and induce activation of adenylate cyclase. Their stability against hydrolysis, however, prevents the GTPase-dependent inactivation (10) and leads to a persistently activated state of $N_s$ and a subsequent persistent activation of the adenylate cyclase.

The model of a GTPase-regulated adenylate cyclase was corroborated by the finding that cholera toxin, which like stable GTP analogs persistently activates adenylate cyclase, inhibits the hormone-stimulated high-affinity GTPase, as shown in membranes of turkey erythrocytes (11), rat pancreas (12) and human platelets (7,13) (Fig. 1). This GTPase inhibition is also reflected in adenylate cyclase studies, showing that cholera toxin largely retards the turn-off of a hormone-activated adenylate cyclase (8). The effect of cholera toxin on the adenylate cyclase system is apparently due to an ADP-ribosylation of the subunit of the regulatory $N_s$ protein, which binds guanine nucleotides and which interacts with the catalytic moiety of the adenylate cyclase (14,15).

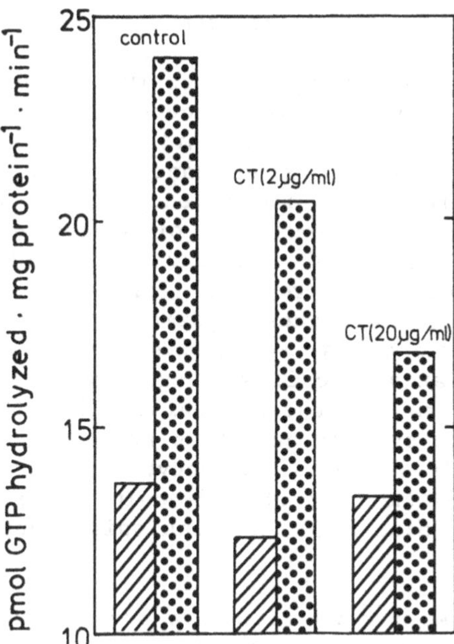


Fig. 1. Influence of cholera toxin (CT) on stimulation of GTP hydrolysis by $PGE_1$ in human platelet membranes. Hatched bars: basal; dotted bars: 1 µM $PGE_1$. Data are taken from ref. 13.

## STIMULATION OF GTP HYDROLYSIS BY ADENYLATE CYCLASE INHIBITORY HORMONES

Similar to hormonal stimulation, hormone-induced inhibition of adenylate cyclase exhibits an absolute GTP requirement. In further analogy to the stimulatory regulation, hormones that inhibit adenylate cyclase also stimulate GTP hydrolysis in membrane preparations. This was reported for prostaglandins from the E-type, nicotinic acid, the adenosine derivative, $N^6$-phenylisopropyladenosine (PIA) and α-adrenergic agonists in adipocytes (16), for various opiates in neuroblastoma x glioma hybrid cells (17,18), for α-adrenergic agonists in human platelets (19) and for somatostatin in S49 lymphoma cells (20). Stimulation of GTP hydrolysis by these hormonal factors exhibited numerous characteristics which were also found for hormone-induced inhibition of adenylate cyclase.

(i) There is compelling evidence that identical hormone receptors are involved in hormone-induced inhibition of adenylate cyclase and the concomitant stimulation of GTP hydrolysis. For instance, in human platelets, inhibition of adenylate cyclase by α-adrenergic agonists is mediated via $\alpha_2$-adrenoreceptors (21). Stimulation of GTP hydrolysis by adrenergic agonists in these membranes also exhibits the typical pharmacological spectrum of an $\alpha_2$-adrenoreceptor-mediated process. The $\alpha_2$-adrenoreceptor-selective antagonist, yohimbine, potently inhibits the epinephrine-induced GTPase stimulation, whereas prazosin, which predominantly blocks $\alpha_1$-adrenoreceptors, is far less effective. The potency orders of α-adrenergic agonists are identical for stimulation of GTPase and inhibition of adenylate cyclase (unpublished observations). Identical potency orders of agonists were also reported for various prostaglandins (16) and opiates (18) in membranes of hamster adipocytes and neuroblastoma x glioma hybrid cells, respectively. The potencies of the prostaglandins and opiates also closely agree with the affinities of the respective receptors for these agonists (18,22). Furthermore, inhibition of adenylate cyclase and stimulation of GTP hydrolysis by the adenosine derivative, PIA, in adipocytes and by opiates in the neuronal tissue are blocked by the adenosine receptor antagonist, 3-isobutyl-1-methylxanthine (23), and the opiate antagonist, naloxone (17), respectively.

(ii) Not only the hormone receptors are identical for inhibition of adenylate cyclase and concomitant stimulation of GTP hydrolysis, but also the concentration-response curves of agonists are almost superimposable for both reactions. This finding indicates that there is a close connection between these two hormonal effects.

(iii) The tissue-specific, maximally inhibitory effects of hormones on adenylate cyclase correlate well with the corresponding stimulation of GTP hydrolysis (16,18). For instance, in hamster adipocyte membranes, $PGE_1$ and nicotinic acid inhibit adenylate cyclase by maximally 60 to 80%. These compounds stimulate GTP hydrolysis in these membranes by maximally 70 to 100%. Inhibitory hormonal factors with lower efficacies in this tissue such as the adenosine derivative, PIA, and epinephrine inhibit adenylate cyclase by maximally 40 to 50%. Correspondingly, these agonists stimulate GTPase activity to a smaller extent (30 to 40%) than $PGE_1$ and nicotinic acid.

(iv) The GTPase stimulated by α-adrenergic agonists, opiates, and $PGE_1$ in membranes of human platelets (19), neuroblastoma x glioma hybrid cells (17) and hamster adipocytes (16), respectively, exhibits an apparent $K_m$ value of 0.1 to 0.3 μM GTP. Similar concentrations of GTP are required for half-maximal inhibition of adenylate cyclase by these hormonal factors. This finding suggests that only one GTP-binding site is involved in these two reactions.

(v) Particularly in adipocytes and neuroblastoma x glioma hybrid cells, but also in membranes of various other tissues, hormone-induced inhibition of adenylate cyclase is largely amplified by sodium ions (24,25). Similarly, hormonal stimulation of GTP hydrolysis in these tissues depends on the presence of sodium ions. Furthermore, the half-maximally and maximally permissive effects of sodium on adenylate cyclase inhibition and GTPase stimulation occur at a very similar concentration of this cation (16,18).

The above summarized findings indicate undoubtedly that stimulation of a GTPase is somehow intimately related to hormonal inhibition of adneylate cyclase. Based on the excellent correlation obtained between adenylate cyclase inhibition and GTPase stimulation, it has been suggested (16,18) that the stimulation of GTP hydrolysis is causally related to the hormone-induced adenylate cyclase inhibition. According to this suggestion, inhibitory hormonal factors after interaction with their receptors should cause a direct stimulation of the GTPase and, thereby, inactivate the adenylate cyclase, whereas hormonal factors that stimulate adenylate cyclase increase GTP hydrolysis by increasing the substrate availability of the enzyme.

## INVOLVEMENT OF DIFFERENT GTPases IN THE BIDIRECTIONAL REGULATION OF THE ADENYLATE CYCLASE

The above mentioned suggestion implied that adenylate cyclase stimulatory and inhibitory hormones increase the activity of the same GTPase. However, recently obtained data clearly indicate that two distinct guanine nucleotide-binding regulatory components, i.e., $N_s$ and $N_i$, and apparently also two GTP-hydrolyzing systems, are involved in the bidirectional control of adenylate cyclase activity by hormonal factors.

As mentioned before, cholera toxin activates adenylate cyclase apparently by ADP-ribosylating a subunit of the stimulatory coupling protein, $N_s$, and, thereby, inhibiting the $N_s$-associated GTPase. From these findings, it follows that hormone-induced inhibition of adenylate cyclase should be inhibited or abolished by cholera toxin, if there is only one GTPase and if stimulation of this GTPase is causally related to hormone-induced adenylate cyclase inhibition. However, as reported for various tissues and hormonal factors, cholera toxin does not impair or prevent adenylate cyclase inhibition by hormones (13,26-29) (Fig. 2).

In analogy to these findings on adenylate cyclase, it was observed that cholera toxin fails to impair GTPase stimulation by adenylate cyclase inhibitory hormones, wherease GTPase stimulation by hormones that stimulate adenylate cyclase is inhibited by this toxin (13,29). For instance, in human platelet membranes, stimulation of GTP hydrolysis by $PGE_1$, which stimulates adenylate cyclase in this tissue, is inhibited by cholera toxin (see Fig. 1). In contrast, stimulation of GTP hydrolysis by epinephrine, which inhibits platelet adenylate cyclase, is not affected by the toxin.

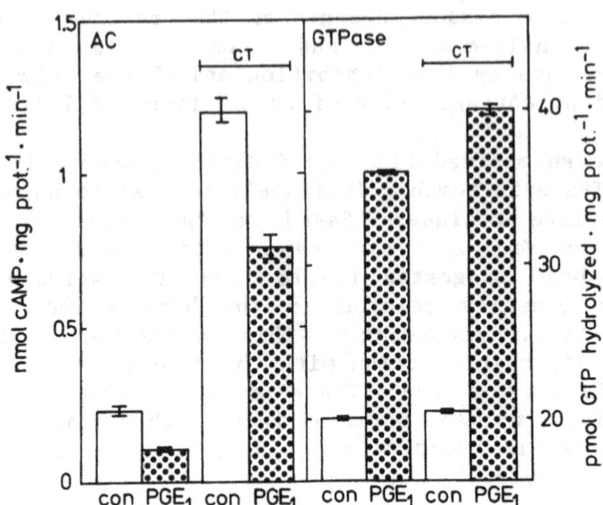


Fig. 2. Influence of cholera toxin (CT) on adenylate cyclase (AC) inhibition and GTPase stimulation by $PGE_1$ in hamster adipocyte membranes. Data are taken from ref. 29.

Furthermore, cholera toxin retards the turn-off reaction of $PGE_1$-stimulated adenylate cyclase but does not affect the turn-off reaction of the epinephrine-inhibited platelet adenylate cyclase (30).

The hypothesis of two hormone-sensitive GTPases is furthermore strengthened by data obtained in combination studies. While combinations of adenylate cyclase inhibitory hormones increase GTPase activity in a non-additive manner, the combination of a hormone that stimulates adenylate cyclase with an adenylate cyclase inhibitory hormone causes an additive increase in GTP hydrolysis (19).

Further indications for two distinct coupling components and two GTPases involved in hormonal stimulation and inhibition of adenylate cyclase came from studies with proteases and the S-alkylating agent, N-ethylmaleimide (NEM). Trypsin and a recently found bovine sperm protease and NEM rather selectively impair or abolish hormone-induced inhibition of adenylate cyclase, whereas hormone-induced stimulation of this enzyme is only minimally affected or even increased (31-34). Concomitantly, these compounds impair or prevent stimulation of GTPase by adenylate cyclase inhibitory hormonal factors, whereas stimulation of GTP hydrolysis by hormones that stimulate adenylate cyclase is largely unaffected by proteases and NEM (33).

The evidence for the existence of two structurally independent

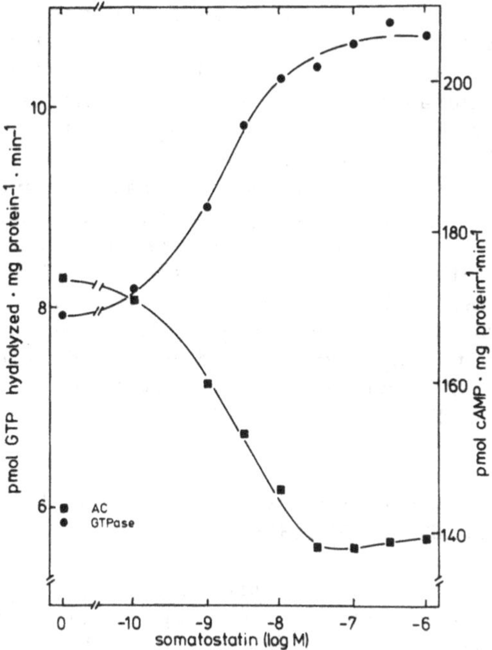


Fig. 3. Inhibition of adenylate cyclase (AC) and stimulation of GTP hydrolysis by somatostatin in cyc$^-$ membranes. Data are taken from ref. 47.

regulatory components was corroborated by studies with cyc$^-$ variants of S49 lymphoma cells. These variants have been shown to lack a functional $N_s$ protein. Accordingly, these variants have little Mg-dependent adenylate cyclase activity, which is not stimulated by hormonal factors, GTP analogs, cholera toxin or fluoride (35). However, as recently shown (36,37), GTP and stable GTP analogs inhibit the cyc$^-$ adenylate cyclase. The inhibition is observed with the unstimulated enzyme and with the enzyme stimulated by forskolin or by purified, preactivated $N_s$ protein. The characteristics of the guanine nucleotide inhibition of the cyc$^-$ adenylate cyclase are identical to those found in complete membrane systems. Furthermore, this inhibition of the cyc$^-$ adenylate cyclase is abolished by limited proteolysis with trypsin or a sperm protease and by NEM, which also impair or abolish hormone and guanine nucleotide inhibition of adenylate cyclase in complete membrane systems.

We have recently reported that this $N_s$-deficient cyc$^-$ adenylate cyclase is inhibited by the peptide hormone, somatostatin, in a strictly GTP-dependent manner (38). Furthermore, somatostatin accelerates the stable GTP analog-induced inhibition of the cyc$^-$ enzyme, as also observed for norepinephrine in human platelet membranes (39). Similar to findings for other inhibitory hormonal factors in complete systems, somatostatin stimulates GTP hydrolysis in $cyc^-$ membranes (20) (Fig. 3). The somatostatin-induced GTPase stimulation in membranes of $cyc^-$ variants exhibits identical characteristics as hormone-stimulated GTP hydrolysis in other cell types, which additionally contain $N_s$. Thus, membranes of $cyc^-$ variants, although lacking a functional $N_s$, contain a guanine nucleotide-binding regulatory component, $N_i$, which mediates the hormone and guanine nucleotide inhibition of the adenylate cyclase and which is apparently associated with a GTPase activity. This component, which is probably present in all hormone-sensitive adenylate cyclase systems, appears to be activated and inactivated by similar mechanisms as is the $N_s$ component, i.e., activation by the binding of GTP or stable GTP analogs and inactivation by the hydrolysis of $N_i$-bound GTP by an $N_i$-associated GTPase (2).

## INFLUENCE OF THE PERTUSSIS TOXIN, ISLET-ACTIVATING PROTEIN, ON HORMONE-STIMULATED GTP HYDROLYSIS

Recently, islet-activating protein (IAP), a *Bordetella pertussis* toxin, has been reported to impair or abolish selectively hormone-induced inhibition of adenylate cyclase in various tissues (40,41). The modulation of the inhibitory regulation of adenylate cyclase by IAP is apparently due to an ADP-ribosylation of a 40K dalton polypeptide (42). This component, which is thought to be a subunit of the inhibitory regulatory component, $N_i$, has been separated from $N_s$ (43), which is the target of the cholera toxin-catalyzed ADP-ribosylation.

The pertussis toxin not only impairs hormone-induced inhibition of adenylate cyclase, but the concomitant stimulation of GTP hydrolysis is also attenuated. Fig. 4 shows such an example of IAP's action. Both inhibition of adenylate cyclase and stimulation of the high-affinity GTPase by epinephrine are largely attenuated when human platelet membranes are pretreated with IAP (44). In contrast, stimulation of GTP hydrolysis by $PGE_1$, which stimulates platelet adenylate cyclase, is not affected by IAP but inhibited by cholera toxin. In membranes of rat adipocytes (45), neuroblastoma x glioma hybrid cells (46) and $cyc^-$ variants (47), inhibition of adenylate cyclase and concomitant stimulation of the low $K_m$ GTPase by nicotinic acid, enkephalin and somatostatin, respectively, are also attenuated or abolished after treatment of intact cells or membranes with the pertussis toxin.

Although the inhibitory effects of hormones and GTP on the adenylate cyclase are blocked after IAP treatment, inhibition of the enzyme by stable GTP analogs is still observed (45,47,48). Recent studies performed in IAP-treated $cyc^-$ variants and rat adipocytes showed that the pertussis toxin treatment largely retards the

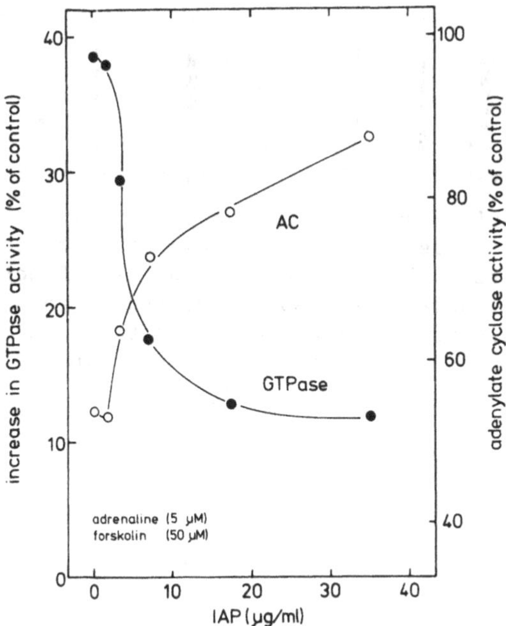


Fig. 4. Influence of IAP on epinephrine-induced adenylate cyclase (AC) inhibition and GTPase stimulation in human platelet membranes. Data are taken from ref. 44.

inhibitory effects of stable GTP analogs on the adenylate cyclase (49). These data suggest that IAP does not affect the inactivation of $N_i$, but that the toxin inhibits the activation process of this inhibitory coupling component. Considering a principally unaltered GTPase activity, the fraction of active $N_i$ will be minimal in the presence of GTP and inhibitory hormones after IAP treatment. Accordingly, GTPase stimulation by adenylate cyclase inhibitory hormones is not seen, since the measured GTP hydrolysis is apparently only a reflection of the turn-over of GTP-bound to GDP-bound $N_i$. In contrast, with stable GTP analogs, active $N_i$ can accumulate, although delayed, and, thus, can cause adenylate cyclase inhibition, even after IAP treatment.

## ROLE OF HORMONE-SENSITIVE GTPases IN REGULATION OF ADENYLATE CYCLASE

The data discussed above indicate that adenylate cyclase activity is controlled via two distinct guanine nucleotide-binding regulatory components, i.e., $N_s$ and $N_i$, which are involved in the stimulatory and inhibitory hormonal regulation of the adenylate cyclase, respectively. Hormone-induced stimulation of the adenylate cyclase is apparently caused by $N_s$ in its GTP-bound form. This activated $N_s$ is inactivated by hydrolysis of the bound GTP to GDP

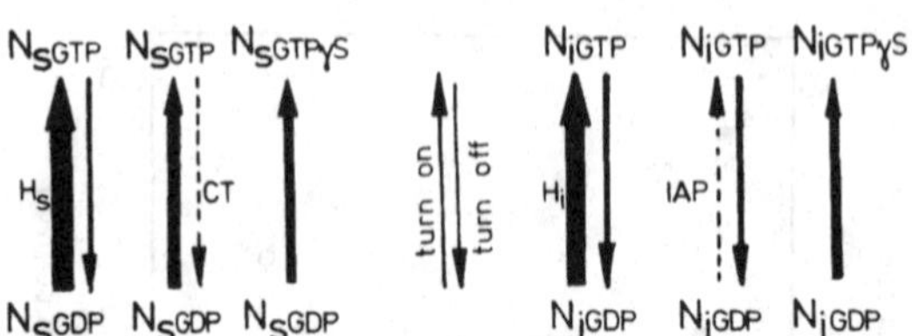


Fig. 5. Hypothetical model of activation and inactivation of the guanine nucleotide-binding regulatory components of the adenylate cyclase system, $N_s$ and $N_i$. Functions of the guanine nucleotides, GDP, GTP, and GTPγS, stimulatory ($H_s$) and inhibitory ($H_i$) hormones and the bacterial toxins, cholera toxin (CT) and pertussis toxin (IAP).

and $P_i$ due to the $N_s$-associated GTPase. Therefore, inhibition of the $N_s$-associated GTPase by cholera toxin or binding of stable GTP analogs, e.g., GTPγS, to $N_s$ causes a persistently activated $N_s$ and a subsequent persistently activated adenylate cyclase (Fig. 5).

The inhibitory coupling component, $N_i$, appears also to be active in its GTP-bound state and inactive after hydrolysis of the bound GTP to GDP and $P_i$. This hydrolysis is due to an $N_i$-associated GTPase, which is not inhibited by cholera toxin. The pertussis toxin, IAP, largely retards the turn-on reaction of $N_i$ and, thereby, this toxin impairs $N_i$-mediated inhibition of adenylate cyclase by GTP and hormones and the hormone-stimulated GTP hydrolysis. In analogy to $N_s$, $N_i$ can also be activated by stable GTP analogs, e.g., GTPγS, which results in a persistently activated $N_i$ and a subsequent persistently inhibited adenylate cyclase.

This model of activation and inactivation of the guanine nucleotide-binding regulatory components implies that the GTPase reaction is not causally related to hormone-induced inhibition of the adenylate cyclase. In contrast, hydrolysis of N-bound GTP turns off either activated $N_s$ or $N_i$ and, thereby, either $N_s$-mediated stimulation or $N_i$-mediated inhibition of the adenylate cyclase. Most striking evidence for this hypothesis that GTP hydrolysis is not causal for hormone-induced inhibition of adenylate cyclase comes from the findings that stable GTP analogs inhibit adenylate cyclase via $N_i$ and that this reaction is potentiated by inhibitory hormonal factors just as $N_s$-mediated adenylate cyclase stimulation by these analogs is increased by stimulatory hormonal factors.

## ROLE OF ADENYLATE CYCLASE IN REGULATION OF HORMONE-SENSITIVE GTPases

For stimulation of GTP hydrolysis by hormones, an active catalytic subunit of the adenylate cyclase system is apparently not required. For instance, the $N_s$-associated GTPase can be stimulated by hormones even at high concentrations (10 mM) of NEM, which almost completely abolish the adenylate cyclase activity (3,7,13). Furthermore, calcium ions are known to inactivate the adenylate cyclase. However, these ions (up to 20 mM) have no effect on GTPase stimulation by either stimulatory or inhibitory hormonal factors, as observed in human platelet and hamster adipocyte membranes. An effective coupling between the regulatory components and the adenylate cyclase is apparently also not required for GTPase stimulation. $Mn^{2+}$ ($\geq$ 1 mM) and $Mg^{2+}$ (> 10 mM) impair or prevent $N_i$-mediated adenylate cyclase inhibition (2). However, at these concentrations of the divalent cations, the $N_i$-associated GTPase stimulation by adenylate cyclase inhibitory hormones is not affected (unpublished observations). For stimulation of GTP hydrolysis by hormones, apparently only the hormone receptor and the N protein are required. As recently reported (50), β-adrenergic agonists stimulate GTP hydrolysis when purified $N_s$ and partially purified β-adrenoceptors are reconstituted in artificial membranes.

## CONCLUSIONS

The data available indicate that two distinct GTPase cycles are involved in the hormonal control of adenylate cyclase via the regulatory components, $N_s$ and $N_i$, and that in both cycles GTP hydrolysis is responsible for the inactivation of the active state of either $N_s$ or $N_i$. From the very limited number of studies on the regulation of hormone-sensitive GTP hydrolysis, it appears that both GTPases are largely independent of the catalytic subunit of the adenylate cyclase. Further studies with the purified regulatory components, $N_s$ and $N_i$, and purified stimulatory and inhibitory hormone receptors are required to clarify the precise regulation of the hormone-sensitive GTPases.

## ACKNOWLEDGEMENTS

The authors' studies reported herein were supported by the Deutsche Forschungsgemeinschaft.

## REFERENCES

1. E.M. Ross and A.G Gilman, Biochemical properties of hormone-sensitive adenylate cyclase, Ann. Rev. Biochem. 49:533 (1980).

2. K.H. Jakobs, K. Aktories, and G. Schultz, Mechanisms and components involved in adenylate cyclase inhibition by hormones, Adv. Cycl. Nucl. Res. in press.
3. D. Cassel and Z. Selinger, Catecholamine-stimulated GTPase activity in turkey erythrocyte membranes, Biochim. Biophys. Acta 452:538 (1976).
4. L.J. Pike and R.J. Lefkowitz, Activation and desensitization of β-adrenergic receptor-coupled GTPase and adenylate cyclase of frog and turkey erythrocyte membranes, J. Biol. Chem. 255:6860 (1980).
5. M. Lambert, M. Svoboda, and J. Christophe, Hormone-stimulated GTPase activity in rat pancreatic plasma membranes, FEBS Lett. 99:303 (1979).
6. N. Kimura and N. Shimada, Glucagon-stimulated GTP hydrolysis in rat liver plasma membranes, FEBS Lett. 117:172 (1980).
7. H.A. Lester, M.L. Steer, and A. Levitzki, Prostaglandin-stimulated GTP hydrolysis associated with activation of adenylate cyclase in human platelet membranes, Proc. Natl. Acad. Sci. USA 79:719 (1982).
8. D. Cassel, H. Levkovitz, and Z. Selinger, The regulatory GTPase cycle of turkey erythrocyte adenylate cyclase, J. Cycl. Nucl. Res. 3:393 (1977).
9. D. Cassel and Z. Selinger, Mechanism of adenylate cyclase activation through the β-adrenergic receptor: Catecholamine-induced displacement of bound GDP by GTP, Proc. Natl. Acad. Sci. USA 75:4155 (1978).
10. D. Cassel and Z. Selinger, Activation of turkey erythrocyte adenylate cyclase and blocking of the catecholamine-stimulated GTPase by guanosine 5'-(γ-thio)triphosphate, Biochem. Biophys. Res. Commun. 77:868 (1977).
11. D. Cassel and Z. Selinger, Mechanism of adenylate cyclase activation by cholera toxin: Inhibition of GTP hydrolysis at the regulatory site, Proc. Natl. Acad. Sci. USA 74:3307 (1977).
12. M. Svoboda, M. Lambert, and J. Schristophe, Distinct effects of the C-terminal octapeptide of cholecystokinin and of a cholera toxin pretreatment on the kinetics of rat pancreatic adenylate cyclase activity, Biochim. Biophys. Acta 675:46 (1981).
13. K. Aktories, G. Schultz, and K.H. Jakobs, Cholera toxin inhibits prostaglandin $E_1$ but not adrenaline-induced stimulation of GTP hydrolysis in human platelet membranes, FEBS Lett. 146:65 (1982).
14. D. Cassel and T. Pfeuffer, Mechanism of cholera toxin action: Covalent modification of the guanyl nucleotide-binding protein of the adenylate cyclase system, Proc. Natl. Acad. Sci. USA 75:2669 (1978).
15. T. Pfeuffer, Guanine nucleotide-controlled interactions between components of adenylate cyclase, FEBS Lett. 101:85 (1979).
16. K. Aktories, G. Schultz, and K.H. Jakobs, Stimulation of a low $K_m$ GTPase by inhibitors of adipocyte adenylate cyclase, Mol. Pharmacol. 21:336 (1982).

17. G. Koski and W.A. Klee, Opiates inhibit adenylate cyclase by stimulating GTP hydrolysis, Proc. Natl. Acad. Sci. USA 78:4185 (1981).
18. G. Koski, R.A. Streaty, and W.A. Klee, Modulation of sodium-sensitive GTPase by partial opiate agonists, J. Biol. Chem. 257:14035 (1982).
19. K. Aktories and K.H. Jakobs, Epinephrine inhibits adenylate cyclase and stimulates a GTPase in human platelet membranes via α-adrenoceptors, FEBS Lett. 130:235 (1981).
20. K. Aktories, G. Schultz, and K.H. Jakobs, Somatostatin-induced stimulation of a high affinity GTPase in membranes of S49 lymphoma cyc⁻ and H21a variants, Mol. Pharmacol. in press.
21. P. Lasch and K.H. Jakobs, Agonistic and antagonistic properties of various α-adrenergic agonists in human platelets, Naunyn-Schmiedeberg's Arch. Pharmacol. 306:119 (1979).
22. R. Grandt, K. Aktories, and K.H. Jakobs, Guanine nucleotides and monovalent cations increase agonist affinity of prostaglandin $E_2$ receptors in hamster adipocytes, Mol. Pharmacol. 22:320 (1982).
23. K. Aktories, G. Schultz, and K.H. Jakobs, Adenosine receptor-mediated stimulation of GTP hydrolysis in adipocyte membranes, Life Sci. 30:269 (1982).
24. K. Aktories, G. Schultz, and K.H. Jakobs, Inhibition of hamster fat cell adenylate cyclase by prostaglandin $E_1$ and epinephrine: Requirement for GTP and sodium ions, FEBS Lett. 107:100 (1979).
25. A.J. Blume, D. Lichtshtein, and G. Boone, Coupling of opiate receptors to adenylate cyclase: Requirement for $Na^+$ and GTP, Proc. Natl. Acad. Sci. USA 76:5626 (1979).
26. K.H. Jakobs and G. Schultz, Different inhibitory effect of adrenaline on platelet adenylate cyclase in the presence of GTP plus cholera toxin and of stable GTP analogues, Naunyn-Schmiedeberg's Arch. Pharmacol. 310:121 (1979).
27. F. Probst and B. Hamprecht, Opioids, noradrenaline and GTP analogs inhibit cholera toxin-activated adenylate cyclase in neuroblastoma x glioma hybrid cells, J. Neurochem. 36:580 (1981).
28. T.E. Cote, C.W. Grewe, J.C. Tsuruta, R.L. Eskay, and J.W. Kebabian, D-2 dopamine receptor-mediated inhibition of adenylate cyclase activity in the intermediate lobe of the rat pituitary gland requires guanosine 5'-triphosphate, Endocrinology 110:812 (1982).
29. K. Aktories, G. Schultz, and K.H. Jakobs, Cholera toxin does not impair hormonal inhibition of adenylate cyclase and concomitant stimulation of a GTPase in adipocyte membranes, Biochim. Biophys. Acta 719:58 (1982).
30. K.H. Jakobs, Determination of the turn-off reaction for the epinephrine-inhibited human platelet adenylate cyclase, Eur. J. Biochem. 132:125 (1983).
31. G.L. Stiles and R.J. Lefkowitz, Hormone-sensitive adenylate cyclase: Delineation of a trypsin-sensitive site in the path-

way of receptor-mediated inhibition, J. Biol. Chem.257:6287 (1982).

32. K.H. Jakobs, R.A. Johnson, and G. Schultz, Activation of human platelet adenylate cyclase by a bovine sperm component, Biochim. Biophys. Acta 756:369 (1983).
33. K.H. Jakobs, P. Lasch, M. Minuth, K. Aktories, and G. Schultz, Uncoupling of α-adrenoceptor-mediated inhibition of human platelet adenylate cyclase by N-ethylmaleimide, J. Biol. Chem. 257:2829 (1982).
34. K. Aktories, G. Schultz, and K.H. Jakobs, Inactivation of the guanine nucleotide regulatory site mediating inhibition of the adenylate cyclase in hamster adipocytes, Naunyn-Schmiedeberg's Arch. Pharmacol. 321:247 (1982).
35. G.L. Johnson, H.R. Kaslow, Z. Farfel, and H.R. Bourne, Genetic analysis of hormone-sensitive adenylate cyclase, Adv. Cycl. Nucl. Res. 13:1 (1980).
36. J.D. Hildebrandt, J. Hanoune, and L. Birnbaumer, Guanine nucleotide inhibition of cyc⁻ S49 mouse lymphoma cell membrane adenylyl cyclase, J. Biol. Chem. 257:14723 (1982).
37. K.H. Jakobs, U. Gehring, B. Gaugler, T. Pfeuffer, and G. Schultz, Occurrence of an inhibitory guanine nucleotide-binding regulatory component of the adenylate cyclase system in cyc⁻ variants of S49 lymphoma cells, Eur. J. Biochem. 130:605 (1983).
38. K.H. Jakobs and G. Schultz, Occurrence of a hormone-sensitive inhibitory coupling component of the adenylate cyclase in S49 lymphoma cyc⁻ variants, Proc. Natl. Acad. Sci. USA 80:3899 (1983).
39. K.H. Jakobs and K. Aktories, Synergistic inhibition of human platelet adenylate cyclase by stable GTP analogs and epinephrine, Biochim. Biophys. Acta in press.
40. T. Katada and M. Ui, Islet-activating protein: A modifier of receptor-mediated regulation of rat islet adenylate cyclase, J. Biol. Chem. 256:8310 (1981).
41. H. Kurose, T. Katada, T. Amano, and M. Ui, Specific uncoupling by islet-activating protein, pertussis toxin, of negative signal transduction via α-adrenergic, cholinergic, and opiate receptors in neuroblastoma x glioma hybrid cells, J. Biol. Chem. 258:4870 (1983).
42. T. Katada and M. Ui, ADP ribosylation of the specific membrane protein of C6 cells by islet-activating protein associated with modification of adenylate cyclase activity, J. Biol. Chem. 257:7210 (1982).
43. G.M. Bokoch, T. Katada, J.K. Northup, E.L. Hewlett, and A.G. Gilman, Identification of the predominant substrate for ADP-ribosylation by islet activating protein, J. Biol. Chem 258: 2072 (1983).
44. K. Aktories, G. Schultz, and K.H. Jakobs, Islet-activating protein impairs $\alpha_2$-adrenoceptor-mediated inhibitory regulation of human platelet adenylate cyclase, Naunyn-Schmiedeberg's Arch. Pharmacol. in press.

45. K. Aktories, G. Schultz, and K.H. Jakobs, Islet-activating protein prevents nicotinic acid-induced GTPase stimulation and GTP but not GTPγS-induced adenylate cyclase inhibition in rat adipocytes, FEBS Lett. 156:88 (1983).
46. D.L. Burns, E.L. Hewlett, J. Moss, and M. Vaughan, Pertussis toxin inhibits enkephalin stimulation of GTPase of NG108-15 cells, J. Biol. Chem. 258:1435 (1983).
47. K. Aktories, G. Schultz, and K.H. Jakobs, Adenylate cyclase inhibition and GTPase stimulation by somatostatin in S49 lymphoma $cyc^-$ variants are prevented by islet-activating protein, FEBS Lett. in press.
48. J.D. Hildebrandt, R.D. Sekura, J. Codina, R. Iyengar, C.R. Manclark, and L. Birnbaumer, Stimulation and inhibition of adenylyl cyclases mediated by distinct regulatory proteins, Nature 302:706 (1983).
49. K.H. Jakobs, K. Aktories, and G. Schultz, Mechanism of pertussis toxin action on the adenylate cyclase: Inhibition of the turn-on reaction of the inhibitory regulatory site, submitted for publication.
50. E.M. Ross, T. Asano, D.R. Brandt, and S.E. Pedersen, Interaction of β-adrenergic receptors and the stimulatory GTP-binding protein of adenylate cyclase in reconstituted vesicles, 15th FEBS Meeting, Brussels, Abstr. vol. p. 47 (1983).

# RECONSTITUTION OF THE REGULATORY FUNCTIONS OF β-ADRENERGIC RECEPTORS

Elliott M. Ross, Tomiko Asano, Steen E. Pedersen, and Douglas R. Brandt

Department of Pharmacology
University of Texas Health Science Center at Dallas
Dallas, Texas

The hormone-sensitive adenylate cyclase system acts as the intracellular effector for numerous neurotransmitters and hormones whose receptors are located on the cell surface. Multiple inhibitory and stimulatory receptors with specificities for different ligands can act simultaneously on a single target cell to modulate the activity of adenylate cyclase on the inner face of the plasma membrane. In addition to acute control, several different mechanisms exist for the long-term regulation of the enzyme, of individual receptors, and of the coupling process. Inhibitory control of adenylate cyclase is discussed by Drs. Ui and Aktories in this volume, and Harden (1) has thoroughly reviewed refractoriness and other modes of chronic regulation of adenylate cyclase recently. Here, we will discuss the biochemical events that couple transmitter binding to its receptor with the stimulation of adenylate cyclase.

Two findings have helped to establish our understanding of the regulation of adenylate cyclase. One is that GTP or a similar purine nucleotide is required for hormonal regulation of the enzyme. The other is that hormone-sensitive adenylate cyclase is a system composed of up to six plasma membrane-bound proteins. In 1971, Rodbell and colleagues (2,3) showed that GTP is required for the activation of hepatic adenylate cyclase by glucagon and that GTP regulates the binding of glucagon to its receptor. They also showed that non-hydrolyzable analogs of GTP, such as GTPγS and Gpp(NH)p, cause the persistent stimulation of the enzyme (4). In the case of these analogs, hormones generally increase the rate at which activation occurs but not the extent of activation. Based on these results and on related studies (5-7), it was suggested that the hydrolysis of bound GTP might be important in the regulation of adenylate cyclase. In 1976, Cassel and Selinger (8) first demonstrated β-adrenergic catecholamine-stimulated GTPase activity in turkey erythrocyte membranes, and showed that this activity was related to the regulation of adenylate cyclase. They suggested that the binding of GTP activates the cyclase and that hydrolysis of GTP to GDP by the GTPase causes deactivation. They also proposed that the release of the bound GDP product limits the rate of reactivation (8-10). While GDP release may not be rate-limiting for the activation of all adenylate cyclases, the idea of a hormone-regulated GTPase cycle has set the framework for recent analyses of the activation and deactivation of the enzyme (ref. 11, for review).

The above studies and most other work done prior to 1977 treated adenylate cyclase as if it were a single protein, despite previous hints to the contrary. By 1977 it had been demonstrated that β-adrenergic receptors and adenylate cyclase are distinct and separable proteins (12,13). There had also been speculation that adenylate cyclase might itself be composed of several components. In 1977, Pfeuffer (14) and Ross and Gilman (15) demonstrated that Gpp(NH)p-sensitive adenylate cyclase could be resolved into two separate components, the catalytic unit itself and a GTP-binding protein that mediates regulation by guanine nucleotides and hormone receptors. Pfeuffer (14) used affinity chromatography on GTP-agarose to separate the two proteins. Ross and Gilman (15) showed that the cyc$^-$ S49 lymphoma cell, that had previously been identified as phenotypically deficient in adenylate cyclase activity (16), actually retained the catalytic unit of the cyclase but was deficient in the GTP-binding regulatory protein. Reconstitution of cyc$^-$ membranes with cyclase-free fractions from wild-type plasma membranes could restore hormone-sensitive and Gpp(NH)p-sensitive activities. Further studies showed that the regulatory protein, referred to as $G_s$, mediates the stimulation caused by guanine nucleotides and by fluoride, another activator of the cyclase, and acts as a GTP-dependent coupling factor between stimulatory receptors and the catalytic unit (15,17). These investigators developed an assay for $G_s$ based on its ability to reconstitute the sensitivity of the catalytic unit in cyc$^-$ to guanine nucleotides or fluoride. The ability to assay $G_s$ allowed its purification, accomplished by Gilman and associates (18,19). Based on these studies (see also refs. 11,20), it became clear that $G_s$ is the proximal physiological stimulator of the adenylate cyclase catalytic unit, and that $G_s$ must itself be "activated" by binding GTP before it is competent to activate the catalyst. The role of stimulatory hormone receptors is to increase the rate at which GTP binds to $G_s$.

Even if one ignores calmodulin and the inhibitory effects of guanine nucleotides and hormones, the stimulatory pathway of the adenylate cyclase system is still quite complex, involving three proteins, two regulatory ligands (hormone and nucleotide), as well as substrate and regulatory divalent cations. Furthermore, the concentrations of the proteins in membranes are generally not known. To gain a detailed biochemical understanding of the regulation of adenylate cyclase, we have tried to simplify the system as much as possible. First, since the stimulation of adenylate cyclase can be thought of as the sequential interaction of receptor with $G_s$ followed by that of $G_s$ with C, we have dealt with each two-protein system separately. Our work on the $G_s$-C couple is summarized elsewhere (21,22), and we will discuss here only the interaction of $G_s$ with β-adrenergic receptors. Second, we have tried to study this interaction not in native membranes, but rather in reconstituted model systems where purified or partially purified preparations of both proteins are reinserted into large, unilamellar vesicles composed of pure lipids. This approach is necessary because receptor and $G_s$ can interact only in relatively unperturbed bilayers. It is also attractive because it allows us to control the lipid environment and to study its effect on $G_s$, receptor, and the coupling process.

## INCORPORATION OF β-ADRENERGIC RECEPTORS INTO PHOSPHOLIPID VESICLES

Our initial goal was to reincorporate crude, detergent-solubilized β-adrenergic receptors into phospholipid vesicles (23). Receptors were solubilized from rat erythrocyte membranes using digitonin and the extract was mixed with excess dimyristoylphosphatidylcholine (DMPC). This pure, synthetic lipid is exceptionally well characterized in most of its physical properties. Digitonin was removed from the mixture to allow the formation of vesicles, first by gel filtration and then by sucrose density gradient centrifugation. Receptors sedimented in a discrete band with some lipid and protein, but well-separated from residual digitonin. This preparation was characterized by freeze-fracture and negative stain electron microscopy and shown to consist of unilamellar vesicles of about 50-100 nm diameter. Ligand binding activity in the vesicles that was characteristic of β-adrenergic receptors could be trapped on glass fiber filters, suggesting that the receptors had reassociated with the vesicle bilayer. The thermal stability of the reconstituted receptors was similar to that of native, membrane-bound receptors, substantially enhanced relative to receptors in detergent solution. More importantly, the reconstituted receptors regained the ability to bind the specific β-adrenergic antagonist [$^{125}$I]iodohydroxybenzylpindolol (IHYP) (Table 1). While soluble receptors can bind [$^{3}$H]dihydroalprenolol (DHA) (24), IHYP binding activity is completely lost upon solubilization (12,23). Since up to 80% of the binding sites for DHA in the reconstituted preparation could also be detected using IHYP, it seemed that our initial reconstitution was effective.

Table 1. Reconstitution of IHYP binding activity in β-adrenergic receptor vesicles.

| | Assayable Receptors (fmol/mg) | |
|---|---|---|
| Preparation | DHA | IHYP |
| Plasma membranes | 338 | 350 (103%) |
| Digitonin extract | 818 | 36 (4%) |
| Receptor vesicles | 1200 | 770 (64%) |

Rat erythrocyte plasma membranes were solubilized and receptors were reconstituted according to Fleming and Ross (23). Binding activity was measured at saturation using both ligands. IHYP binding is also expressed as a percentage of DHA binding.

This protocol was also successful in reconstituting crude, deoxycholate-solubilized receptors from turkey erythrocytes (25). However, as we began to use increasingly purified receptors that had been depleted of membrane lipids, we found that DMPC was no longer effective as the sole added lipid for the reconstitution. When partially purified receptors were used, reconstitution required a mixture of erythrocyte phospholipids in addition to DMPC (26). The basis of this requirement is unclear, but the use of erythrocyte lipids increases the recovery of receptors during gel filtration. This finding is similar to those of Cerione et al. (27) or Kirilovsky and Schramm (28). More recently, we have used the gel filtration procedure to reconstitute receptors that were purified by affinity chromatography (29) to about 40% of theoretical homogeneity (8 nmol/mg protein; about 15,000 fold purified). The reconstitution of purified receptors in our hands requires both phosphatidylserine and phosphatidylethanolamine, but not phosphatidylcholine. The addition of cholesterol (or cholesteryl-hemisuccinate) stabilizes the reconstituted receptors, but is is probably not important for receptor binding activity (S.E. Pedersen and E.M Ross, unpublished data; contrast with Kirilovsky and Schramm, 28).

## RECEPTOR-$G_s$ COUPLING IN PHOSPHOLIPID VESICLES

The coupling of vesicle-bound receptors to $G_s$ is the most sensitive and significant criterion of faithful reconstitution, and its mechanism is the focus of our interest. In our initial procedure for recoupling receptors and $G_s$, we reincorporated turkey erythrocyte β-adrenergic receptors into unilamellar DMPC vesicles, again using gel filtration to remove detergent. We then added pure rabbit hepatic $G_s$ to the preformed receptor vesicles. $G_s$ bound to the vesicles, and several control experiments indicated that bound $G_s$ did not exchange measurably among vesicles. Using these procedures, both receptors and $G_s$ could be reconstituted with a high yield, and the receptor-$G_s$ vesicles displayed the appropriate behavior of a well-coupled β-adrenergic system (25). The interaction of receptor and $G_s$ in the vesicles was probed by measuring the ability of agonist to increase the rate at which $G_s$ could be activated by GTPγS. Activated $G_s$ was measured according to its ability to stimulate the catalytic unit of adenylate cyclase in $cyc^-$ membranes (17). This procedure is conceptually similar to the semiquantitative assay used by Citri and Schramm (30,31). However, we used purified hepatic $G_s$, which has a known reconstitutive specific activity when assayed under appropriate conditions (18,19). Using this assay, therefore, we could quantitate the number of $G_s$ molecules that became activated in each experiment as well as the total concentration of $G_s$ that was present in the vesicles. The concentration of receptors and their fractional saturation by agonists were measured directly using IHYP binding. In these experiments, the initial rate of activation of $G_s$ by GTPγS in the receptor-$G_s$ vesicles was increased up to four-fold by the addition of β-adrenergic agonists. Furthermore, the $G_s$:receptor ratio indicated that a single agonist-liganded receptor could promote the rapid activation of 6-10 molecules of $G_s$. This "catalytic" action of receptors, their ability to interact sequentially with multiple $G_s$ molecules, indicates restoration of the highly efficient coupling that is characteristic of native plasma membranes. This process forms the molecular basis of the pharmacologic phenomenon of "spare receptors". Thus, the dose-response curve for an agonist yielded an EC50 for $G_s$ activation that was significantly lower than the equilibrium $K_d$ for the same drug. The major limitation of our first receptor-$G_s$ vesicle preparation was that while all vesicles contained 5-10 molecules of $G_s$, relatively few of the vesicles contained a receptor molecule. This was a consequence of our using a crude preparation of receptors. Thus, while reconstituted receptors functioned with high efficiency, only a small fraction of the $G_s$ molecules -- those that happened to be on receptor-containing vesicles -- could be stimulated by the receptor-agonist complex.

The problem of low receptor concentrations has been alleviated by our development of an improved receptor-$G_s$ vesicle preparation composed of pure $G_s$ plus β-adrenergic receptors that have been purified at least 1000-fold and, recently, up to 15,000-fold. Receptors were purified by affinity chromatography (29), mixed with pure $G_s$ and phospholipid, and reconstituted using gel filtration to remove detergent (26). The purified receptors can be reconstituted to a concentration greater than one per vesicle, so that essentially all the $G_s$ can be activated via the hormone-stimulated pathway. The efficiency of coupling in these vesicles remains high, and the purity of the preparation allows experiments that were impossible using the earlier preparations. First, since nonspecific binding of guanine nucleotides is minimal, high affinity GTPγS binding is measurable directly as a molecular correlate of $G_s$ activation (Fig. 1). The receptor-$G_s$ vesicles display efficient receptor-catalyzed binding of [$^{35}$S]GTPγS and activation of $G_s$ by GTPγS, and the two processes are parallel as predicted (18).[s] The hormone-stimulated rates of both processes are increased about five-fold over basal activation at short times under the conditions shown here. Stimulation by isoproterenol or other agonists and the blockade of stimulation by antagonists displayed appropriate β-adrenergic specificity and were specific for the (-)isomers.

We have now used the receptor-$G_s$ vesicle system to probe the coupling process by investigating the affinity, the initial rate and the extent of reconstituted, hormone-stimulated GTPγS binding. The binding experiments were designed such that the nucleotide concentration did not change significantly over the course of the hormone-stimulated association with $G_s$. Thus, data could be analyzed as a

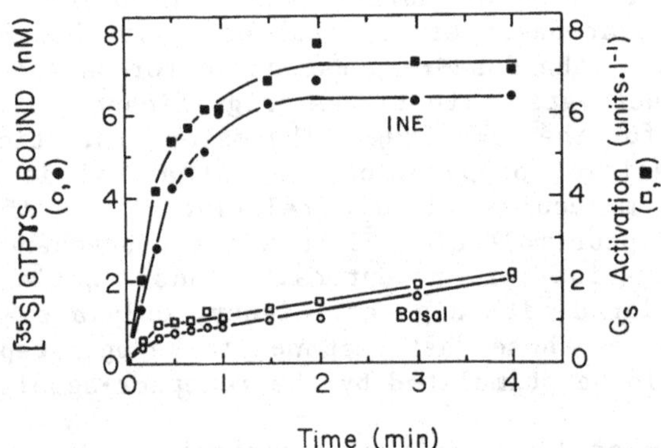


Fig. 1. [$^{35}$S]GTPγS binding and activation of $G_s$ in receptor-$G_s$ vesicles. Vesicles were incubated at 30° C in a binding assay mixture containing 1 mM free $Mg^{2+}$, 0.1 μM [$^{35}$S]GTPγS, and either 1 μM (-)isoproterenol (original data not shown) or 100 nM (-)propranolol (O,□). At the times indicated, aliquots were assayed both for [$^{35}$S]GTPγS bound (O,●) (18) and for $G_s$ activation (□,■) (25). The incremental effects due to isoproterenol on binding (●) and activation (■) are shown. Taken from a manuscript submitted by Asano and Ross.

potentially pseudo-first order reaction. Such an analysis of the rate of isoproterenol-stimulated nucleotide binding is shown in Fig. 2. These data reveal an apparent first-order rate constant (k') for agonist-stimulated GTPγS binding of 2.9 $min^{-1}$ under the conditions shown (average k' = 2.4 $min^{-1}$ in 10 experiments; range = 1.7-3.4 $min^{-1}$).

This rate of agonist-stimulated GTPγS binding in the reconstituted vesicles is about 100-fold greater than the rate of $Mg^{2+}$-stimulated binding to Lubrol-solubilized $G_s$ (18). Those authors found that binding to soluble $G_s$ was apparently first order over a wide range of initial GTPγS concentrations, and that the first-order rate constant changed less than two-fold between $10^{-8}$M

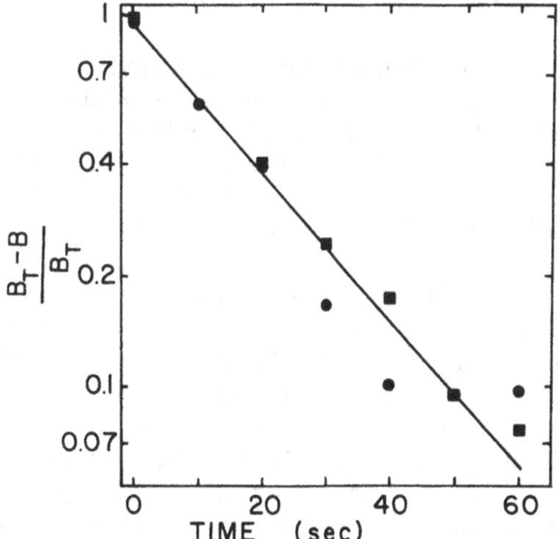


Fig. 2. First-order replot of the rate of hormone-stimulated binding of [$^{35}$S]GTPγS to receptor-$G_s$ vesicles. Data are from two experiments using two different vesicle preparations (●,■). Vesicles were incubated at 30° as described in the legend to Fig. 1. [$^{35}$S]GTPγS was allowed to associate with the vesicles for the times shown, as well as for 90, 120 and 180 s. Bound [$^{35}$S]GTPγS at 90, 120 and 180 s were averaged for each experiment to yield total binding, $B_T$, and used to normalize binding at shorter times. The data shown are the incremental amounts of binding due to isoproterenol. Taken from Brandt et al.(26).

and $10^{-4}$M GTPγS. To explain this finding, they proposed that the rate of high-affinity binding of nucleotides is limited by the initial dissociation of its two subunits. $G_s$ is an 80,000 dalton heterodimer of a 45,000 (or 52,000) dalton $\alpha^s$ subunit and a 35,000 dalton β subunit (32). Northup and coworkers (18,33) showed that guanine nucleotides (and fluoride) bind to the α subunit exclusively, and that isolated, liganded α subunit is able to activate the catalytic unit of the cyclase in the absence of β. GTPγS-activated $G_s$ behaves hydrodynamically as an isolated α subunit, and conditions that favor the dissociation of the αβ dimer (dilution, addition of $Cl^-$, high concentrations of $Mg^{2+}$) also increase the rate of nucleotide binding. Their model for the activation of $G_s$ is depicted in Eqn.1. Activation of $G_s$ may occur by the slow dissociation of its subunits followed by the tight binding of guanine nucleotide (N) to the α subunit (reactions 2 and 3). Alternatively, low affinity binding of nucleotide to $G_s$ may facilitate the dissociation of the β subunit, again leading to α·N (reactions 1 and 4). If the latter pathway is significant at high GTPγS concentrations, then the forward rates of reactions 2 and 4 must be similar, because the first-order rate of binding varies slightly if at all with increasing concentration of $G_s$ (18).

The kinetics of agonist-stimulated GTPγS binding to $G_s$ in the vesicles differs markedly from the data obtained using soluble $G_s$. First, the apparent affinity of GTPγS binding to vesicle-bound $G_s$ is 10-50 fold greater than that observed with Lubrol-solubilized $G_s$. The apparent affinity in Lubrol solution is actually a kinetic phenomenon that reflects the stability of the free α subunit of $G_s$ (33). $G_s$ is much more stable to denaturation in vesicles than it is in solution and the apparent affinity for nucleotide is consequently increased. Second, the first-order rate constant for agonist-stimulated GTPγS binding is strongly dependent on the concentration of GTPγS (Fig. 3). Because the response to GTPγS is clearly nonlinear, these data are inconsistent with a simple second-order binding reaction between $G_s$ and GTPγS. We interpret these findings in terms of the scheme of Smigel et al. (33) (Eqn. 1) by assuming that reaction 1 is in rapid equilibrium, that reaction 3 is fast and essentially irreversible (18), and that the receptor-hormone complex markedly increases the forward rate of reaction 4 but not that of reaction 2. GTPγS increases the rate of the binding reaction by

$$\begin{array}{ccc} \alpha\beta & \overset{K_2}{\rightleftharpoons} & \alpha \\ K_1 \updownarrow & & \updownarrow K_3 \\ \alpha\beta\cdot N & \underset{K_4}{\rightleftharpoons} & \alpha\cdot N \end{array} \qquad \text{(Eqn. 1)}$$

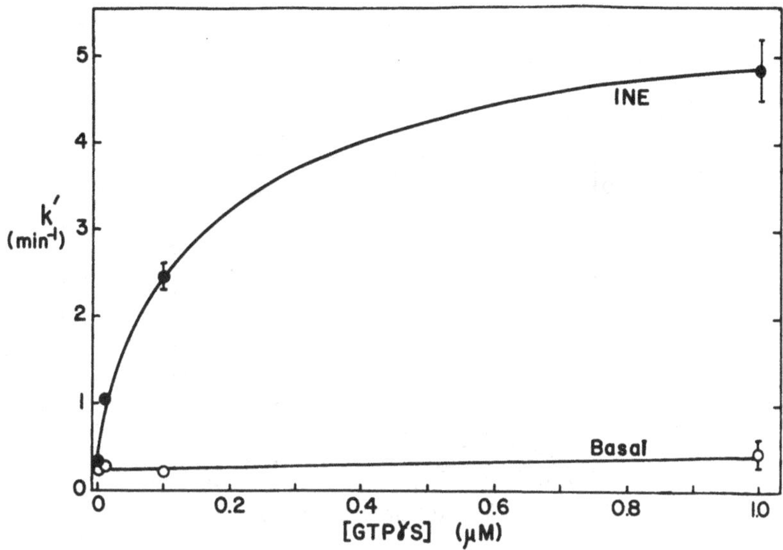


Fig. 3. Influence of GTPγS concentration on the apparent first-order rate constant for the binding of GTPγS to receptor-$G_s$ vesicles. Binding was measured as described in the legend to Fig. 1 except that the GTPγS concentration was varied as indicated. The apparent first-order association constant (k') was determined as outlined in Fig. 2. Incubations were performed in the presence of 1 μM (-)isoproterenol (●) or 0.1 μM (-)propranolol (O). The standard deviation of the mean is indicated by brackets where significant.

shifting the equilibrium of reaction 1 to the hormone-stimulated pathway. The data shown in Figure 4 further support this concept. At low nucleotide concentrations, where flux should be primarily through reactions 2 and 3, very little effect of hormone is seen. At high nucleotide concentrations, where the formation of αβ·N and flux through reactions 1 and 4 should be favored, at least a 10-fold effect of hormone is noted.

Determining whether subunit dissociation really occurs in vesicles as it does in detergent solution will require a detailed analysis of the dependence of these binding rates on the concentration of each subunit of $G_s$ in the vesicles. The observed first-order binding reaction may represent a conformational isomerization of $G_s$ rather than subunit dissociation. Nevertheless,

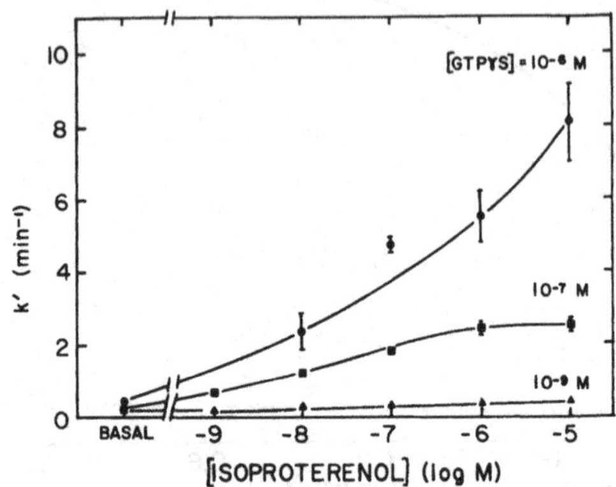


Fig. 4. Influence of isoproterenol concentration on the apparent first-order rate constant for GTPγS binding to receptor-$G_s$ vesicles. Binding was measured as described in the legend to Fig. 1 except that the (-)isoproterenol concentrations were varied as indicated, and the GTPγS concentration was 1 μM, (●), 100 nM (■), or 1 nM (▲). The rate constant (k') for GTPγS binding was determined as outlined in Fig. 2. The basal value was determined in the presence of 0.1 μM (-) propranolol. The standard deviation of the mean is indicated by brackets where significant. From a submitted manuscript by Asano and Ross.

our findings so far begin to indicate the pathway that represents hormone-stimulated activation of $G_s$, and suggest that the reconstituted receptor-$G_s$ vesicles provide a system in which this pathway may be studied in greater detail.

## HORMONE-STIMULATED GTPase IN RECEPTOR-$G_s$ VESICLES

Since the discovery of the catecholamine-stimulated GTPase by Cassel and Selinger (8) and the resolution of $G_s$ in 1977, it was widely assumed that the GTPase is localized on $G_s$. However, Northup et al. (18) found that pure, Lubrol-solubilized $G_s$ does not display significant GTPase activity, and speculated that either receptor or the catalytic unit of adenylate cyclase was required for its expression. Therefore, when we prepared receptor-$G_s$ vesicles that lacked both the adenylate cyclase catalytic unit and spurious nucleoside triphosphatase activity, it was again of interest to look for a hormone-stimulated GTPase. Such activity was in fact observed when receptor-$G_s$ vesicles were incubated with [$\gamma$-$^{32}$P]GTP and isoproterenol (Fig. 5; see also Brandt et al., 26). Isoproterenol stimulated GTP hydrolysis as much as 15-fold over activity measured in the absence of agonist, the "basal" rate. Appropriate controls indicated that it is $G_s$ that catalyzes the hydrolysis of GTP in the vesicles and that it is the interaction with receptor that stimulates the rate of the reaction. There is no evidence to suggest that the GTPase activity is stimulated by the catalytic unit of adenylate cyclase, as has been widely assumed. We speculate that

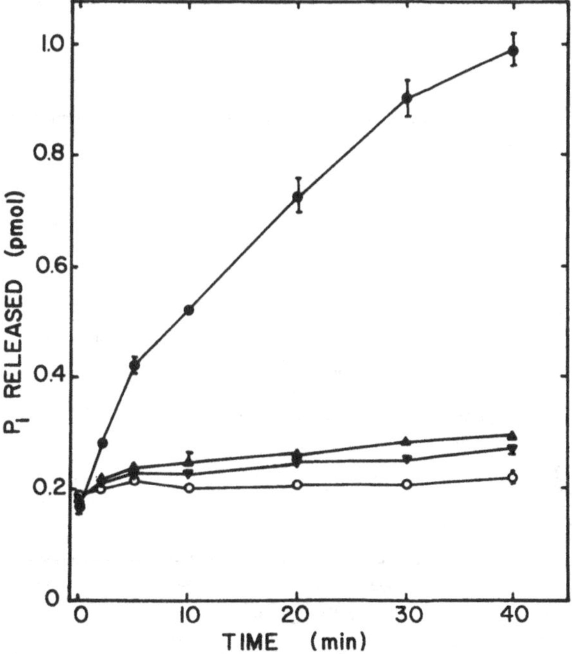


Fig. 5. GTPase activity in receptor-$G_s$ vesicles. Vesicles containing 74 fmol of $G_s$ were incubated at 30° for the times shown. The concentration of [$\gamma$-$^{32}$P]GTP was 0.1 μM. Assays contained 1 μM (-)isoproterenol (●); isoproterenol plus 10 μM (-)propranolol (▲), or 0.1 μM propranolol alone (▼). Hydrolysis in the absence of vesicles is also shown (O). The zero-time, zero-protein value represents [$^{32}$P]$P_i$ contaminating the [$\gamma$-$^{32}$P]GTP. The standard deviations of triplicate determinations are shown. From Brandt et al. (26).

C actually may inhibit steady-state turnover of the GTPase by stabilizing the $G_s$-GTP intermediate.

While there is a large, specifically β-adrenergic stimulation of GTPase activity in the receptor-$G_s$ vesicles, even the stimulated rate is quite low. By dividing the GTPase rate by the number of $G_s$ molecules in the vesicles (determined by GTPγS binding), we can calculate that the reconstituted GTPase has a maximum observed turnover number of 2 mol GTP hydrolyzed / min / mol $G_s$. While this would be very slow for a typical metabolic enzyme, it is not unreasonable in the case of $G_s$. First, the rate of the reconstituted GTPase is similar to that observed in native membranes (8-10,26) and to that displayed by transducin, a GTP-binding protein from the rod outer segment that is homologous to $G_s$ (34). Moreover, the turnover number is similar to the rate constant for the binding of GTPγS (Figs. 2-4). The rate of GTPγS binding should be a good estimate of the rate of binding of the GTP substrate since the rate-limiting reaction, 2 and 4, are nucleotide-independent (Eqn.1).

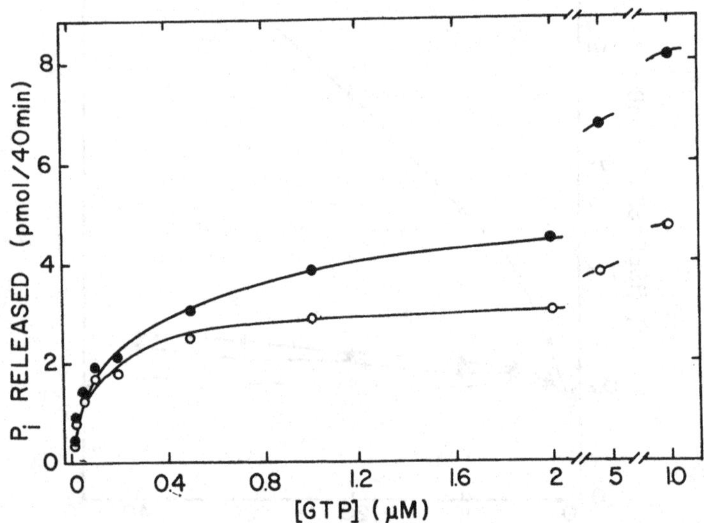


Fig. 6. Substrate dependence of GTPase activity in receptor-$G_s$ vesicles. Each assay contained 230 fmol of $G_s$. Assays were carried out in the presence of 50 μM (-)isoproterenol (●) or 0.1 μM (-)propranolol (not shown). The increase in activity caused by isoproterenol is also shown (O). From Brandt et al. (26).

The similarity of the rates of binding and hydrolysis suggests that agonist stimulates the GTPase by stimulating the binding of the GTP substrate to $G_s$. The substrate dependence of the reconstituted hormone-stimulated GTPase is in fact biphasic, and activity is not saturated below 10 μM GTP (26). If receptor acts by stimulating the binding of GTP, this pattern of substrate dependence may indicate that receptor promotes binding of GTP to the α subunit of $G_s$ via the αβ·GTP intermediate, as was observed in our studies of GTPγS binding (see Eqn. 1).

## CONCLUSIONS

It has been difficult to define the molecular mechanisms of hormonal regulation of adenylate cyclase primarily because of the intrinsic complexity of a system that includes three membrane-bound proteins, one with dissociable subunits, and at least two regulatory ligands. The use of receptor-$G_s$ vesicles limits the number of these interactions in a simplified model system where the coupling of the two proteins can be investigated in detail. Our current understanding of the system suggests that hormone-liganded receptor acts to increase the rate of GTP binding to $G_s$, thereby activating it and allowing it to further activate C. The lifetime of $G_s$·GTP determines the extent of steady-state activation of adenylate cyclase, and the hydrolysis of GTP to GDP terminates activation. While this lifetime may be prolonged by the association of $G_s$-GTP with C, control of the intrinsic rate of GTP hydrolysis on the surface of $G_s$ remains obscure. It is still uncertain if the receptor is directly responsible for the release of GDP from $G_s$, as suggested by Cassel and Selinger (9,10), or if rebinding of the β subunit is more directly responsible for removal of GDP. Regardless, the reconstituted receptor-$G_s$ vesicles are an excellent model system in which to pursue these studies. The concentrations of both receptor and $G_s$ can be controlled and quantitated, proteins from different tissues may be used, and the lipid:protein ratio and the lipid composition can be manipulated easily. Thus, kinetic and thermodynamic parameters can be determined for both ligand-protein and protein-protein interactions. Determination of these parameters will enable us to assess proposed mechanisms for receptor-$G_s$ coupling and to ascertain the dependence of coupling on the composition and state of the membrane bilayer.

REFERENCES

1. T. K. Harden, Agonist-induced desensitization of the β-adrenergic receptor-linked adenylate cyclase, Pharmacol. Rev. 35:5 (1983).
2. M. Rodbell, L. Birnbaumer, S. L. Pohl, and H. M. J. Krans, The glucagon-sensitive adenyl cyclase system in plasma membranes of rat liver. V. An obligatory role of guanyl nucleotides in glucagon action, J. Biol. Chem. 246:1877 (1971).
3. M. Rodbell, H. M. J. Krans, S. L. Pohl, and L. Birnbaumer, The glucagon-sensitive adenyl cyclase system in plasma membranes of rat liver. IV. Effects of guanyl nucleotides on binding of [$^{125}$I]glucagon, J. Biol. Chem. 246:1872 (1971).
4. M. Schramm and M. Rodbell, A persistent active state of the adenylate cyclase system produced by the combined actions of isoproterenol and guanylyl imidodiphosphate in frog erythrocyte membranes, J. Biol. Chem. 250:2232 (1975).
5. A. J. Blume and C. J. Foster, Neuroblastoma adenylate cyclase: role of 2-chloroadenosine, prostaglandin E, and guanine nucleotides in the regulation of activity, J. Biol. Chem. 251:3399 (1976).
6. E. M. Ross, M. E. Maguire, T. W. Sturgill, R. L. Biltonen, and A. G. Gilman, Relationship between the β-adrenergic receptor and adenylate cyclase. Studies of ligand binding and enzyme activity in purified membranes of S49 lymphoma cells, J. Biol. Chem. 252:5761 (1977).
7. G. L. Johnson, T. K. Harden, and J. P. Perkins, Regulation of adenosine 3':5'-monophosphate content of Rous sarcoma virus-transformed human astrocytoma cells, J. Biol. Chem. 253:1465 (1978).
8. D. Cassel and Z. Selinger, Catecholamine-stimulated GTPase activity in turkey erythrocytes, Biochim. Biophys. Acta 452:538 (1976).
9. D. Cassel and Z. Selinger, Mechanism of adenylate cyclase activation by cholera toxin: an inhibition of GTP hydrolysis at the regulatory sites, Proc. Natl. Acad. Sci. USA 74:3307 (1977).
10. D. Cassel and Z. Selinger, Mechanism of adenylate cyclase activation through the β-adrenergic receptor: catecholamine-induced displacement of bound GDP by GTP, Proc. Natl. Acad. Sci. USA 75:4155 (1978).
11. E. M. Ross and A. G. Gilman, Biochemical properties of hormone-sensitive adenylate cyclase, Ann. Rev. Biochem. 49:533 (1980).
12. T. Haga, K. Haga, and A. G. Gilman, Hydrodynamic properties of the β-adrenergic receptor and adenylate cyclase from wild type and variant S49 lymphoma cells, J. Biol. Chem. 252:5776 (1977).

13. L. E. Limbird, and R. J. Lefkowitz, Resolution of β-adrenergic receptor binding and adenylate cyclase activity by gel exclusion chromatography, J. Biol. Chem. 252:799 (1977).
14. T. Pfeuffer, GTP-binding proteins in membranes and the control of adenylate cyclase activity, J. Biol. Chem. 252:7224 (1977).
15. E. M. Ross and A. G. Gilman, Resolution of some components of adenylate cyclase necessary for catalytic activity, J. Biol. Chem. 252:6966 (1977).
16. H. R. Bourne, P. Coffino, and G. M. Tomkins, Selection of a variant lymphoma cell deficient in adenylate cyclase, Science 187:750 (1975).
17. E. M. Ross, A. C. Howlett, K. M. Ferguson, and A. G. Gilman, Reconstitution of hormone-sensitive adenylate cyclase activity with resolved components of the enzyme, J. Biol. Chem. 253:6401 (1978).
18. J. K. Northup, M. D. Smigel, and A. G. Gilman, The guanine nucleotide activating site of the regulatory component of adenylate cyclase: identification by ligand binding, J. Biol. Chem. 257:11416 (1982).
19. P. C. Sternweis, J. K. Northup, M. D. Smigel, and A. G. Gilman, The regulatory component of adenylate cyclase: purification and properties, J. Biol. Chem. 256:11517 (1981).
20. P. C. Sternweis and A. G. Gilman, Reconstitution of catecholamine-sensitive adenylate cyclase. Reconstitution of the uncoupled variant of the S49 lymphoma cell, J. Biol. Chem. 254:3333 (1979).
21. E. M. Ross, S. E. Pedersen, and V.A. Florio, Hormone-sensitive adenylate cyclase: Identity, function, and regulation of protein components, Curr. Top. Membranes Transport 18:109 (1982).
22. E. M. Ross, Phosphatidylcholine-promoted interaction of the catalytic and regulatory proteins of adenylate cyclase, J. Biol. Chem. 257:10751 (1982).
23. J. W. Fleming and E. M. Ross, Reconstitution of β-adrenergic receptors into phospholipid vesicles: restoration of [$^{125}$I]iodohydroxybenzylpindolol binding to digitonin-solubilized receptors, J. Cyclic Nucleotide Res. 6:407 (1980).
24. M. G. Caron and R. J. Lefkowitz, Solubilization and characterization of the β-adrenergic receptor binding sites of frog erythrocytes, J. Biol. Chem. 251:2374 (1976).
25. S. E. Pedersen and E. M. Ross, Functional reconstitution of β-adrenergic receptors and the stimulatory GTP-binding protein of adenylate cyclase, Proc. Natl. Acad. Sci. USA 79:7228 (1982).
26. D. R. Brandt, T. Asano, S. E. Pedersen, and E. M. Ross, Reconstitution of catecholamine-stimulated GTPase activity, Biochemistry 22:4357 (1983).

27. R. A. Cerione, B. Strulovici, J. L. Benovic, C. D. Strader, M. G. Caron, and R. J. Lefkowitz, Reconstitution of β-adrenergic receptors in lipid vesicles: Affinity chromatography purified receptors confer catecholamine responsiveness on a heterologous adenylate cyclase system, Proc. Natl. Acad. Sci. USA 80:4899 (1983).

28. J. Kirilovsky and M. Schramm, Delipidation of a β-adrenergic receptor preparation and reconstitution by specific lipids, J. Biol. Chem. 258:6841 (1983).

29. R. G. L. Shorr, M. W. Strohsacker, T. N. Lavin, R.J. Lefkowitz, and M. G. Caron, The β-adrenergic receptor of the turkey erythrocyte. Molecular heterogeneity revealed by purification and photoaffinity labeling, J. Biol. Chem. 257:12341 (1982).

30. Y. Citri and M. Schramm, Resolution, reconstitution, and kinetics of the primary action of a hormone receptor, Nature (London) 287:297 (1980).

31. Y. Citri, and M. Schramm, Probing the coupling sites of the β-adrenergic receptor. Competition between different forms of the guanyl nucleotide binding protein for interaction with the receptor, J. Biol. Chem. 257:13257 (1982).

32. J. K. Northup, P. C. Sternweis, M. D. Smigel, L. S. Schleifer, E. M. Ross, and A.G. Gilman, Purification of the regulatory component of adenylate cyclase, Proc. Natl. Acad. Sci. USA 77:6516 (1980).

33. M. D. Smigel, J. K. Northup, and A. G. Gilman, Characteristics of the guanine nucleotide-binding regulatory component of adenylate cyclase, Recent Prog. Horm. Res. 38:601 (1982).

34. W. Baehr, E. A. Morita, R. J. Swanson, and M. L. Applebury, Characteristics of bovine outer segment G protein, J. Biol. Chem. 257:6452 (1982).

# REGULATION OF THE CATALYTIC UNIT OF ADENYLATE CYCLASE SYSTEM IN RAT BRAIN

S. Ishibashi, T. Kurokawa, K. Higashi and T. Dan'ura

Department of Physiological Chemistry
Hiroshima University School of Medicine
1-2-3, Kasumi, Minami-ku, Hiroshima 734, Japan

## INTRODUCTION

It is well known that the adenylate cyclase system in the cell membrane consists of the ligand receptor, the guanine nucleotide-binding regulatory unit (G/F), and the catalytic unit which catalyzes the formation of cyclic AMP from ATP (1,2). As compared with the former two, the catalytic unit has been less characterized, principally because of instability of this protein after solubilization with the use of detergents. Nevertheless, many efforts have gradually clarified the nature of this protein. Among them, success in separation of the catalytic unit from G/F (3) and use of the cyc$^-$ mutant of S49 lymphoma cell which apparently lacks a functional G/F protein (4) have facilitated the clarification. As a result, the molecular size of the catalytic unit has been estimated to be 170,000-220,000 in several tissues (5,6).

As to the regulation of the catalytic unit, it has been observed that the catalytic unit is regulated not only by G/F but also by various factors independent from G/F. Divalent cations, $Mg^{2+}$ and $Mn^{2+}$, activate the catalytic unit in addition to their involvement in the formation of a metal-ATP complex, the actual substrate of the catalytic unit (7,8). For the activation, a much higher concentration of the divalent cation is required than that needed to form the complex. $Mn^{2+}$ is more potent than $Mg^{2+}$ in activating the catalytic unit, and the former enhances further the activity of the unit which has been maximally activated by the latter. The finding suggests a special role of $Mn^{2+}$ in the regulation of the function of the catalytic unit. It is thus postulated that the binding of $Mn^{2+}$ to the specific site for divalent cations on the catalytic unit molecule induces a conformational change in the molecule resulting in activation of the unit.

On the other hand, in recent years, it has been reported that forskolin, a hypotensive diterpene from the root of Coleus forskohlii, stimulates both membrane-bound and detergent solubilized adenylate cyclases (9). Since the activation by forskolin can be observed in cyc$^-$ S49 lymphoma cells (10), the catalytic unit itself is assumed to be the target of the diterpene. In line with the finding, forskolin-Sepharose affinity chromatography is used for separation of the catalytic unit from G/F (11). In addition to these factors, it is known that calmodulin is involved in the regulation of adenylate cyclase activity, especially of that in the brain (12). Though the effects of these regulatory factors on the function of the catalytic unit have been recognized in various experimental systems, the mechanisms of their action and interaction of their effects have not yet been fully clarified.

In this paper, we present results of our studies on the direct regulatory effects of $Mn^{2+}$ and forskolin on the catalytic unit of adenylate cyclase prepared from rat brain with the use of sodium cholate and ammonium sulfate. Additional data obtained from similar studies on bovine brain adenylate cyclase will also be presented for the confirmation of the results obtained for the rat brain enzyme.

## SEPARATION OF THE CATALYTIC UNIT OF RAT BRAIN ADENYLATE CYCLASE FROM OTHER FUNCTIONAL PROTEINS

### Preparation of the Catalytic Unit of Rat Brain Adenylate Cyclase

Frozen and pooled rat brains were homogenized in 5 volumes of ice-chilled 250 mM sucrose solution containing 1 mM EDTA and 0.1 % ethanol, ph 7.4, using a Potter-Elvehjem homogenizer. The homogenate was centrifuged at 1,000 x g for 10 min to remove cell nuclei and debris. Then, the post-nuclear supernatant was centrifuged at 1,000 x g for 30 min. These procedures as well as the following procedures, when needed, were all carried out at 4° C.

Adenylate cyclase was solubilized from this precipitate by suspending it in 50 mM Tris-HCl buffer containing 250 mM sucrose, 15 mM $MgCl_2$ and 1 mM dithiothreitol, pH 8.0, to which 0.7 % sodium cholate and 0.6 M ammonium sulfate were added as reported by Ross (4). After standing for 20 min, the suspension was centrifuged at 100,000 x g for 30 min. To the supernatant, solid ammonium sulfate was added with stirring to 35 % saturation, and the mixture was centrifuged at 20,000 x g for 30 min. The pellet was suspended in 50 mM Tris-HCl buffer, pH 8.0, to which 8.0 mg/ml lecithin was added for the stabilization of the catalytic unit thus obtained.

Table 1. Absence of G/F and Calmodulin in the Catalytic Unit of Adenylate Cyclase Prepared from Rat Brain

| Additions | Adenylate cyclase activity (pmol/min/mg protein) |
|---|---|
| Experiment 1 (4)[a] | |
| none | 14.0 ± 0.8[b] |
| Gpp(NH)p 50 μM | 17.2 ± 1.8 |
| Experiment 2 (3) | |
| none | 10.1 ± 0.1 |
| trifluoperazine 50 μM | 9.5 ± 0.7 |
| EGTA 0.5 mM | 11.4 ± 0.5 |

[a]Numbers in the parentheses represent the number of experiments performed for the different preparations of the catalytic unit.
[b]Mean ± S.E.

## Measurement of Adenylate Cyclase Activity

Adenylate cyclase activity was measured according to the method of Salomon et al. (13) with some modifications (14). The standard assay mixture consisted of 25 mM Tris-HCl, 1 mM cyclic AMP, 10 mM theophylline, 14.5 mM phosphocreatine, 250 units/ml creatine phosphokinase and 0.03-1.25 mM [$^{3}$H]ATP ($10^6$ cpm), pH 8.0, in a final volume of 400 μl. Standard incubation was performed at 30° C for 20 min.

## Absence of G/F and Calmodulin in the Catalytic Unit Preparation

The catalytic unit of rat brain adenylate cyclase prepared as mentioned above was not stimulated by the addition of 50 μM guanosine 5'-(β, γ-imino)triphosphate (Gpp(NH)p) (Table 1). Combination of 5 mM sodium fluoride and 0.05 mM aluminum chloride, which is known to activate G/F protein, also failed to increase the activity (data not shown). Furthermore, addition of G/F protein (separately prepared from rat brain synaptic membranes by Ultrogel AcA34 and hydroxylapatite column chromatography) to the catalytic unit caused the stimulation of adenylate cyclase activity in the presence of the same concentration of sodium fluoride and aluminum chloride. The stimulation was dependent on the concentration of the G/F protein, and the addition of 15 μg/ml G/F caused about a 2.5 fold increase in the activity. The findings indicate that the catalytic unit preparation is free from the functional G/F protein,

Table 2. Apparent Km Values of the Catalytic Unit of Rat Brain Adenylate Cyclase

| | MgATP | MnATP | $Mg^{2+}$ | $Mn^{2+}$ |
|---|---|---|---|---|
| Km(app.) (μM) | 80 | 80 | 3,800 | 650 |

and that the site on the catalytic unit molecule which interacts with G/F is retained normally.

On the other hand, neither addition of 50 μM trifluoperazine, a calmodulin inhibitor, nor 0.5 mM EGTA changed adenylate cyclase activity of the catalytic unit preparation (Table 1), suggesting the absence of calmodulin in this preparation. Though the preparation was still far from the purified state, it was used in the following analyses.

## ACTIVATION OF THE CATALYTIC UNIT OF RAT BRAIN ADENYLATE CYCLASE BY DIVALENT CATIONS

### Changes in the Catalytic Unit of Adenylate Cyclase Activity by $Mg^{2+}$ and $Mn^{2+}$

Adenylate cyclase activity of the catalytic unit thus prepared from rat brain was stimulated by either $Mg^{2+}$ or $Mn^{2+}$. The Vmax was increased about 3 and 4.5 times by 10 mM $Mg^{2+}$ and 1 mM $Mn^{2+}$, respectively. The mode of the activation was noncompetitive with respect to metal-ATP complex as the substrate. From double reciprocal plots of activity vs. substrate concentration under various divalent ion concentrations as well as from their secondary plots, apparent Km values for the metal-ATP complexes and divalent cations were calculated as shown in Table 2. The values for divalent cations are much higher than those for the metal-ATP complexes, and the value for $Mn^{2+}$ is lower than that for $Mg^{2+}$, agreeing with results reported previously (15).

The results for the divalent cations indicate that the catalytic unit has an allosteric site for these cations other than a catalytic site for the metal-ATP complexes, and that the affinity of the divalent cation site is higher for $Mn^{2+}$ than for $Mg^{2+}$. This site is assumed to be common for divalent cations, since it has been reported that the rate of activation by $Mn^{2+}$ is decreased by addition of $Mg^{2+}$ in a dose-dependent manner (16), and $Ca^{2+}$ competitively inhibits the activation by both $Mg^{2+}$ and $Mn^{2+}$ (7).

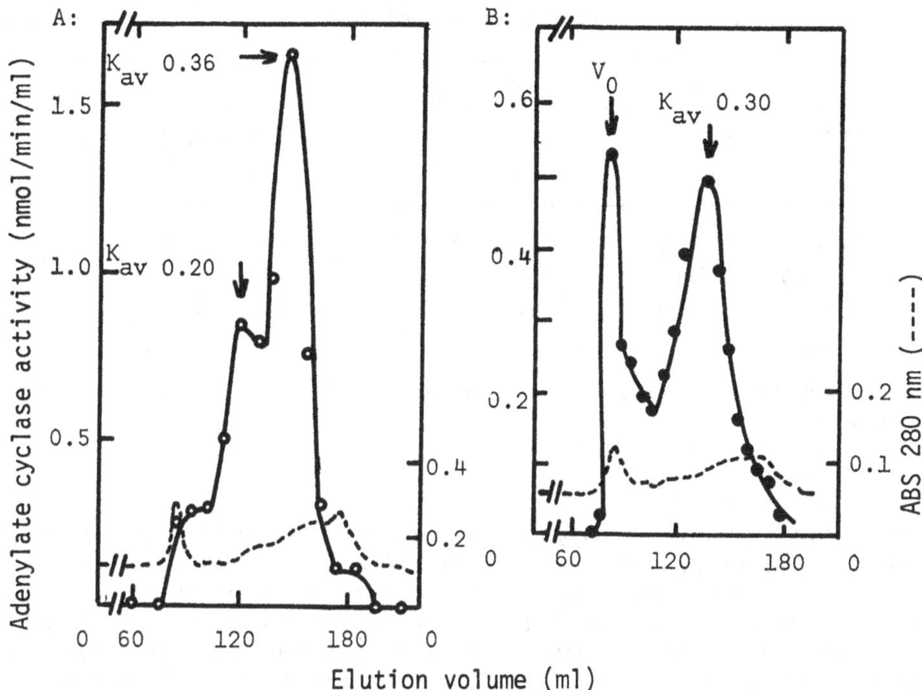


Fig. 1. Sepharose 6B Gel Filtration Pattern of the Catalytic Unit of Rat Brain Adenylate Cyclase. A (left): The gel was eluted with 50 mM Tris-HCl buffer containing 0.1% Lubrol PX, pH 8.0, in the absence of $Mn^{2+}$. B (right): The experimental conditions were the same as for A, except for the presence of 1 mM $Mn^{2+}$ in the sample and elution buffer.

## Change in Gel Filtration Behavior of the Catalytic Unit by $Mn^{2+}$

When the catalytic unit of rat brain adenylate cyclase was solubilized in 1.0% Lubrol-PX and chromatographed on DEAE-cellulose with a linear KCl gradient from 0 to 0.2 M, the enzyme activity was eluted as three peaks. The major peak eluting at 0.05 M KCl was then fractionated by Sepharose 6B gel filtration. The majority of adenylate cyclase activity was eluted with 50 mM Tris-HCl buffer containing 0.1% Lubrol-PX, pH 8.0, in a position with $K_{av}$ of 0.36 (Fig. 1A). However, when the Sepharose 6B gel filtration was performed in the same way but in the presence of 1 mM $Mn^{2+}$, adenylate cyclase activity was eluted at a position with $K_{av}$ of 0.30 and in the void volume (Fig. 1B). The change in the elution pattern was reproducible, and seemed to be specific for $Mn^{2+}$, because the Tris

Table 3. Effect of Forskolin on Apparent Km Value for ATP and Vmax of the Catalytic Unit of Rat Brain Adenylate Cyclase

| Forskolin concentration (μM) | | Km(app.) for ATP (μM) | Vmax (pmol/min/mg protein) |
|---|---|---|---|
| 0 | (4)[a] | 83 ± 2[b] | 22 ± 3 |
| 1.0 | (4) | 217 ± 3 | 134 ± 33 |
| 10.0 | (4) | 246 ± 8 | 337 ± 38 |

[a]Numbers in the parentheses represent the number of the experiments.
[b]Mean ± S.E.

buffer for the elution always contained 15 mM $Mg^{2+}$, as mentioned earlier. The change in the gel filtration profile was also specific for the catalytic unit, since the elution pattern of the bulk of proteins was not different between the presence and absence of $Mn^{2+}$. Though the relationship between the activation of the catalytic unit and the change in the gel filtration pattern by $Mn^{2+}$ is not clear, similar findings for a change in the profiles of separation by isoelectric focusing (17) and gel filtration (18) were reported for adenylate cyclase activity in mouse brain and rat liver, respectively.

## ACTIVATION OF THE CATALYTIC UNIT OF RAT BRAIN ADENYLATE CYCLASE BY FORSKOLIN

### Mode of Activation of the Catalytic Unit by Forskolin

Addition of forskolin to the assay system for adenylate cyclase activity caused an increase in Vmax of the catalytic unit preparation of rat brain adenylate cyclase in a concentration-dependent manner as far as we examined up to 10 μM forskolin. Table 3 shows the values at several concentrations. It was of interest that the apparent Km value for ATP, which was about 80 μM without forskolin, was also increased in accordance with the increase in Vmax at a certain range of the concentration of forskolin. The apparent Km value, however, converged gradually on about 200 μM as the concentration of forskolin exceeded 1.0 μM. Thus, the activation of the catalytic unit by forskolin seems to be accompanied by an apparent decrease in the affinity of the catalytic site for metal-ATP complex in a certain range of forskolin concentration, provided that the change in the Km value can be regarded as a reflection of the

Table 4. Activation of the Catalytic Unit of Rat Brain Adenylate Cyclase by Forskolin with Different ATP Concentrations

| Forskolin concentration (μM) | Adenylate cyclase activity (pmol/min/mg protein) with 125 μM ATP | with 1.25 mM ATP |
|---|---|---|
| 0 | 12.9 | 19.9 |
| 2.5 | 65.6 | 131 |
| 5.0 | 79.4 | 185 |
| 10.0 | 82.4 | 239 |

affinity for the substrate.

### Dependence on ATP Concentration of Activation of the Catalytic Unit by Forskolin

In relation to the increase in the apparent Km value for ATP, activation of the catalytic unit was examined at two different concentrations of ATP (Table 4). In these experiments, the concentration of $Mg^{2+}$ was kept at 15 mM as in other experiments. When the ATP concentration was 125 μM, the activation almost saturated at 5.0 μM forskolin, whereas the activation was still increasing with 1.25 mM ATP up to 10.0 μM forskolin. The rate of activation with 10.0 μM forskolin is about 12 fold with 1.25 mM ATP, while is about 6.5 fold with 125 μM ATP. It may be inferred that this finding is related to the above mentioned change in the apparent Km value for ATP by forskolin, i.e., higher concentrations of ATP would be favorable for the forskolin activation of the catalytic unit. It may also be inferred that the binding of forskolin to the catalytic unit is dependent on ATP concentration.

### Modulation of Forskolin-induced Activation of the Catalytic Unit by $Mn^{2+}$

The relationship between the activating effect of $Mn^{2+}$ and that of forskolin on the catalytic unit of rat brain adenylate cyclase was examined, since both factors seem to have specific binding sites on the catalytic unit molecule. As shown in Table 5, synergism was observed in the activation between $Mn^{2+}$ and forskolin, agreeing with the results reported by Seamon et al. (19).

Table 5. Synergism between $Mn^{2+}$ and Forskolin in Activating the Catalytic Unit of Rat Brain Adenylate Cyclase

| Additions | Adenylate cyclase activity (pmol/min/mg protein) |
|---|---|
| none | 44 |
| 1 mM $Mn^{2+}$ | 239 |
| 50 μM forskolin | 369 |
| 1 mM $Mn^{2+}$ + 50 μM forskolin | 1910 |

It was of particular interest that the increase in the apparent Km value of the catalytic unit for ATP was not observed any more when $Mn^{2+}$ was added to the assay system for adenylate cyclase activity, in spite of the presence of forskolin (Table 6). It is deduced from these findings that $Mn^{2+}$ not only activates the catalytic unit preparation of rat brain adenylate cylase directly but also modulates the forskolin-induced change in the affinity of the catalytic site for ATP. The table shows that the effect of 10 μM forskolin on the two kinetic parameters of the catalytic unit is gradually modified when the concentration of $Mn^{2+}$ is increased in the range of 0.01-1.0 mM, although 0.01 mM $Mn^{2+}$ only increases the Vmax without changing the apparent Km value for ATP increased by forskolin.

The modulation of the forskolin induced change in the Km (app.) for ATP of the catalytic unit by addition of $Mn^{2+}$ above 0.1 mM may be related to the synergistic activation of the unit by these two effectors. Since all of these experiments have been carried out in

Table 6. Apparent Km Values for ATP and Vmax of the Catalytic Unit of Rat Brain Adenylate Cyclase in the Presence of 10 μM Forskolin and Various Concentrations of $Mn^{2+}$

| $Mn^{2+}$ concentration (mM) | Km (app.) for ATP (μM) | Vmax (pmol/min/mg protein) |
|---|---|---|
| 0 | 250 | 155 |
| 0.01 | 250 | 200 |
| 0.1 | 180 | 300 |
| 0.5 | 100 | 455 |
| 1.0 | 80 | 555 |

Table 7. Activation of Adenylate Cyclase in Synaptic Membranes Prepared from Bovine Cerebrum by Forskolin and $Mn^{2+}$

| Additions | Km (app.) for ATP (μM) | Vmax (pmol/min/mg protein) |
|---|---|---|
| A[a] | | |
| none | 80 | 120 |
| 10 μM forskolin | 200 | 670 |
| 10 μM forskolin + 1.0 mM $Mn^{2+}$ | 80 | 1140 |
| B[b] | | |
| none | 90 | 510 |
| 10 μM forskolin | 200 | 1180 |
| 10 μM forskolin + 1.0 mM $Mn^{2+}$ | 90 | 1560 |

[a]Control synaptic membranes.
[b]Synaptic membranes preactivated by incubating with 0.2 mM Gpp(NH)p at 30° C for 30 min.

the presence of much excess $Mg^{2+}$, the effect of $Mn^{2+}$ to modulate the effect of forskolin on the apparent Km value may be specific for this cation. It seems to be unlikely that these results are due to fluctuation of the ATP concentration by other ATP-utilizing enzymes, because adenylate cyclase activity has always been measured in the presence of an ATP-regenerating system.

Recently, Awad et al. (20) reported that forskolin caused an increase in the apparent Km value for MgATP of human platelet adenylate cyclase, agreeing with our results. They also reported that such an increase in the apparent Km value was not observed for solubilized rat brain enzyme when the activity was measured with MnATP and in the presence of 2 mM $Mn^{2+}$. Their results on the rat brain adenylate cyclase are also in line with ours, and may be interpreted to mean that the effect of forskolin on the apparent Km value is modulated by $Mn^{2+}$ at concentrations on the order of 2 mM.

## ACTIVATION OF BOVINE CEREBRAL ADENYLATE CYCLASE BY FORSKOLIN AND ITS MODULATION BY MANGANESE ION AND G/F

Synaptic membranes were prepared from bovine cerebrum (21), and activation of adenylate cyclase in the membranes was examined for comparison with the above-mentioned results for rat brain enzyme. As shown in Table 7A, addition of 10 μM forskolin induced increases in both the apparent Km value for ATP and in Vmax, confirming the results obtained for the catalytic unit of rat brain adenylate cyclase. It was also confirmed that 1.0 mM $Mn^{2+}$ enhanced the activating effect of forskolin on adenylate cyclase, as manifested by an increase in Vmax, and modulated the augmenting effect of forskolin on the apparent Km value for ATP. The findings indicate that the mode of activation of adenylate cyclase by forskolin and $Mn^{2+}$ and their interaction in the activation are not specific for rat brain but rather common, though these two experimental systems are different.

Similar studies were then performed using the synaptic membranes which had been preactivated by incubating in the presence of 0.2 mM Gpp(NH)p at 30° C for 30 min, expecting to see the effect of functional G/F. Addition of the same concentration of forskolin increased adenylate cyclase activity in the Gpp(NH)p-treated membrane (Table 7B), but the rate of the activation by forskolin was less in the Gpp(NH)p-treated membrane than in the membrane without the treatment. The finding seems to suggest that the association of a functional G/F unit to the catalytic unit interferes the forskolin action. However, the increase in the apparent Km value for ATP by forskolin was unchanged by the preactivation with Gpp(NH)p. When both 10 μM forskolin and 1.0 mM $Mn^{2+}$ were added to the Gpp(NH)p-treated membrane, the increase in Vmax was not so marked as in the membrane without the treatment. But, the apparent Km value for ATP was modulated as was observed in the control membrane. In other words, the modulating effect of $Mn^{2+}$ on the forskolin-induced change in the apparent Km value for ATP can be observed irrespective of the association of the functional G/F unit to the catalytic unit of bovine brain adenylate cyclase.

## CONCLUSION

Divalent cations, forskolin and a functional G/F unit activate the catalytic unit of rat and bovine brain adenylate cyclase probably by binding to different sites on the catalytic unit molecule. However, they do not regulate the function of the catalytic unit independently, but interact with each other in activating the unit, as manifested, for instance, by modulation of the forskolin-induced increase in the apparent Km value for ATP by $Mn^{2+}$.

ACKNOWLEDGEMENT

We are grateful for the co-operation of Misses J. Kamegashira and A. Miyawaki.

REFERENCES

1. M. Rodbell, The role of hormone receptors and GTP-regulatory proteins in membrane transduction, Nature 284:17 (1980).
2. L. E. Limbird, Activation and attenuation of adenylate cyclase, Biochem. J. 195:1 (1981).
3. S. Strittmatter and E. J. Neer, Properties of the separated catalytic and regulatory units of brain adenylate cyclase, Proc. Natl. Acad. Sci. USA 77:6377 (1980).
4. E. M. Ross, Physical separation of the catalytic and regulatory proteins of hepatic adenylate cyclase, J. Biol. Chem. 256:1949 (1981).
5. T. Haga, K. Haga, and A. G. Gilman, Hydrodynamic properties of the β-adrenergic receptor and adenylate cyclase from wild type and variant S49 lymphoma cells, J. Biol. Chem. 252:5776 (1977).
6. E. J. Near and R. S. Salter, Modification of adenylate cyclase structure and function by ammonium sulfate, J. Biol. Chem. 256:5497 (1981).
7. R. D. Lasker, R. W. Downs, Jr., and G. D. Aurbach, Calcium inhibition of adenylate cyclase: Studies in turkey erythrocytes and S49 $cyc^-$ cell membrane, Arch. Biochem. Biophys. 216:345 (1982).
8. S. G. Somkuti, J. D. Hildebrandt, J. T. Herberg, and R. Iyengar, Divalent cation regulation of adenylyl cyclase, J. Biol. Chem. 257:6387 (1982).
9. K. B. Seamon and J. W. Daly, Forskolin: A Unique diterpene activator of cyclic AMP-generating systems, J. Cyclic Nucleotide Res. 7:201 (1981).
10. K. Seamon and J. W. Daly, Activation of adenylate cyclase by the diterpene forskolin does not require the guanine nucleotide regulatory protein, J. Biol. Chem. 256:9799 (1981).
11. T. Pfeuffer and H. Metzger, 7-O-Hemisuccinyl-deacetyl forskolin-Sepharose: a novel affinity support for purification of adenylate cyclase, FEBS Lett. 146:369 (1982).
12. R. S. Salter, M. H. Krinks, C. B. Klee, and E. J. Neer, Calmodulin activates the isolated catalytic unit of brain adenylate cyclase, J. Biol. Chem. 256:9830 (1981).
13. Y. Salomon, C. Londos, and M. Rodbell, A highly sensitive adenylate cyclase assay, Anal. Biochem. 58:541 (1974).

14. T. Kurokawa, M. Kurokawa, and S. Ishibashi, Anti-microtubular agents as inhibitors of desensitization to catecholamine stimulation of adenylate cyclase in Ehrlich ascites tumor cells, Biochim. Biophys. Acta 583:467 (1979).
15. D. Garbers and R. A. Johnson, Metal and metal-ATP interactions with brain and cardiac adenylate cyclases, J. Biol. Chem. 250:8499 (1975).
16. E. J. Near, Interaction of soluble brain adenylate cyclase with manganese, J. Biol. Chem. 254:2089 (1979).
17. D. F. Malamud, C. C. DiRusso, and J.T. Aprille, Multiple forms of brain adenylate cyclase: Stimulation by $Mn^{2+}$, Biochim. Biophys. Acta 485:243 (1977).
18. C. Londos, P. M. Lad, T. B. Nielsen, and M. Rodbell, Solubilization and conversion of hepatic adeylate cyclase to a form requiring MnATP as substrate, J. Supramol. Struct. 10:31 (1979).
19. K. B. Seamon, W. Padgett, and J. W. Daly, Forskolin: Unique diterpene activator of adenylate cyclase in membranes and in intact cells, Proc. Natl. Acad. Sci. USA 78:3363 (1981).
20. J. A. Awad, R. A. Johnson, K. H. Jakobs, and G. Schultz, Interaction of forskolin and adenylate cyclase, J. Biol. Chem. 258:2960 (1983).
21. V.P. Whittaker, I. A. Michaelson, and R. J. Kirkland, The separation of synaptic vesicles from nerve-ending particles ('Synaptosomes'), Biochem. J. 90:293 (1964).

# MOLECULAR REGULATORY MECHANISM OF $D_2$ DOPAMINE RECEPTOR IN THE BOVINE STRIATUM

Chikako Tanaka, Takayoshi Kuno, Osamu Shirakawa, Kiyofumi Saijoh and Nobuo Kubo

Department of Pharmacology
Kobe University School of Medicine, Kobe, Japan

## INTRODUCTION

There is considerable evidence for the existence of multiple types of dopamine receptor in the brain. $D_1$ receptors have been defined by their ability to elicit an increase in adenylate cyclase activity while $D_2$ receptors do not activate this enzyme (1). Recently, in experiments using homogenates of the intermediate lobe of the rat pituitary gland, it has been shown that stimulation of the $D_2$ receptors decreases the responsiveness of the beta-adrenoceptor by reducing adenylate cyclase activity, and that guanine nucleotides are obligatory for receptor-mediated stimulation and for inhibition of adenylate cyclase (2). However, although it has been demonstrated that guanine nucleotides decrease the affinity of agonists but not antagonists for the brain $D_2$ receptor and that stimulation of the $D_2$ receptor results in a reduction in cAMP efflux from rat striatal slices, the inhibition of adenylate cyclase activity mediated by the $D_2$ receptor has not been directly determined (3-6). Islet-activating protein (IAP), a pertussis toxin, inhibits the inhibition of adenylate cyclase activity mediated by $alpha_2$-adrenergic, cholinergic, and opiate receptors in neuroblastoma x glioma hybrid cells, and modulates the binding affinities of these receptors, presumably as a result of ADP-ribosylation of one of the subunits of the inhibitory guanine nucleotide regulatory protein, designated $N_i$ (3-6). Thus, we attempted to determine whether IAP modulates the GTP-sensitive adenylate cyclase activity and the binding affinity of the bovine striatal $D_2$ receptor associated with ADP-ribosylation of a subunit of $N_i$.

## MATERIALS AND METHODS

### Materials

IAP purified from the 3-day culture supernatant of _B. pertussis_ cells (Tohama strain, phase I) according to the procedure described elsewhere (7) was generously provided by Dr. M. Yajima (Research Laboratories of Kakenyaku Kako Co., Japan). The stock solution was prepared by dissolving 1 mg of IAP in 1 ml of the vehicle consisting of 0.1 M potassium phosphate buffer, pH 7.0, and 2 M urea, and storing at 4° C until use. The vehicle alone was used as control. Sulpiride (Fujisawa, Japan), (-)- and (+)-butaclamol (Ayerst, Canada), ketanserin (Kyowa Hakko, Japan) and haloperidol (Yoshitomi, Japan) were gifts from the respective companies. [$^3$H]Spiperone (31.7 Ci/mmol), [$^3$H]_n_-propylapomorphine (NPA, 58.8 Ci/mmol) and [$^{32}$P]NAD (52 Ci/mmol) were purchased from New England Nuclear. The cAMP radioimmunoassay kit was purchased from Yamasa, Japan. ATP, GTP, NAD, thymidine, dithiothreitol, dopamine hydrochloride and apomorphine hydrochloride were purchased from Sigma.

### Membrane Preparation and Radiation Inactivation

Crude synaptic membrane preparations from bovine striatum were prepared as described previously (8,9). The membrane preparations (2-5 mg protein) in 1 ml of 50 mM Tris-HCl buffer, pH 7.4, were lyophilized in ampules and sealed in _vacuo_. Samples were irradiated with gamma ray from the source of $^{60}$Co (34,400 Ci) at a dose of approximately 5.0 Mrad/hr at 0-10° C. Irradiated samples were rehydrated and assayed for $D_2$ receptor binding and acetylcholinesterase activity. Size of the target was calculated from the following formula, as described by Schlegel et al. (10): molecular size (daltons) = $6.4 \times 10^{11}/D_{37}$, where $D_{37}$ is the radiation dose (rad) giving a residual binding activity of 37%.

### Treatment of Membranes with IAP

The membrane preparation (15-20 mg protein) was incubated with 125 µg of preactivated IAP for 15 min at 37° C in 5 ml of 25 mM Tris-HCl buffer, pH 7.4, containing 2.5 mM $MgCl_2$, 1 mM ATP, 0.2 mM GTP, 10 mM thymidine, 5 mM dithiothreitol and 1 mM NAD. Incubation was terminated by cooling the reaction tube followed by centrifugation (10,000 g, 10 min) at 4° C. The pellet was washed three times with ice cold 50 mM Tris-HCl buffer, pH 7.4, by repeated dilution and centrifugation. For radiolabeling, 0.01 mM [$^{32}$P]NAD was used instead of nonradioactive NAD. IAP was preactivated by incubation in 20 mM dithiothreitol for 15 min at 37° C.

## $D_2$ Receptor Binding Assay

Assays of [$^3$H]NPA binding and [$^3$H]spiperone binding were carried out by the method described previously (11). Stereospecific binding was defined as the difference in the binding obtained with incubation in the presence of $10^{-6}$ M (-)- and (+)-butaclamol. $D_2$ specific binding was defined as that which occurred in the presence of $10^{-7}$ M ketanserin (to occlude $S_2$ serotonergic sites) but which was displaced by $10^{-6}$ M (+)-butaclamol for [$^3$H]spiperone binding, or 5 X $10^{-8}$ M spiperone for [$^3$H]NPA binding. The difference between the [$^3$H]spiperone binding, with and without $10^{-4}$ M GTP at $10^{-5}$ M dopamine, was taken as the value for the GTP effect.

## Polyacrylamide Gel Electrophoresis and Autoradiography

Discontinuous buffered SDS-polyacrylamide slab gel electrophoresis was used to analyze the labeled membrane preparation, as described previously (12). The molecular weights of bands of the gels were estimated by the method of Weber and Osborn (13).

## Adenylate Cyclase and Acetylcholinesterase Assays

Adenylate cyclase assay was carried out by the method of Katada et al. (3-5) using a cAMP radioimmunoassay kit. Acetylcholinesterase activity was determined by a decrease in the amount of acetylcholine which had been added to the enzymic reaction mixture as substrate. The enzymic reaction was performed by addition of 0.1 ml of the rehydrated lyophilized membrane preparations into a test tube containing 1 ml of acetylcholine-buffer-salt mixture reagent. This reagent was composed of 8 volumes of 1/15 M phosphate buffer and 1 volume each of 0.5 M acetylcholine and a salt mixture containing 4.2 g $MgCl_2$ and 0.2 g of KCl per 100 ml. After incubation of the sample for 1 hr at 37° C, the content of acetylcholine remaining in the samples was determined colorimetrically.

# RESULTS AND DISCUSSION

Membrane preparations from bovine striatum and frontal cortex were incubated with [$^{32}$P]NAD in the presence or absence of IAP and then polyacrylamide gel analysis of radioactive products was carried out. IAP treatment of membranes resulted in a labeling of a protein with a molecular weight about 40,000 daltons (8), this finding being consistent with observations made using cultured cells (3-6,14). In addition, IAP treatment produced a three-fold increase in adenylate cyclase activity, when it was assayed in the presence of $10^{-5}$ M GTP. Thus, the mammalian brain membranes also seem to contain a pertussis toxin substrate, probably one of the subunits of $N_i$.

To examine the effect of IAP on the $D_2$ dopamine receptor, $D_2$ receptor specific binding of [$^3$H]spiperone to IAP-treated and non-treated bovine striatal membranes was measured in the presence of increasing concentrations of [$^3$H]spiperone. Scatchard analysis of both membrane preparations gave a linear relationship indicating a single class of specific binding sites. The numbers of binding sites (Bmax) and the dissociation constants ($K_D$) for binding were 210 ± 20 fmol/mg protein, 0.34 ± 0.03 nM in IAP-treated membranes and 200 ± 10 fmol/mg protein, 0.32 ± 0.02 nM in nontreated membranes, respectively. This indicates that neither the density of $D_2$ receptors nor the affinity for [$^3$H]spiperone was affected by IAP treatment.

$IC_{50}$ concentrations for dopamine and apomorphine with control membranes were 40 ± 3 and 2.1 ± 0.2 μM and were significantly increased to 120 ± 10 and 4.8 ± 0.4 μM with IAP treatment (8). The values for sulpiride, haloperidol and (+)-butaclamol were not significantly changed. Addition of GTP to nontreated membranes also resulted in a selective decrease in the affinity of the $D_2$ receptor agonists, the $IC_{50}$ values for dopamine and apomorphine being 120 ± 10 and 5.0 ± 0.4 μM, respectively. However, there was no further reduction in affinity for the agonists with addition of GTP to the IAP-treated membranes. These findings suggest that the guanine nucleotide regulatory protein is not coupled to the $D_2$ dopmine receptor in IAP-treated membranes. On the other hand, there was no change in the affinity of $D_2$ receptors for agonists or antagonists with cholera toxin-treated membranes, and IAP-induced modulation of the $D_2$ receptor was not observed in the membranes treated with IAP in the absence of NAD. Our results suggest that ADP-ribosylation of one of the subunits of $N_i$ leads to a conformational change of the $D_2$ receptor protein, through the molecular interaction between these two proteins and that bidirectional coupling of $N_i$ to adenylate cyclase and to the $D_2$ receptor recognition protein exists, as shown in Fig. 1.

In the next series of experiments, we used target size analysis to evaluate the functional molecular size of the $D_2$ receptor relative to other membrane macromolecules (11). The inclusion of $10^{-7}$ M ketanserin in all incubation tubes and the definition of nonspecific binding by $10^{-6}$ M (+)-butaclamol for [$^3$H]spiperone binding and by 5 X $10^{-8}$ M spiperone for [$^3$H]NPA binding made feasible labeling of only the $D_2$ receptor, as validated by drug displacement experiments. When the striatal membrane preparations were lyophilized before binding assays, no loss of binding of either ligand occurred and the same $K_d$ and Bmax values were obtained upon resuspension in the binding assay medium. We carried out radiation inactivation of an enzyme of known molecular weight, acetylcholinesterase, in the same lyophilized membranes, in order to verify the appropriate conditions of the radiation inactivation. As shown in Fig. 2, acetylcholinesterase activity decreased linearly on a semilogarithmic scale with

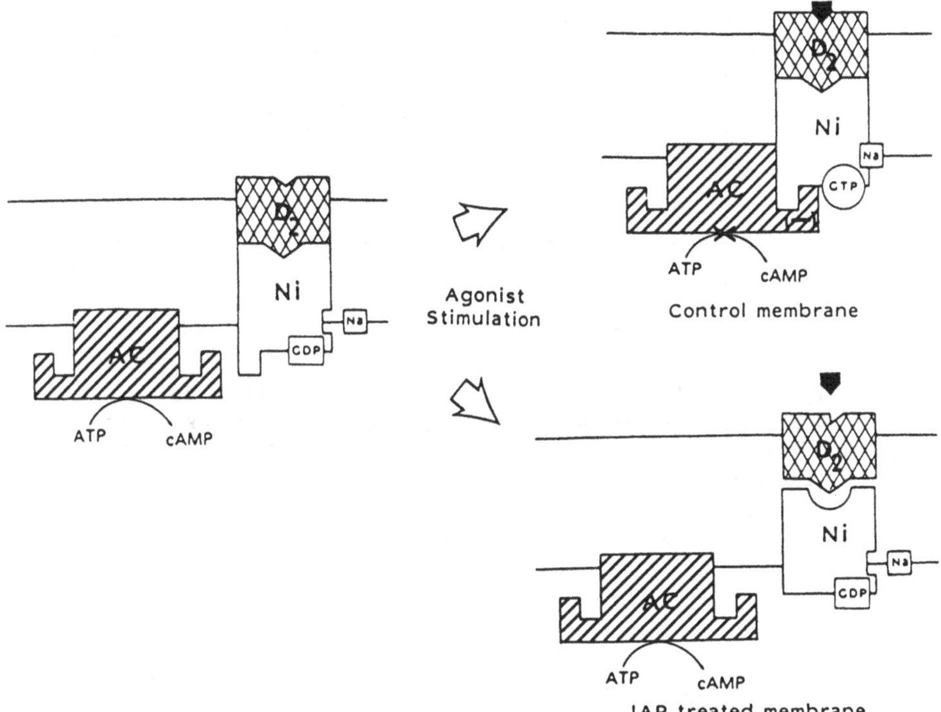


Fig. 1. Schematic representation of the working hypothesis of bidirectional coupling of inhibitory guanine nucleotide regulatory protein ($N_i$) to the $D_2$ receptor recognition protein ($D_2$) and to adenylate cyclase (AC) in the bovine striatal membrane. A dopaminergic agonist occupies a $D_2$ receptor interacting with a $N_i$; GTP interacts with $N_i$ to inhibit adenylate cyclase activity. Pertussis toxin (IAP) selectively acts at $N_i$ and produces a decrease in the $D_2$ receptor affinity for agonist but not for antagonist, as a result of ADP-ribosylation of one of the subunits of $N_i$, which leads to the conformational change of the $D_2$ receptor recognition protein. (The figure does not indicate stoichiometry of various components.)

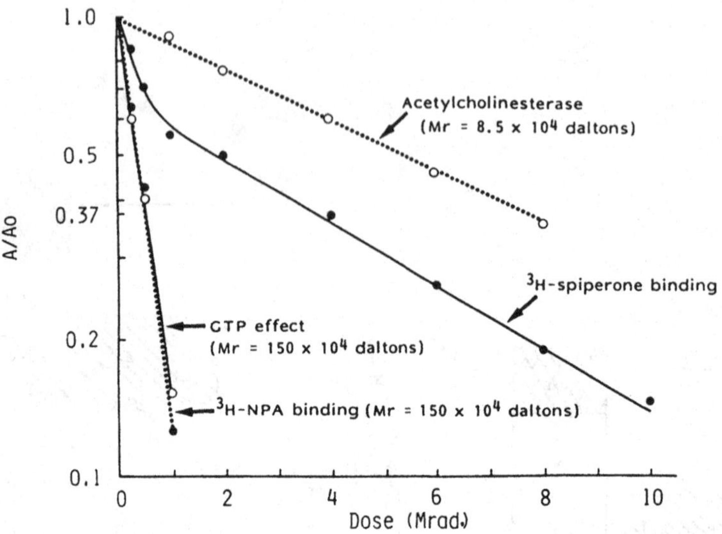


Fig. 2. Representative radiation inactivation curves for [$^3$H]NPA binding (●······●), [$^3$H]spiperone binding (●——●), GTP effect (O——O) and acetylcholinesterase (O·····O). (From Ref. 11).

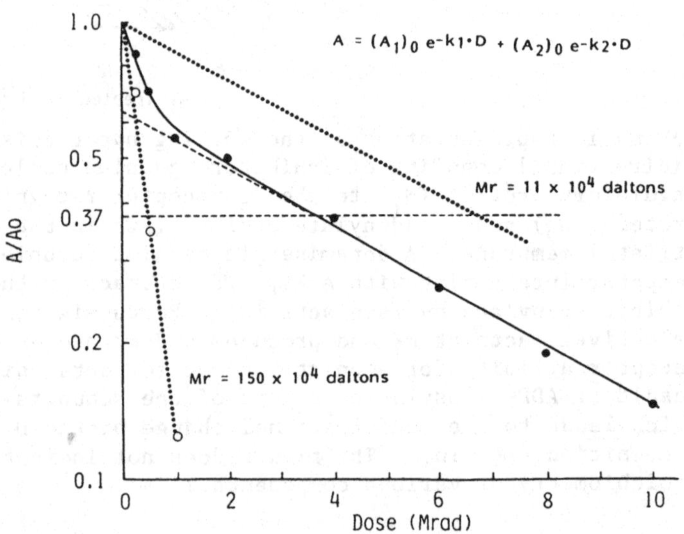


Fig. 3. Analysis of the nonlinear radiation inactivation curve of [$^3$H]spiperone binding. $K_1$ represents the shallower slope of the small target, $K_2$ the slope of the larger target and D the radiation dose.

an increase in the dose of irradiation, and the molecular size was calculated to be 85,000 daltons, which is all but identical to 80,000 daltons determined by gel filtration of acetylcholinesterase in calf brain (15).

The GTP effect and specific [$^3$H]NPA binding decayed linearly with the radiation dose. Their calculated target sizes averaged 150 X $10^4$ daltons. On the other hand, the radiation inactivation curve for the specific [$^3$H]spiperone binding was unexpectedly nonlinear. For analysis of this nonlinear curve, we assumed two target sizes which were fitted graphically to the equation shown in Fig. 3. The calculated size of the larger target was 150 X $10^4$ daltons, whereas that of the smaller target was 11 X $10^4$ daltons. Thus, in brain membranes, there are two distinct sub-populations of $D_2$ receptors labeled by [$^3$H]spiperone, and the larger sized sub-population may correspond to the receptor labeled by [$^3$H]NPA and the functional unit of the GTP effect on [$^3$H]spiperone binding. Lilly et al. (16) showed that the molecular size of the $D_2$ receptor in canine and human striatum was calculated to be 123,000 daltons by target size analysis using high-energy electrons, which is consistent with that of small sized $D_2$ receptor in our study.

Irradiation of the lyophilized striatal membranes at 3 Mrad is the minimal dose required to induce a selective destruction of the large sized sub-population of the $D_2$ receptor. Drug displacement experiments of [$^3$H]spiperone binding in the presence or absence of 0.1 mM GTP revealed that $IC_{50}$ values of dopamine and apomorphine (32 ± 4 and 1.9 ± 0.2 μM) were increased to 110 ± 10 and 4.0 ± 0.4 μM, while that of sulpiride and (+)-butaclamol were not significantly changed by irradiation at 3 Mrad. These results indicate that the small size sub-population of the $D_2$ receptor has a low affinity for agonists, and the large sized sub-population of the $D_2$ receptor has a high affinity for agonists, that is, [$^3$H]NPA binding sites and the [$^3$H]spiperone binding sites in the guanine nucleotide sensitive state. These results are also consistent with our previous findings that the zwitter-ionic detergent-solubilized $D_2$ receptor in a form sensitive to guanine nucleotides has a much larger Stokes radius compared with that of digitonin-solubilized $D_2$ receptor in a form insensitive to guanine nucleotides (9).

Fig. 4 is a schematic illustration of the correspondence between different affinity states and target size of the bovine striatal $D_2$ dopamine receptor. Large size (150 X $10^4$ daltons) appears to represent the minimal assembly of units required for a high-affinity agonist binding and the GTP sensitive state of the $D_2$ receptor.

Rodbell proposed that neurotransmitter and hormone receptors are regulated by guanine nucleotides and form oligomeric complexes with the guanine nucleotide regulatory protein in the cell membrane

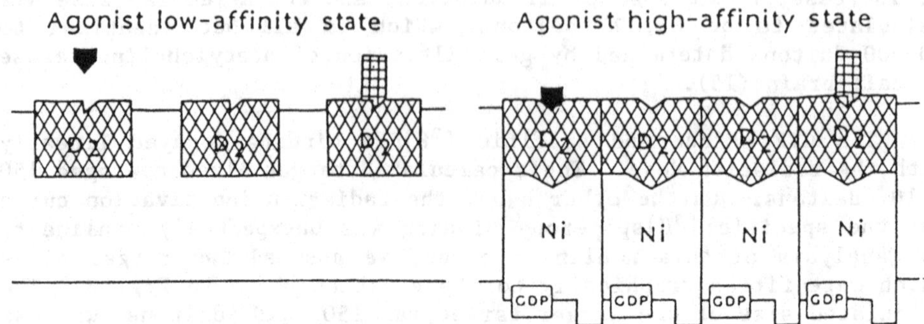


Fig. 4. Schematic representation of two different affinity states of the $D_2$ receptor. Left: agonist low-affinity state corresponds to the small sized sub-population of the $D_2$ receptor (Mr = 11 X $10^4$ daltons); right: agonist high-affinity state corresponds to the large sized sub-population of the $D_2$ receptor (Mr = 150 X $10^4$ daltons). See text for detailed description.

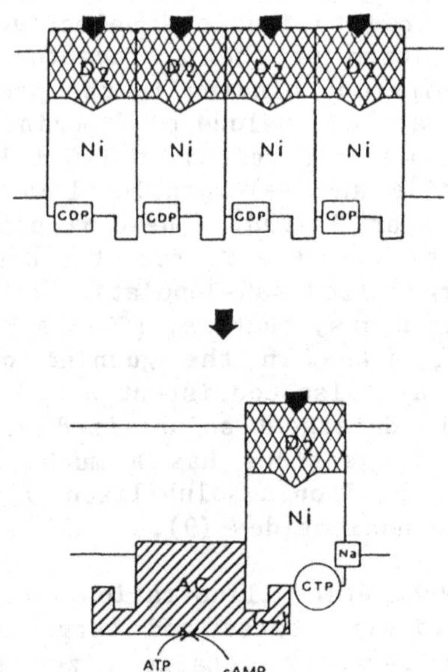


Fig. 5. Schematic representation of the working hypothesis of the "disaggregation coupling model" of the $D_2$ receptor. Binding of the agonist to the oligomeric complex may result in disaggregation of the complex and the coupling of $N_i$ to adenylate cyclase (AC). (The figure does not indicate stoichiometry of various components.)

based on target size analysis of the receptor-adenylate cyclase system (17). If we consider the $D_2$ receptor mediated inhibition of adenylate cyclase in the framework of a "disaggregation coupling model" which was proposed to account for the receptor mediated activation of adenylate cyclase, our findings in the case of the brain $D_2$ receptor suggest that the agonist high-affinity state of the $D_2$ receptor has the large target size which is an oligomeric complex consisting of the $D_2$ receptor recognition protein and $N_i$. The agonist low-affinity state of the $D_2$ receptor has a small target size which may be the monomeric or dimeric receptor recognition protein and can bind with antagonists. Binding of agonist to this oligomeric complex may result in disaggregation of the complex and the coupling of $N_i$ to adenylate cyclase as shown in Fig. 5.

ACKNOWLEDGEMENTS

This work was supported by grants from the Ministry of Education, Science and Culture and Ministry of Health and Welfare, Japan. We thank M. Ohara of Kyushu University for helping us prepare this article.

REFERENCES

1. J. W. Kebabian and D. B. Calne, Multiple receptors for dopamine, Nature 277:93 (1979).
2. T. E. Cote, C. W. Grewe, K. Tsuruta, J. C. Stoof, R. L. Eskay, and J. W. Kebabian, D-2 dopamine receptor-mediated inhibition of adenylate cyclase activity in the intermediate lobe of the rat pituitary gland requires guanosine 5'-triphosphate, Endocrinol. 110:812 (1982).
3. T. Katada and M. Ui, Islet-activating protein, a modifier of receptor-mediated regulation of rat islet adenylate cyclase, J. Biol. Chem. 256:8310 (1981).
4. T. Katada, T. Amano, and M. Ui, Modulation by islet-activating protein of adenylate cyclase activity in C6 glioma cells, J. Biol. Chem. 257:3739 (1982).
5. T. Katada and M. Ui, ADP-ribosylation of the specific membrane protein of C6 cells by islet-activating protein associated with modification adenylate cyclase activity, J. Biol. Chem. 257:7210 (1982).
6. H. Kurose, T. Katada, T. Amano, and M. Ui, Specific uncoupling by islet-activating protein, pertussis toxin, of negative signal transduction by a alpha-adrenergic, cholinergic, and opiate receptors in neuroblastoma x glioma hybrid cells, J. Biol. Chem. 258:4870 (1983).

7. M. Yajima, K. Hosoda, Y. Kanbayashi, T. Nakamura, Y. Nogimori, Y. Nakase, and M. Ui, Islets-activating protein (IAP) in Bordetella pertussis that potentiates insulin secretory responses of rats: purification and characterization, J. Biochem. 83:295 (1978).

8. T. Kuno, O. Shirakawa, and C. Tanaka, Selective decrease in the affinity of $D_2$ dopamine receptor for agonist induced by islet activating protein, pertussis toxin, associated with ADP-ribosylation of the specific membrane protein of bovine striatum, Biochem. Biophys. Res. Commun. 115:325 (1983).

9. T. Kuno, K. Saijoh, and C. Tanaka, Solubilization of $D_2$ dopamine receptor coupled to guanine nucleotide regulatory protein from bovine striatum, J. Neurochem. 41:841 (1983).

10. W. Schlegel, E. S. Kempner, and M. Rodbell, Activation of adenylate cyclase in hepatic membranes involves interactions of the catalytic unit with multimeric complexes of regulatory proteins, J. Biol. Chem. 254:5168 (1979).

11. T. Kuno and C. Tanaka, Correspondency between different affinity states and target size of the bovine striatal $D_2$ dopamine receptor, in preparation.

12. T. Kuno and C. Tanaka, Identification of the D-1 dopamine receptor subunit in rat striatum after photoaffinity labeling, Brain Res. 230:417 (1981).

13. K. Weber and M. Osborn, The reliability of molecular weight determinations by dodecyl sulfate polyacrylamide gel electrophoresis, J. Biol. Chem. 244:4406 (1969).

14. J. D. Hildebrandt, R. D. Sekura, J. Codina, R. Iyengar, C.R. Manclark, and L. Birnbaumer, Stimulation and inhibition of adenylyl cyclases mediated by distinct regulatory proteins, Nature 302:706 (1983).

15. E. G. Hollunger and D. H. Niklasson, The release and molecular state of mammalian brain acetylcholinesterase, J. Neurochem. 20:821 (1973).

16. L. Lilly, C. M. Fraser, C. Y. Jung, P. Seeman, and J. C. Venter, Molecular size of the canine and human brain $D_2$ dopamine receptor as determined by radiation inactivation, Mol. Pharmacol. 24:10 (1983).

17. M. Rodbell, The role of hormone receptors and GTP-regulatory proteins in membrane transduction, Nature 284:17 (1980).

# GENETIC AND FUNCTIONAL STUDIES OF GUANINE NUCLEOTIDE-BINDING REGULATORY PROTEINS

Henry R. Bourne, Cornelis Van Dop, and
Gerald F. Casperson

Department of Pharmacology and Medicine and the
Cardiovascular Research Institute, University
of California, San Francisco, CA 94143

## INTRODUCTION

Recent research in many laboratories is rapidly elucidating the structure and functions of a family of GTP-binding regulatory proteins, each of which carries information across a membrane from a receptor protein to an effector enzyme. The family includes three known members: $N_s$, $N_i$, and retinal transducin. Adenylate cyclase serves as the effector for two of these proteins, which stimulate ($N_s$) or inhibit ($N_i$) the enzyme when triggered by specific receptors for hormones or neurotransmitters. Transducin, a protein found in high abundance in retinal rod outer segments, couples photoexcitation of rhodopsin to stimulation of a specific cGMP phosphodiesterase (PDE).

Homologies among the three proteins (1-3) include similar subunit structures and amino acid compositions, and the presence of a guanine nucleotide binding site in the α subunit of each protein. In each case, the excited signal detector (hormone receptor or rhodopsin) activates the coupling protein by catalyzing exchange of GTP for GDP at the guanine nucleotide binding site, associated with separation of the coupling protein's α and β subunits. For transducin and $N_s$, the active GTP-bound α subunit stimulates the effector enzyme, adenylate cyclase or PDE. Finally, each of the three proteins has a GTPase activity, which serves to terminate the active state. Light or specific hormones maintain activity of the cognate coupling protein by promoting repeated replacement of GDP by GTP, a process that results in light- or hormone-stimulated GTP hydrolysis (1,2,4).

The present review will summarize studies from this laboratory in two areas: (a) genetic investigations of $N_s$ in mouse, man, and yeast; (b) the use of bacterial toxins to probe the structure and function of transducin. In both areas, we will emphasize questions that should be tackled by new experiments.

## GENETIC INVESTIGATIONS OF $N_s$

### Mutations in S49 Mouse Lymphoma Cells

The ability of cAMP and agents that stimulate cAMP synthesis to kill S49 lymphoma cells allowed isolation of S49 mutants with lesions affecting hormone-sensitive adenylate cyclase (5). Adenylate cyclase in the *cyc⁻* mutant failed to respond to effectors that stimulate cAMP synthesis in wild type S49 cells, including β-adrenergic amines and cholera toxin (6). In a series of elegant experiments, Ross, Gilman, and their colleagues used the *cyc⁻* mutant to discover, purify, and characterize $N_s$ (4,7). They discovered the functional absence of $N_s$ in *cyc⁻*, and used biochemical complementation of the defect of mutant membranes as an assay in purifying the protein. $N_s$ is an oligomer of two subunits, the larger of which (α, a 42-kDa polypeptide) contains the guanine nucleotide binding site and the site that is ADP-ribosylated by the exotoxin of *Vibrio cholerae* (8-10). The α subunit is functionally absent in *cyc⁻*, which apparently contains normal amounts of the smaller β subunit.

Two other S49 mutants display phenotypes (11-13) that result from defective function of the α subunit of $N_s$. $N_s$ in the *unc* mutant is specifically uncoupled from hormone receptors, while the regulatory protein of the H21a mutant is uncoupled from the catalytic unit of adenylate cyclase. α Subunits of both the *unc* and H21a proteins can be ADP-ribosylated by cholera toxin, although this is not true in *cyc⁻*. The *unc* polypeptide differs in electrical charge from the corresponding wild type polypeptide (14); this change probably reflects a mutation that resulted in substitution of an amino acid. All three phenotypes represent mutations in a single gene (presumably the gene coding for the α subunit of $N_s$), as shown by the failure of each to correct defects of the other phenotypes in hybrid S49 cells (13). We have not isolated phenotypes that result from mutations in genes that encode catalytic adenylate cyclase or the β subunit of $N_s$.

Eventual isolation of the gene that encodes the α subunit of $N_s$ will make it possible to pinpoint the changes in DNA base sequence -- and therefore of amino acid sequence -- that produced the three phenotypes described above. Thus, these and other mutants will eventually allow identification of the domains of the coupling proteins that interact with receptors, and those that interact with the catalytic unit.

## Inherited N Deficiency in Man

$N_s$ activity is reduced in erythrocytes and other cells and tissues of patients with Type I-pseudohypoparathyroidism (PHP-I), an inherited syndrome of resistance to parathyroid hormone and other hormones that act *via* stimulating cAMP synthesis (15, 16). Many PHP-I patients exhibit hypocalcemia and its sequelae, including tetany and convulsions. The hypocalcemia is associated with elevated serum parathyroid hormone (PTH) and resistance to the metabolic effects of PTH. Most of these patients also exhibit a typical body habitus called "Albright's hereditary osteodystrophy" (AHO), which includes round face, short stature, and short fingers and toes. (For references, see 17, 18.)

The resistance of many PHP-I patients to hormones other than PTH, including thyrotropin, suggested that the biochemical lesion should be generalized and distal to the hormone receptor. To test this prediction, two assays were used to measure $N_s$ activity in cells from PHP-I patients: 1. Assessment of the ability of membrane extracts to complement the *cyc*$^-$ defect *in vitro*; 2. Quantitation of incorporation of radioactive ADP-ribose from $NAD^+$ into the $N_s$ protein's α subunit, a reaction catalyzed by cholera toxin. Using these assays, $N_s$ activity was found to be reduced by about 50% in erythrocytes (15,16), platelets (19), cultured skin fibroblasts (20), and virus-transformed lymphoblasts (21) obtained from PHP-I patients. The observation of $N_s$ deficiency in cultured fibroblasts and lymphoblasts indicates that $N_s$ deficiency is coded directly in the genome of individual cells of PHP-I patients. None of the cell types we tested, however, exhibits any obvious endocrine abnormality in PHP-I.

Downs et al. (22) recently reported measurement of decreased $N_s$ activity in an important PTH target tissue, the kidney. Their patient's erythrocytes also showed decreased $N_s$ activity. For obvious reasons, confirmation of $N_s$ deficiency in kidneys of additional PHP-I patients will be difficult. Nonetheless, it is reasonable to conclude that the $N_s$ deficiency of PHP-I not only involves erythrocytes, platelets, fibroblasts, and lymphoblasts, but also includes endocrine target cells in the organs that exhibit resistance to cAMP-stimulating hormones *in vivo*. These include kidney, thyroid, liver, gonad, and probably bone and anterior pituitary (17, 18).

We have found (23) both autosomal dominant and autosomal recessive inheritance of $N_s$ deficiency and the PHP-I phenotype in different kindreds. This genetic heterogeneity surely indicates that more than one kind of molecular defect can cause $N_s$ deficiency. The autosomal dominant disorder probably results from mutational loss of one of two alleles that encode a polypeptide subunit of the $N_s$ protein. This would be consistent with a 50% loss of $N_s$

activity. We suspect that the affected subunit is α, because both the $cyc^-$ complementation and ADP-ribosylation assays measure α activities. The recessive pattern of inheritance, of course, cannot be explained in this way.

From the PHP-I phenotype we can infer that normal regulation of calcium by PTH requires all or most of the available $N_s$ protein in important target tissues, because a 50% deficiency causes a clinically obvious disturbance of calcium homeostasis. By the same reasoning, $N_s$ activity is not limiting -- in most patients -- for functionally adequate regulation of target cells by other hormones. Sufficiently sensitive and subtle endocrine testing can detect subnormal responsiveness to certain hormones in PHP-I, even when no clinical abnormality is obvious (for references see 17, 18). In addition, some $N_s$-deficient PHP-I patients do not suffer from hypocalcemia and have only minimally elevated serum PTH, but nonetheless exhibit decreased urinary cAMP responses to PTH. Thus, some individuals apparently do not require all or most of the normal complement of $N_s$ to regulate calcium homeostasis.

Finally, patients in some families with PHP-I do not have detectable $N_s$ deficiency (17). These patients have a lower incidence of AHO and endocrine disorders such as hypothyroidism. At present the molecular basis of this variant of PHP-I is unknown. Isolation of the genes that encode polypeptide subunits of the $N_s$ protein and the catalytic unit of adenylate cyclase will eventually contribute to understanding the molecular pathophysiology of the various subtypes of PHP-I.

## Mutations in Simple Eukaryotes

As objects for genetic investigation, simple eukaryotes such as the yeast *Saccharomyces cerevisiae* possess distinct advantages over mouse cells in tissue culture or human patients in an endocrine clinic. Yeast cells have a much smaller genome, can express recessive mutations in the haploid stage of their life cycle, can be efficently transformed with exogenous DNA, and have a generation time of 2 hours -- as opposed to 17 hours for S49 cells or 20+ years for human beings.

We recently found (24) that adenylate cyclase in yeast membranes responds to stimulation by guanine nucleotides in a pattern that closely resembles responses observed in vertebrate cells: GTP and GTP analogs, such as Gpp(NH)p and GTPγS, stimulate yeast adenylate cyclase, and this stimulation is competitively inhibited by a GDP analog, GDPβS. Just as in vertebrate systems, regulation by guanine nucleotides is most prominent in the presence of $Mg^{2+}$, rather than $Mn^{2+}$. Furthermore, yeast membranes appear to contain distinct components analogous to the $N_s$ and C units of vertebrate adenylate cyclase. The putative C activity (i.e., the activity detectable in the presence of $Mn^{2+}$) is thermostable; in contrast, the guanine nucleotide-regulated $Mg^{2+}$-dependent activity is very thermolabile, unless protected by the presence of a guanine nucleotide during heat treatment.

Matsumoto et al. (25) recently isolated an adenylate cyclase-deficient yeast mutant, _cyr1_, which required exogenous cAMP for growth. The same group (T. Ishikawa, personal communication) has shown that _CYR1_ is the structural gene for catalytic adenylate cyclase. Presently available genetic techniques allow transformation of yeast cells with yeast DNA incorporated into plasmids that can replicate both in yeast and in _E. coli_. It is likely that the yeast C gene will be isolated in the near future.

Recent experiments with the _cyr1_ mutant (Casperson et al., unpublished) provide very strong evidence that yeast adenylate cyclase contains distinct gene products that are cognate to the components of vertebrate adenylate cyclase. Particulate extracts of _cyr1_ lack detectable adenylate cyclase activity (24,25), whether assayed in the presence of $Mn^{2+}$ or $Mg^{2+}$. Heat-treated wild type particulate extracts contain almost no guanine nucleotide-sensitive, $Mg^{2+}$-dependent adenylate cyclase activity, but retain $Mn^{2+}$-dependent activity. Mixture of the _cyr1_ extracts with heat-treated wild type extracts reconstitutes the $Mg^{2+}$-dependent activity. This result strongly suggests that _cyr1_ membranes lack C activity but can supply functional $N_s$ to wild type membranes in which C remains intact, while $N_s$ has been thermally inactivated. Rosenberg and Pall recently reported (26,27) quite similar findings in the adenylate cyclase system of _Neurospora crassa_.

These findings in fungi raise a host of interesting questions. What is the role of a guanine nucleotide-binding signal-coupling protein in these organisms? What hormone-like signal is transduced? Do these cells contain proteins cognate to hormone receptors? Can genes encoding fungal cyclase components be isolated and used to isolate cognate genes of vertebrate cells?

## BACTERIAL TOXINS AND RETINAL TRANSDUCIN

In this section we shall summarize recent work utilizing two bacterial toxins as probes to study the structure and function of transducin, the GTP-binding regulatory protein of retinal rod outer segments (ROS).

First, a word about the toxins: Cholera toxin played a useful part in the early characterization of $N_s$, because toxin-catalyzed ADP-ribosylation allowed the protein's α subunit to be tagged with $[^{32}P]$ADP-ribose (8-10). More recently a toxin produced by another bacterial pathogen, _Bordetella pertussis_, led to identification of

$N_i$, a membrane protein closely homologous to $N_s$. Ui and his colleagues (as reviewed elsewhere in this volume) showed that treatment with pertussis toxin caused attenuation of the inhibition of adenylate cyclase produced by inhibitory ligands, including α-adrenergic, muscarinic, and opiate agonists. They also showed that the toxin has an A-B structure similar to that of cholera toxin, and that it acts by catalyzing ADP-ribosylation of a 41-kDa polypeptide that is bound to the plasma membrane (28-30). Later work in several laboratories (31,32) has shown that the 41-kDa polypeptide is in fact the α subunit of $N_i$.

Recent investigations of the biochemistry of signal transduction in ROS have characterized a set of proteins that exhibit striking functional homologies with hormone-sensitive adenylate cyclase (1-3,33). In ROS the flow of information (Fig. 1) is from photon through rhodopsin and transducin to activation of the hydrolysis of cGMP by a specific phosphodiesterase (PDE). The retinal system allows enormous amplification of the external signal: a single photon may activate hundreds of transducin and PDE molecules and stimulate hydrolysis of millions of cGMP molecules (1). The sequence of reactions by which transducin couples photoexcitation of rhodopsin to stimulation of the PDE (Fig. 1, top) is similar or identical to the mechanism by which $N_s$ is thought to mediate hormonal activation of adenylate cyclase. Both involve receptor-catalyzed exchange of GTP for GDP at a binding site on the α subunit; termination of the activity of the effector enzyme by hydrolysis of the GTP bound to the coupling protein; and eventual re-initiation of the coupling process by another interaction between receptor and coupling protein.

Compared to the amounts of receptor and N proteins obtainable from plasma membranes, ROS contain vast amounts of rhodopsin, transducin, and the effector enzyme, cGMP PDE. The relative abundance of the retinal proteins makes possible some biochemical experiments that would be impossible with hormone-sensitive adenylate cyclase. To the extent that the two systems are analogous, results in one system can be used to help understand the other. This is the rationale of the experiments described in this section.

## Cholera toxin and transducin

Cholera toxin stimulates adenylate cyclase by ADP-ribosylating the α-subunit of $N_s$, a modification that stabilizes $N_s$ in the GTP-bound active conformation (34), resulting in inhibition of hormone-stimulated GTPase and persistent activation of adenylate cyclase. In collaboration with the laboratory of Lubert Stryer, we found (33) that cholera toxin also catalyzes ADP-ribosylation of the α-subunit of transducin in ROS, and thereby inhibits its light-stimulated GTPase activity.

The parallels between cholera toxin-catalyzed ADP-ribosylation of $N_s$ and transducin were quite striking: (a) Both proteins served as substrates for the toxin only in the membrane-bound form; (b) activation of the detector molecule by hormone (35) or light (33) enhanced toxin-catalyzed ADP-ribosylation (indeed, transducin was not ADP-ribosylated at all in dark-adapted ROS); (c) binding of a stable GTP analog, such as guanylyl-5'-imidodiphosphate [Gpp(NH)p], enhanced ADP-ribosylation, both in the adenylate cyclase system (35) and in ROS (33). In sum, it appears in both systems that cholera toxin prefers as a substrate the "active," GTP-bound conformation of the coupling protein; strikingly, the covalent attachment of ADP-ribose stabilizes the coupling protein in the same "active" conformation, or one very similar to it.

The close functional homologies between $N_s$ and transducin accompany structural homologies as well. Thus, both proteins are heterodimers of polypeptides with similar molecular weights and amino acid compositions (3). Because cholera toxin requires such similar conditions for ADP-ribosylating both proteins, and because the functional effects of ADP-ribosylation are also similar, it is likely that this toxin ADP-ribosylates similar sites in the two proteins. Accordingly, we sought to identify the site at which cholera toxin attaches ADP-ribose to transducin. In collaboration with M. Tsubokawa and J. Ramachandran, we modeled our experiments on previous studies (36) of the site on Elongation Factor 2 that is ADP-ribosylated by diptheria toxin. ADP-ribosylated transducin was partially purified and then subjected to tryptic hydrolysis. The ADP-ribosylated fragment, purified on a boronate column followed by high performance liquid chromatography, was sequenced by a micro dansyl-Edman procedure (37). The tetrapeptide we isolated had the sequence SER-ARG-VAL-LYS. Arginine was the ADP-ribosylated amino acid.

This is the first identification of an ADP-ribosylated arginine from a protein that is specifically ADP-ribosylated by cholera toxin. Incubations of cholera toxin with $NAD^+$ and various amino acids had previously indicated that the guanidinium group of arginine can serve as an acceptor for ADP-ribose (38). Like cholera toxin, two other bacterial toxins affect cellular metabolism by mono-ADP-ribosylating specific cellular proteins (30,39). Diptheria toxin ADP-ribosylates diphthamide on Elongation Factor 2 (40) and pertussis toxin (D.R. Manning, B. Fraser, R. Kahn, and A.G. Gilman, personal communication) ADP-ribosylates a different amino acid on retinal transducin (see below). In addition, ADP-ribosylation of an arginine on RNA polymerase has been reported during T4 phage infection of *E. coli* (41). Thus, each of the toxins ADP-ribosylates a unique amino acid on its specific protein substrate.

Because of the close structural and functional similarities between $N_s$ and transducin, we predict that the amino acid sequence around the cholera toxin substrate site on $N_s$ will closely resemble that found in transducin, and that the ADP-ribosylated amino acid will be arginine.

Pertussis Toxin and Transducin

Because of the striking parallels, outlined above, between hormone-sensitive adenylate cyclase and the rod photoreceptor, we imagined that ROS might contain a protein analogous to $N_i$. Results of exposure of photoactivated ROS to pertussis toxin and $[^{32}P]NAD^+$ were quite disappointing. The toxin appeared to catalyze very little labeling of any protein. Eventually we discovered that this disappointing outcome resulted from performing the experiments with photoexcited ROS. We subsequently found that pertussis toxin very rapidly ADP-ribosylates one ROS polypeptide in the dark, and that this reaction is markedly inhibited by photoexcitation of rhodopsin (C. Van Dop, G. Yamanaka, F. Steinberg, R.D. Sekura, C.R. Manclark, L. Stryer, and H.R. Bourne, unpublished).

A second surprise was that the polypeptide ADP-ribosylated by pertussis toxin in ROS is the α subunit of transducin itself. We had initially expected to find a substrate distinct from transducin, just as $N_i$ is distinct from $N_s$. The polypeptide ADP-ribosylated by pertussis toxin, however, was clearly transducin's α subunit. Indeed, transducin serves as an excellent substrate for ADP-ribosylation by the toxin, which can catalyze incorporation of one mole of ADP-ribose per mole of transducin (A.G. Gilman, personal communication).

Further studies indicated that pertussis toxin prefers the inactive GDP-bound form of transducin -- a preference directly opposite to that of cholera toxin for the same protein. Thus, light and exposure to GTP analogs inhibit ADP-ribosylation by pertussis toxin and increase ADP-ribosylation by cholera toxin. Conversely, dark adaptation and GDP enhance ADP-ribosylation by pertussis toxin, and markedly attenuate the reaction catalyzed by cholera toxin. It should be noted that the two toxins ADP-ribosylate quite distinct sites on the α subunit: The amino acid sequence surrounding the substrate site for pertussis toxin (A.G. Gilman, personal communication) bears no resemblance to the sequence we found for cholera toxin.

ADP-ribosylation by pertussis toxin dramatically alters the function of transducin. Firstly, we found that ADP-ribosylated transducin binds much less tightly to photoexcited rhodopsin than does native transducin. In addition, the light-activated GTPase activity of rod outer segments is inhibited. Finally, the signal-coupling capacity of transducin is blocked, thereby preventing photoactivation of phosphodiesterase.

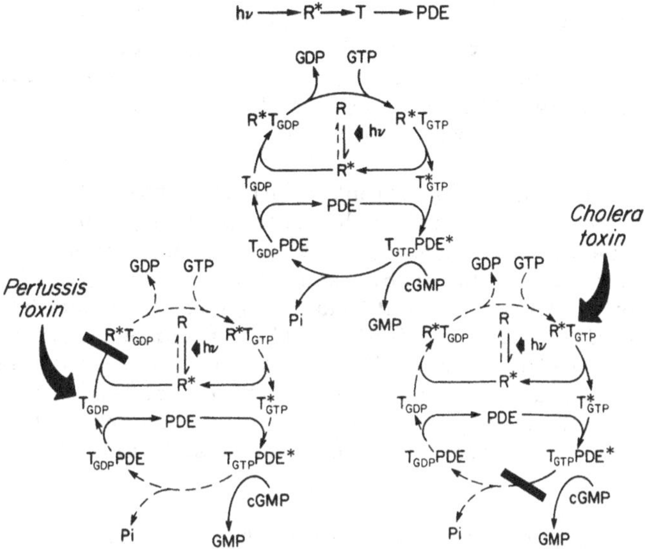


Fig. 1. Effects of pertussis and cholera toxins on signal transmission in rod outer sigments. The flow of information in this system is from light (the signal, analogous to hormone in the adenylate cyclase system) to rhodopsin ($R^*$, analogous to the hormone receptor), through transducin (analogous to the N proteins) to the effector enzyme, a cGMP phosphodiesterase (PDE, analogous to the catalytic unit of adenylate cyclase). The diagram at the top of the figure shows the current view of how transducin couples $R^*$ to activation of PDE, based on work from several laboratories (1-3). In this scheme, photoexcited rhodopsin catalyzes the exchange of GDP for GTP at the guanine nucleotide binding site on transducin. The GTP-bound "active" transducin ($T^*$) then associates with PDE to form a cGMP-hydrolyzing ternary complex. The activity of this complex terminates in association with hydrolysis of GTP to GDP, and the now inactive T dissociates from the PDE. To maintain cGMP hydrolysis, T must be reactivated by the GDP-GTP exchange mechanism involving $R^*$. Our data indicate that pertussis toxin (diagram at lower left) preferentially ADP-ribosylates the GDP-bound, inactive conformation of transducin. This modification blocks the cycle of reactivation by preventing T from interacting with $R^*$ (heavy bar). In contrast, cholera toxin (diagram at lower right) preferentially ADP-ribosylates the active conformation of transducin, probably when it is bound to $R^*$. This modification slows the hydrolysis of GTP by transducin (heavy bar) and thereby prolongs the duration of the active ternary complex of T, GTP, and PDE.

Fig. 1 depicts our interpretation of the effects of the two toxins on photoexcitation in ROS: ADP-ribosylation by cholera toxin inhibits the GTPase activity of transducin by stabilizing the T-GTP state. Fittingly, the T-GTP state is also the preferred substrate conformation for cholera toxin.

Pertussis toxin treatment inhibits the GTPase activity of transducin by a different mechanism, probably through interference with exchange of GTP for GDP. Transducin ADP-ribosylated by pertussis toxin appears to be stabilized in the T-GDP state. Just as with cholera toxin, the conformational state stabilized by ADP-ribosylation is similar or identical to the preferred substrate conformation for the toxin. The stability of the T-GDP state is accompanied by diminished capacity of the modified protein to interact with photoexcited rhodopsin ($R^*$ in Fig. 1), the catalyst for GTP-GDP exchange, as indicated by its decreased affinity for bleached membranes.

This interpretation suggests that pertussis toxin treatment may decrease the affinity of the cognate coupling protein, $N_i$, for hormone receptors that inhibit adenylate cyclase. Indeed, the toxin abolished high affinity binding of α-adrenergic and muscarinic agonists to their respective receptors in a cultured cell line (42). In adenylate cyclase systems, this high affinity binding of agonists (which is decreased or absent in the presence of GTP analogs) is considered an index of receptor coupling to guanine nucleotide-binding regulatory proteins. As a second corollary of the present findings with transducin, we predict that pertussis toxin will ADP-ribosylate $N_i$ better in the presence of stable GDP analogs than in the presence of GTP analogs or of agonists that inhibit adenylate cyclase. Initial experiments in our laboratory are consistent with this prediction.

What is the biological significance of ADP-ribosylation sites on transducin for both pertussis and cholera toxins? Why do toxin-catalyzed modifications at these sites cause parallel functional consequences in transducin and the two N proteins? We imagine that the genes encoding transducin and a primordial N protein diverged in the course of evolution from a common ancestral gene. This divergence presumably occurred before the subsequent separation of genes for the $N_s$ and $N_i$ proteins, each of which has retained the ADP-ribosylation site for only one of the toxins. One is hard put, however, to imagine why any of these proteins serves as a substrate for ADP-ribosylation by a bacterial toxin, or why the ADP-ribosylation sites appear to have been conserved. Preservation of the two substrate sites in transducin suggests that the sites are functionally important. Perhaps they serve as recognition sites for other polypeptides -- e.g., β subunits of the same protein, effector enzymes, signal detectors, or even endogenous ADP-ribosylating enzymes.

## PROSPECTS

Investigation of signal transduction across membranes stands on the threshold of an explosion of new information, to be derived from isolation and characterization of genes that encode protein components of signal-transducing systems. Here we have outlined some of the questions raised by mutations of adenylate cyclase in mouse, man, and yeast, and by biochemical studies of retinal transducin. $N_s$ and $N_i$ belong to a family of GTP-binding signal coupling proteins, which includes transducin in retinal rod outer segments and perhaps other membrane proteins as well. Characterization of the genes that encode polypeptide components of these systems will provide precise information regarding the primary structure of the proteins, will help to define changes in their expression and function during differentiation, and will make possible genetic experiments designed to pinpoint structural domains responsible for interactions among the proteins.

## ACKNOWLEDGEMENTS

This work was supported by grants from the National Institutes of Health (GM 27800 and GM 28310) and from the March of Dimes. C. Van Dop was supported by the Johnson and Johnson Institute for Pediatric Service. H. Bourne is a Burroughs-Wellcome Scholar in Clinical Pharmacology.

## REFERENCES

1. L. Stryer, J. B. Hurley, and B. K.-K. Fung, Transducin: an amplifier protein in vision, Trends Biochem. Sci. 6:245 (1981).
2. M. W. Bitensky, G. L. Wheeler, A. Yamazaki, M. W. Rasenick, and P. J. Stein, Cyclic-nucleotide metabolism in vertebrate photoreceptors: A remarkable analogy and an unraveling enigma, Current Topics Membranes and Transport 15:237 (1981).
3. D. R. Manning and A. G. Gilman, The regulatory components of adenylate cyclase and transducin. A family of structurally homologous guanine nucleotide binding proteins, J. Biol. Chem. 258:7059 (1983).
4. M. D. Smigel, J. K. Northup, and A. G. Gilman, Characteristics of the guanine nucleotide-binding regulatory component of adenylate cyclase, Recent Progr. Hormone Res. 38:601 (1982).
5. G. L. Johnson, H. R. Kaslow, Z. Farfel, and H. R. Bourne, Genetic analysis of hormone-sensitive adenylate cyclase, in: "Advances in Cyclic Nucleotide Research," Vol. 13, P. Greengard and G. A. Robison, eds., Raven Press, New York, p. 1 (1980).

6. H. R. Bourne, P. Coffino, and G. M. Tomkins, Selection of a variant lymphoma cell deficient in adenylate cyclase, Science 187:750 (1975).
7. E. M. Ross and A. G. Gilman, Resolution of some components of adenylate cyclase necessary for catalytic activity, J. Biol. Chem. 252:6966 (1977).
8. D. M. Gill and R. Meren, ADP-ribosylation of membrane proteins catalyzed by cholera toxin: Basis of the activation of adenylate cyclase, Proc. Nat. Acad. Sci. USA 75:3050 (1978).
9. D. Cassel and T. Pfeuffer, Mechanism of cholera toxin action: covalent modification of the guanyl nucleotide-binding protein of the adenylate cyclase system, Proc. Natl. Acad. Sci. USA 75:2669 (1978).
10. G. L. Johnson, H. R. Kaslow, and H. R. Bourne, Genetic evidence that cholera toxin substrates are regulatory components of adenylate cyclase, J. Biol. Chem. 253:7120 (1978).
11. T. Haga, E. M. Ross, H. J. Anderson, and A. G. Gilman, Adenylate cyclase permanently uncoupled from hormone receptors in a novel variant of S49 lymphoma cells, Proc. Natl. Acad. Sci. USA 74:2016 (1977).
12. M. R. Salomon and H. R. Bourne, Novel S49 lymphoma variants with aberrant cyclic AMP metabolism, Mol. Pharmacol. 19:109 (1981).
13. H. R. Bourne, B. Beiderman, F. Steinberg, and V. M. Brothers, Three adenylate cyclase phenotypes in S49 lymphoma cells produced by mutations of one gene, Mol. Pharmacol. 22:204 (1982).
14. L. S. Schleifer, J. C. Garrison, P. C. Sternweis, J. K. Northup, and A. G. Gilman, The regulatory component of adenylate cyclase from uncoupled S49 lymphoma cells differs in charge from the wild type protein, J. Biol. Chem. 255:2641 (1980).
15. Z. Farfel, A. S. Brickman, H. R. Kaslow, V. M. Brothers, and H. R. Bourne, Defect of receptor-cyclase coupling protein in pseudohypoparathyroidism, New Engl. J. Med. 303:237 (1980).
16. M. A. Levine, R. W. Downs, M. Singer, S. J. Marx, G. D. Aurbach and A. M. Spiegel, Deficient activity of guanine nucleotide regulatory protein in erythrocytes from patients with pseudohypoparathyroidism, Biochem. Biophys. Res. Comm. 94:1319 (1980).
17. Z. Farfel, and H. R. Bourne, Pseudohypoparathyroidism: Mutation affecting adenylate cyclase, Mineral Electrolyte Metab. 8:227 (1982).
18. A. M. Spiegel, M. A. Levine, G. D. Aurbach, R. W. Downs, Jr., S. J. Marx, R. D. Lasker, A. M. Moses, and N. A. Breslau, Deficiency of hormone receptor-adenylate cyclase coupling protein: basis for hormone resistance in pseudohypoparathyroidism, Am. J. Physiol. 243:E37 (1982).

19. Z. Farfel and H. R. Bourne, Deficient activity of receptor-cyclase coupling protein in platelets of patients with pseudohypoparathyroidism, J. Clin. Endocrinol. Metab. 51:1202 (1980).
20. H. R. Bourne, H. R. Kaslow, A. S. Brickman, and Z. Fargel, Fibroblast defect in pseudohypoparathyroidism, type I: Reduced activity of receptor-cyclase coupling protein, J. Clin. Endocrinol. Metab. 53:636 (1981).
21. Z. Farfel, M. E. Abood, A. S. Brickman, and H. R. Bourne, Deficient activity of receptor-cyclase coupling protein in transformed lymphoblasts of patients with pseudohypoparathyroidism, Type I, J. Clin. Endocrinol. Metab. 55:113 (1982).
22. R. W. Downs, Jr., M. A. Levine, M. K. Drezner, W. M. Burch, Jr. and A. M. Spiegel, Deficient adenylate cyclase regulatory protein in renal membranes from a patient with pseudohypoparathyroidism, J. Clin. Invest. 71:231 (1983).
23. Z. Farfel, V. M. Brothers, A. S. Brickman, F. Conte, R. Neer, and H. R. Bourne, Pseudohypoparathyroidism: Inheritance of deficient receptor-cyclase coupling activity, Proc. Natl. Acad. Sci. USA 78:3098 (1981).
24. G. F. Casperson, N. Walker, A. R. Brasier, and H. R. Bourne, A guanine nucleotide sensitive adenylate cyclase in the yeast Saccharomyces cerevisiae, J. Biol. Chem. 258:7911 (1983).
25. K. Matsumoto, I. Uno, Y. Oshima, and T. Ishikawa, Isolation and characterization of yeast mutants deficient in adenylate cyclase and cAMP-dependent protein kinase, Proc. Natl. Acad. Sci. USA 79:2355 (1982).
26. G. B. Rosenberg and M. L. Pall, Characterization of an ATP-$Mg^{2+}$-dependent guanine nucleotide-stimulated adenylate cyclase from Neurospora crassa, Arch. Biochem. Biophys. 221:243 (1983).
27. G. B. Rosenberg and M. L. Pall, Reconstitution of adenylate cyclase in Neurospora from two components of the enzyme, Arch. Biochem. Biophys. 221:254 (1983).
28. O. Hazeki and M. Ui, Modification by islet-activating protein of receptor-mediated regulation of cyclic AMP accumulation in isolated rat heart cells, J. Biol. Chem. 256:2856 (1981).
29. T. Katada and M. Ui, Islet-activating protein. A modifier of receptor-mediated regulation of rat islet adenylate cyclase, J. Biol. Chem. 256:8310 (1981).
30. T. Katada and M. Ui, ADP-ribosylation of the specific membrane protein of C6 cells by islet-activating protein associated with modification of adenylate cyclase activity, J. Biol. Chem. 257:7210 (1982).
31. G. M. Bokoch, T. Katada, J. K. Northup, E. L. Hewlett, and A. G. Gilman, Identification of the predominant substrate for ADP-ribosylation by islet-activating protein, J. Biol. Chem. 258:2072 (1983).

32. J. Codina, J. Hildebrandt, R. Iyengar, and L. Birnbaumer, Pertussis toxin substrate, the putative $N_i$ component of adenylate cyclases, in an αβ heterodimer regulated by guanine nucleotide and magnesium, Proc. Natl. Acad. Sci. USA, in press, 1983.
33. M. E. Abood, J. B. Hurley, M.-C. Pappone, H. R. Bourne, and L. Stryer, Functional homology between signal-coupling proteins: Cholera toxin inactivates the GTPase activity of transducin, J. Biol. Chem. 257:10540 (1982).
34. D. Cassel and Z. Selinger, Mechanism of adenylate cyclase activation by cholera toxin: inhibition of GTP hydrolysis at the regulatory site, Proc. Natl. Acad. Sci. USA 74:3307 (1977).
35. K. Enomoto and D. M. Gill, Cholera toxin activation of adenylate cyclase. Roles of nucleoside triphosphates and a macromolecular factor in the ADP ribosylation of the GTP-dependent regulatory component, J. Biol. Chem. 255:1252 (1980).
36. B. A. Brown and J. W. Bodley, Primary structure at the site in beef and wheat elongation factor 2 of ADP-ribosylation by diptheria toxin, FEBS Lett. 103:253 (1979).
37. D. H. Spackman, W. H. Stein and S. Moore, Automatic recording apparatus for use in the chromatography of amino acids, Anal. Chem. 30:1190 (1958).
38. S. Nakaya, J. Moss, and M. Vaughan, Effects of nucleoside triphosphates on choleragen-activated brain adenylate cyclase, Biochemistry 19:4871 (1980).
39. R. J. Collier, Structure and activity of diptheria toxin, in: "ADP-Ribosylation Reactions," Academic Press, New York, p. 575 (1982).
40. B. G. Van Ness, J. B. Howard and J. W. Bodley, ADP-ribosylation of elongation factor 2 by diptheria toxin. NMR spectra and proposed structures of ribosyl-diphthamide and its hydrolysis products, J. Biol. Chem. 255:10710 (1980).
41. C. G. Goff, Chemical structure of a modification of the Escherichia coli ribonucleic acid polymerase α polypeptides induced by bacteriophage T4 infection, J. Biol. Chem. 249:6181 (1974).
42. H. Kurose, T. Katada, T. Amano, and M. Ui, Specific uncoupling by islet-activating protein, pertussis toxin, of negative signal transduction via α-adrenergic, cholinergic, and opiate receptors in neuroblastoma x glioma hybrid cells, J. Biol. Chem. 258:4870 (1983).

# DIFFERENTIAL REGULATION OF PUTATIVE $M_1/M_2$ MUSCARINIC RECEPTORS: IMPLICATIONS FOR DIFFERENT RECEPTOR-EFFECTOR COUPLING MECHANISMS

Thomas W. Vickroy, Mark Watson, Henry I. Yamamura and William R. Roeske

Departments of Pharmacology and Internal Medicine
University of Arizona Health Sciences Center
Tucson, Arizona 85724

## INTRODUCTION

This chapter focuses upon results from our recent studies concerning the drug specificities and regulatory profiles of high-affinity muscarinic agonist binding site subtypes. The data are presented in conjunction with other evidence for muscarinic receptor subtypes and a significant portion of the discussion emphasizes the potential involvement of distinct receptor-effector coupling mechanisms for these subtypes. Other reviews relevant to this topic are currently available (1-4).

The pioneering studies of Dale in 1914 (5) and Loewi in 1921 (6) are now recognized as the first clear demonstration of chemical neurotransmission involving a membrane receptor. Since that early work focused upon the interaction of acetylcholine with muscarinic receptors, it is not surprising that muscarinic receptors have been the subject of extensive biochemical, physiological and pharmacological investigations. Nevertheless, until very recently little credence was given to the concept of distinct muscarinic receptor subtypes. However as outlined below, the recent availability of selective muscarinic agents in conjunction with the increased sophistication of receptor binding techniques have provided firm evidence which appears to be inconsistent with a unitary muscarinic receptor hypothesis.

## EVIDENCE FOR MUSCARINIC RECEPTOR SUBTYPES

### Studies In Vivo

Dale's report in 1914 (5), which categorized the actions of acetylcholine as belonging to the "...depressor, cardioinhibitory 'muscarinic' type..." or the "...pressor action of nicotine type...", was a classical discovery which remained unchallenged for almost 50 years. However, in 1961, a report by Roszkowski (7) concerning the general pharmacology of a new synthetic compound revealed the first solid evidence that Dale's classification may be oversimplified. This compound, whose chemical name is 4-(*m*-chlorophenylcarbamoyloxy)-2-butynyltrimethylammonium chloride (abbreviated McN-A-343), appeared to elevate canine blood pressure by selectively stimulating muscarinic receptors in sympathetic ganglia. The action was called "selective" since McN-A-343 had little or no effect on muscarinic receptors in isolated heart or jejunal preparations. However, while this study clearly provided evidence that muscarinic receptors were dissimilar in some tissues, the concept of muscarinic receptor heterogeneity received little immediate attention.

Then in 1978, a report by Goyal and Rattan (8) rekindled interest in muscarinic receptor subtypes. Based upon their studies in the opposum lower esophageal sphincter, these investigators proposed that distinct muscarinic receptor subtypes ($M_1$ and $M_2$) must exist. Nevertheless, this study like its predecessors suffered from a lack of sufficiently selective drugs which were capable of clearly demonstrating pharmacological differences between these subtypes. Thus if it were not for the immediate discovery of a selective muscarinic drug, interest in this topic might have delined once again. Soon after Goyal and Rattan's report (8), however, the $M_1$ selective muscarinic action of the antagonist pirenzepine was described (9,10) and resulted in more widespread attention being focused upon this topic. Following these reports, several studies *in vivo* clearly demonstrated the selective antimuscarinic actions of pirenzepine (11,12) in paradigms where the classical antimuscarinic atropine revealed no selectivity. For example, in pithed rats pirenzepine and atropine were observed to be roughly equipotent as inhibitors of McN-A-343-induced pressor responses (an effect probably mediated by muscarinic receptors in sympathetic ganglia); by comparison pirenzepine was 50 times weaker than atropine in reversing bradycardia following vagal nerve stimulation (an effect mediated by muscarinic receptors in the heart [11]). Thus unlike the majority of other muscarinic compounds, pirenzepine (11) and McN-A-343 (7) appear to selectively influence a subset of muscarinic receptors which are currently classififed as the $M_1$ subtype.

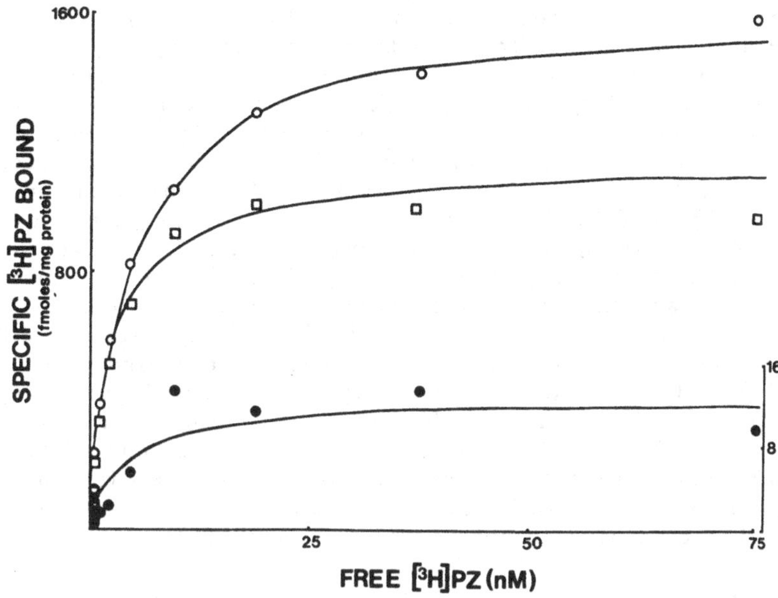


Fig. 1. Regional binding of [$^3$H]PZ. Saturation experiments were carried out in 10 mM sodium-potassium phosphate buffer for rat corpus striatum (O), cerebral cortex (□) and cerebellum (●). The ordinate on the right was used for plotting cerebellar data (reprinted from Watson et al., 13).

## Studies In Vitro

Although still in the early stages, studies of muscarinic receptor subtypes using isolated organs and tissue homogenates have already provided invaluable information. As early as 1976, a report by Barlow and coworkers (14) demonstrated tissue-selective muscarinic effects by the antagonist 4-diphenylacetoxy-N-methylpiperidine methiodide. However, with the discovery of pirenzepine's selectivity via indirect receptor binding studies (10), most subsequent investigations have focused upon this compound (15). Direct studies of [$^3$H]pirenzepine ([$^3$H]PZ) binding have confirmed and extended many of these findings.

The availability of [$^3$H]PZ has recently permitted the direct estimation of pirenzepine binding sites ($M_1$) in tissues which respond to or are relatively insensitive to this selective muscarinic compound. Direct measurements of $[^3H]$PZ binding (13,16) have demonstrated that this drug binds to a saturable population of high-affinity muscarinic sites in many tissues which receive cholinergic innervation (Fig. 1). By itself, this finding is not remarkable since several muscarinic cholinergic ligands are already available. However, when compared with the classical muscarinic antagonist (-)3-quinuclidinyl benzilate (QNB), an important trend becomes apparent (13,17). Tissues in which pirenzepine has potent effects (for example, sympathetic ganglia [11,15]) contain a high-density of high-affinity $[^3H]$PZ binding sites (17). Conversely, tissues which are only weakly affected by pirenzepine (heart) have a very low density of high-affinity $[^3H]$PZ binding sites (Table 1). Thus there appears to be a correlation between the binding of $[^3H]$PZ (13,16,17) and the effects of the unlabelled drug *in vitro* (15) as well as *in vivo* (11,12). In addition, autoradiographic comparisons of $[^3H]$PZ and $[^3H]$(-)QNB binding in the central nervous system (Fig. 2) clearly demonstrate additional differences in the distribution of binding sites for these ligands and may account for the unique effects of pirenzepine in behavioral paradigms (12).

Table 1. Densities of $[^3H]$PZ and $[^3H]$(-)QNB Binding Sites in Various Tissues

| Tissue | $[^3H]$PZ | $[^3H]$(-)QNB | $\%M_1$ |
|---|---|---|---|
| Cerebral Cortex[a] | 80.0 | 100.0 | 80.0 |
| Human Stellate Ganglia[a] | 2.1 | 4.0 | 52.5 |
| Cerebellum[b] | 1.0 | 9.9 | 10.4 |
| Heart[b] | 0.5 | 14.4 | 3.1 |

Values for binding site densities are in fmoles/mg tissue. Portions of this data have been reprinted from Watson et al., (13).

[a]Binding was measured in a physiological salt solution.

[b]Binding was measured in 10mM sodium-potassium phosphate buffer.

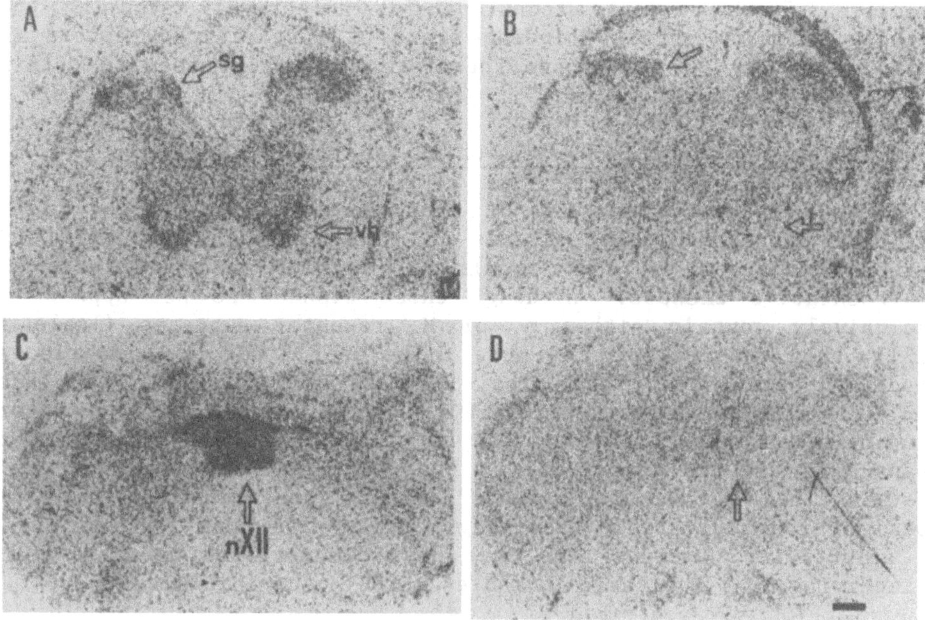


Fig. 2. Differential autoradiographic localization of [$^{3}$H](-)QNB and [$^{3}$H]PZ binding sites in the rat spinal cord (panels A and B) and brainstem (panels C and D) of the rat. Mounted slides were incubated with [$^{3}$H](-)QNB (A and C) or [$^{3}$H]PZ (B and D). Symbols: sg, substantia gelatinosa of the dorsal horn; vh, ventral horn; nXII, hypoglossal nucleus; bar = 500 μM (reprinted by permission from Yamamura et al., 18).

## AGONIST BINDING TO MUSCARINIC SUBTYPES

Direct studies of muscarinic agonist binding to cholinergic receptors have always been subject to many technical and interpretational problems. The complex nature (2,19-23) and low specific binding of [$^{3}$H]-labelled agonists are undoubtedly responsible for the limited number of direct agonist binding studies which have been carried out to date. However, due to interpretational problems which can arise from indirect approaches (agonist/[$^{3}$H] antagonist) (24), it is often essential to study regulator-induced changes of agonist recognition sites by a direct method. In this regard, we have recently described a novel rapid filtration binding assay for [$^{3}$H]cismethyldioxolane or [$^{3}$H]CD (25), a potent muscarinic agonist (26). We have shown that this assay provides a direct and highly-specific (80-90 percent) technique for selectively studying the highest affinity agonist binding state of muscarinic receptors in various tissues (27) and have therefore used this assay to obtain the results described below.

## Pharmacological Comparisons of Agonist-Labelled Subtypes

For comparative studies of drug specificities at the high-affinity agonist-labelled $M_1$ and $M_2$ sites, the hearts ($M_2$) and cerebral cortices ($M_1$) from adult male albino rats were used as representative tissues. Although these tissues do not appear to exclusively contain either $M_1$ or $M_2$ receptor subtypes (Table 1 [13]), they are sufficiently pure to be useful for the comparisons described here. Initial studies with this technique indicated that [$^3$H]CD binds to high-affinity (Kd = 1-2 nM) muscarinic receptor binding sites in both tissues (27) and therein provides an excellent means for directly comparing the pharmacological profiles of agonist-labelled $M_1$ and $M_2$ sites. From our preliminary studies, a total of four muscarinic cholinergic drugs have been found to show $M_1$ selectivity, whereas none have been demonstrated to possess $M_2$ selectivity. As indicated in Table 2, the muscarinic agonists

Table 2. Drugs Possessing $M_1/M_2$ Selectivity In [$^3$H]CD-Labelled Membranes

| Drug Class | Cortex[a] | Heart[a] | Potency Ratio |
|---|---|---|---|
| Agonists | | | |
| Philocarpine | 22.8 | 96.2 | 4.2 |
| McN-A-343 | 93.7 | 208 | 2.2 |
| Antagonists | | | |
| Pirenzepine | 56.2 | 615 | 10.9 |
| Levetimide | 236 | 1084 | 4.6 |
| Other Drugs | | | |
| Gallamine | 237 | 3.5 | 0.014 |
| Physostigmine | 26.0 | 177 | 6.8 |

Homogenates of rat cerebral cortex ($M_1$) or heart ($M_2$) were incubated with 1 nM [$^3$H]CD and 9-16 concentrations of each drug for 2 hr at 25° C in 10 mM sodium-potassium phosphate buffer (pH 7.4).

[a]Numbers in the table are $IC_{50}$ values (in nanomolar) for each drug at the most prevalent class of [$^3$H]CD-labelled sites.

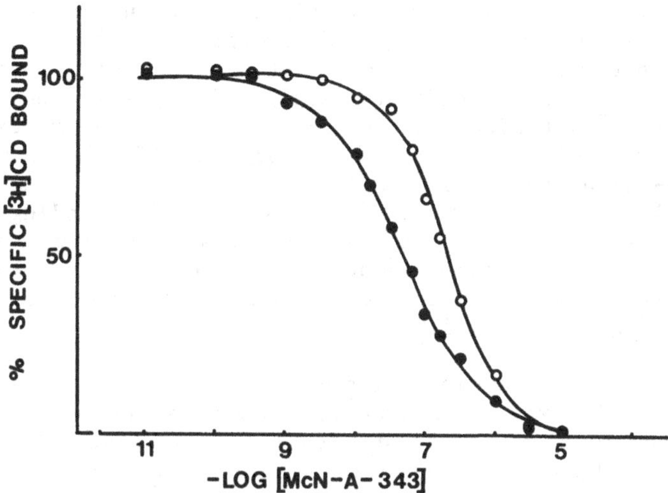


Fig. 3. Inhibition of high-affinity myocardial (O) and cerebral cortical (●) [$^{3}$H]CD (1 nM) binding by McN-A-343. Each point is the mean from 4 - 7 separate experiments (in duplicate) incubated for 2 hr at 25° C in 10 mM sodium-potassium phosphate buffer.

pilocarpine and McN-A-343 and the antagonists pirenzepine and levetimide (inactive isomer of benzetimide) all possess some degree of $M_1$ selectivity under conditions designed to study the highest affinity agonist binding state. By comparision acetylcholine, oxotremorine, carbamylcholine, atropine, scopolamine and dexetimide possess little or no selectivity for these sites in either tissue. Upon closer anlysis of the inihibition data for the "selective" muscarinic compounds, it is apparent that most of these drugs interact with multiple sites in the cerebral cortex and/or heart. For example, inhibition curves for the selective agonist McN-A-343 against [$^{3}$H]CD are shown in Fig. 3. Single-site analyses of these curves suggest that McN-A-343 is four-fold more potent in the cerebral cortex as compared with the heart. However, multiple-site analyses reveal that at least two populations of McN-A-343 binding sites are present in the cerebral cortex ($IC_{50}$ (high) = 2.7 nM (24 percent); $IC_{50}$ (low) = 94 nM (76 percent)) whereas only a single myocardial site is detected ($IC_{50}$ = 186 nM). These results for

McN-A-343 and pirenzopine (Table 2) are especially noteworthy since the apparent selectivity of these compounds has been previously detected by studies in vivo (7,11,12). Interestingly, other cholinergic drugs which are believed to primarily influence nonmuscarinic mechanisms also revealed $M_1/M_2$ selectivity. As shown in Table 2, gallamine (nicotinic cholinergic antagonist) and physostigmine (acetylcholinesterase inhibitor) are both potent inhibitors of high-affinity [$^3$H]CD binding under these conditions. However, while these compounds reveal selective muscarinic effects in this assay (gallamine--$M_2$; physostigmine--$M_1$), their action apparently involved a site which is distinct from the muscarinic drug binding site (28, 29). While the importance of these allosteric drug effects are currently unknown, it should be noted that previous studies in vivo have suggested possible direct muscarinic actions by both physostigmine (30) and gallamine (31,32).

In summary, pharmacological comparisons of the highest affinity agonist-labelled forms of $M_1$ and $M_2$ receptor subtypes indicate that certain muscarinic compounds do possess a moderate degree of $M_1$ selectivity. While these studies clearly point out that greater numbers of selective muscarinic drugs must be uncovered in order to establish distinct pharmacological profiles for muscarinic subtypes, additional studies now suggest that subtypes may also be classified according to nonpharmacological criteria.

## Regulation of Agonist Binding to $M_1$ and $M_2$ Subtypes

For several years it has been widely known that muscarinic agonist binding (and to a much lesser degree antagonist binding) is highly influenced by various endogenous and exogenous regulators. These regulators, which include guanine nucleotides (22-24,33,34), monovalent and divalent cations (2,24,35-37), and sulfhydryl reagents (2,38,39) are believed to act at sites distinct from but somehow coupled to the drug binding site. However, until recently, little or no information was available concerning the effects of these regulators specifically upon $M_1$ or $M_2$ binding site subtypes.

The guanine nucleotide-induced reduction in binding site affinity for muscarinic agonists was first studied by indirect approaches (33,34) and later by direct agonist binding studies (2,22,40). Fig. 4 depicts a typical binding site affinity shift for the muscarinic agonist oxotremorine in the presence of the guanine nucleotide guanyl-5'-yl imidodiphosphate (Gpp(NH)p). As is evident in the heart, the effect of the guanine nucleotide typically involves both a rightward shift (to lower affinity) and a steepening (less heterogeneity) of the agonist competition curve (panel C). However, in the cerebral cortex, a much smaller shift is typically observed in [$^3$H](-)QNB-labelled membranes (panel B) and even less effect (if any) is evident in membranes labelled with the selective antagonist [$^3$H]PZ (panel A). While several complex proposals have

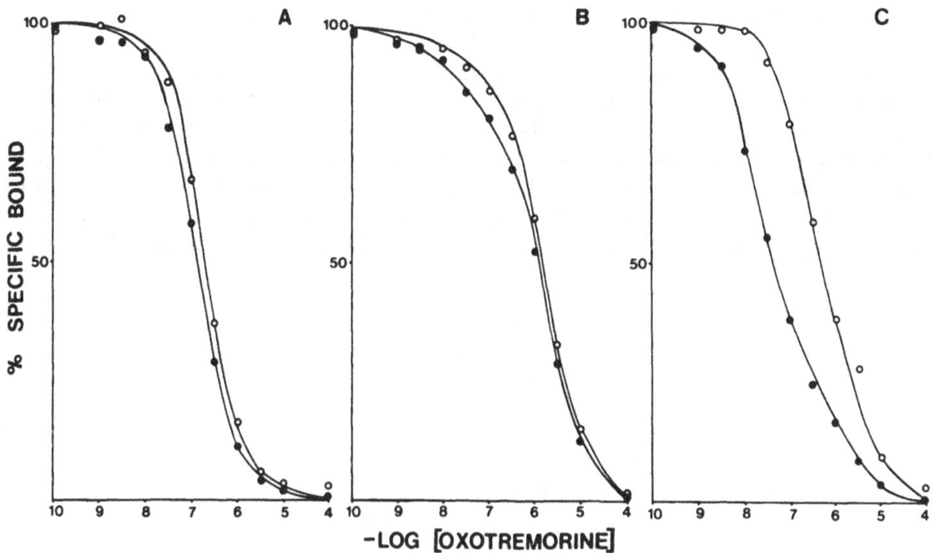


Fig. 4. Gpp(NH)p-induced changes in oxotremorine inhibition curves. Homogenates were incubated with 100 pM [$^3$H]-ligand and various oxotremorine concentrations in 10 mM sodium-potassium phosphate buffer at 25° C with (O) or without (●) 30 μM Gpp(NH)p. $IC_{50}$ values and Hill slopes are given below.
Panel A: [$^3$H]PZ in cerebral cortex. $n_H$ = 0.95, $IC_{50}$ = 130 nM (control); $n_H$ = 1.03, $IC_{50}$ = 196 nM (Gpp(NH)p)
Panel B: [$^3$H](-)QNB in cerebral cortex. $n_H$ = 0.60, $IC_{50}$ = 856 nM (control); $n_H$ = 0.69, $IC_{50}$ = 1320 nM (Gpp(NH)p)
Panel C: [$^3$H](-) QNB in heart. $n_H$ = 0.62, $IC_{50}$ = 52 nM (control); $n_H$ = 0.86, $IC_{50}$ = 520 nM (Gpp(NH)p).

been offered to account for this differential tissue-sensitivity to guanine nucleotides (1,23), we now propose that the guanine nucleotide sensitivity of high-affinity agonist binding is a reflection of muscarinic receptor subtypes in these tissues (27,41). As shown in Fig. 5, the nonhydrolyzable nucleotide Gpp(NH)p is extremely efficacious in reducing $M_2$ high-affinity [$^3$H]CD binding (cerebellum and heart) but is relatively ineffective in the predominantly $M_1$ tissue (cerebral cortex). Kinetic studies not only confirm this guanine nucleotide specificity (Fig. 6) but also support an allosteric mechanism of action. Thus the effect of guanine nucleotides on muscarinic agonist binding clearly involves a nucleotide specific site which may be selectively coupled to and regulate the high-affinity agonist binding site for $M_2$ but not $M_1$ receptors.

In view of the previous work by other investigators concerning the hormonal (or neurotransitter) regulation of adenylate cyclase, it seems highly probable that there is at least some involvement by this membrane-bound enzyme in the effect of guanine nucleotides on muscarinic agonist binding sites. As previously reviewed by Rodbell (42) and Ross and Gilman (43), adenylate cyclase appears to consist of three major components: a catalytic unit, a nucleotide binding

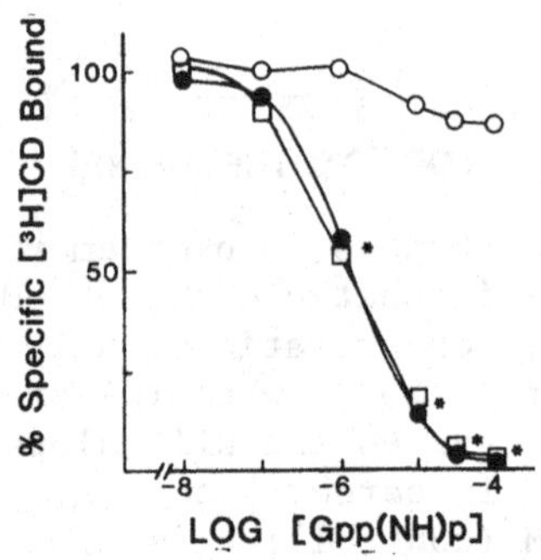


Fig. 5. Differential effects of Gpp(NH)p on high-affinity [$^3$H]CD binding. Homogenates of rat cerebral cortex (O), heart (●) and cerebellum (□) were incubated with 1 nM [$^3$H]CD and Gpp(NH)p in 10 mM sodium-potassium phosphate buffer (reprinted by permission from Vickroy et al., 41). * Significantly different from control ($p < 0.05$).

component ($N_s$), and the hormone receptor. Indeed, previous work has shown that muscarinic receptors are coupled with and influence the activity of adenylate cyclase in certain tissues (44-48). For example, in the heart, an $M_2$ tissue (low density of [$^3$H]PZ sites) in which guanine nucleotides allosterically reduce high-affinity muscarinic agonist binding (Figs. 5 and 6), muscarinic receptor stimulation inhibits β-adrenergic-enhanced adenylate cyclase activity. GTP plays a permissive role in this response (46) and thereby provides a link between guanine nucleotide-sensitive muscarinic agonist binding and the receptor-mediated control of adenylate cyclase activity.

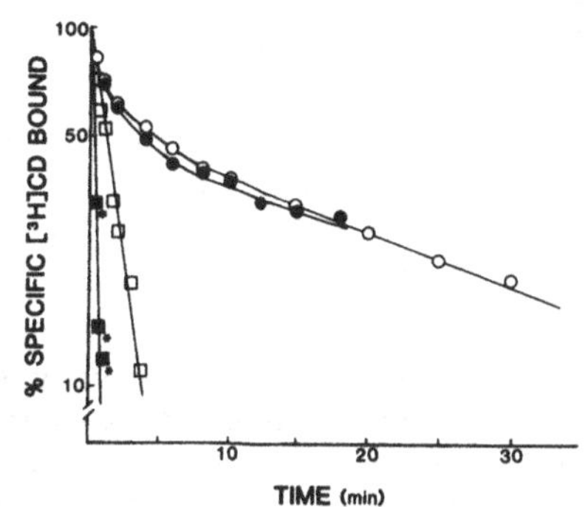


Fig. 6. Differential effects of Gpp(NH)p on [$^3$H]CD dissociation rates. Homogenates of rat cerebral cortex (O,●) and heart (□,■) were equilibrated with 1 nM [$^3$H]CD for 2 hr prior to the addition of 1 μM atropine (open symbols) or 1 μM atropine containing 30 μM Gpp(NH)p (closed symbols).
* Significant enhancement of [$^3$H]CD dissociation by Gpp(NH)p ($p < 0.05$).

However, recent studies of ADP-ribosylation in the presence of pertussis toxin (IAP) indicate that the nucleotide binding protein coupled to muscarinic receptors ($N_i$) differs from the protein coupled to β-adrenergic receptors (49). Thus it is possible that the guanine nucleotide-sensitivity of agonist binding is a direct indicator of the effector (adenylate cyclase) or the regulator protein ($N_i$) which is coupled to the $M_2$ muscarinic receptor in various tissues.

In addition to guanine nucleotides, magnesium ions (and possibly other divalent cations) and the sulfhydryl reagent N-ethylmaleimide (NEM) also demonstrate $M_1/M_2$ selective regulatory effects (27). While previous studies by Wei and Sulhake (36,37) demonstrated that magnesium ions enhance the affinity of myocardial muscarinic sites for agonists, our direct studies with [$^3$H]CD binding have demonstrated this to be an allosteric effect which is evident only in the $M_2$ tissues (27). Again, the basis of this important observtion may be related to the $M_2$ receptor-effector coupling mechanism or the presence of $N_i$, since magnesium ions and NEM are known to somehow influence these proteins.

## CONCLUDING REMARKS

In summary, we have presented an overview of the current evidence which supports the existence of muscarinic receptor subtypes ($M_1$ and $M_2$) in various tissues. Criteria such as high-affinity [$^3$H]PZ binding and the differential regulation of high-affinity agonist binding by several regulators provide strong support for these distinct binding site subtypes. Furthermore, it is possible that our recent findings concerning the $M_2$ selective allosteric effects of magnesium ions, NEM and guanine nucleotides at high-affinity agonist binding sites are directly related to the $M_2$ muscarinic effector (adenylate cyclase) or a regulatory nucleotide binding protein ($N_i$) which is coupled to the receptor. While little data are available concerning the $M_1$ effector, consideration has been given to the potential involvement of phosphatidyl inositol (PI) turnover. This observation is indeed noteworthy since it implies that the effector linkage may be an important factor in determining the drug specificity of the receptor binding site and underscores the importance of understanding the receptor-effector linkage as well as the nature of the receptor binding site. In order to more fully address this problem, however, it will be necessary to ascertain the precise nature of both $M_1$ and $M_2$ muscarinic receptor-effector linkages in a wide variety of tissues where the pharmacological specifities of the binding sites can be clearly characterized. In this way, we should more fully understand the molecular basis of muscarinic receptor subtypes and gain greater insight into the biochemical responses mediated by these subtypes.

## ACKNOWLEDGEMENTS

The authors wish to acknowledge the excellent secretarial assistance of Pat Gonzalez and technical assistance of Martha Ackerman.

REFERENCES

1. N. J. M. Birdsall, E. C. Hulme, R. Hammer, and J. S. Stockton, Subclasses of muscarinic receptors, in: "Psychopharmacology and Biochemistry of Neurotransmitter Receptors," H. I. Yamamura, R. W. Olsen, and E. Usdin, eds., Elsevier/North-Holland, New York, p. 97 (1980).
2. F. J. Ehlert, W. R. Roeske, and H. I. Yamamura, Muscarinic receptor: regulation by guanine nucleotides, ions, and N-ethyl-maleimide, Fed. Proc. 40:153 (1981).
3. W. R. Roeske, F. J. Ehlert, D. S. Barritt, K. Yamanaka, L. B. Rosenberger, S. Yamada, S. Yamamura, and H. I. Yamamura, Recent advances in muscarinic receptor heterogeneity and regulation, in: "Molecular Pharmacology of Neurotransmitter Receptors," T. Segawa, H. I. Yamamura, and K. Kuriyama, eds., Raven Press, New York (1983).
4. F. J. Ehlert, W. R. Roeske, and H. I. Yamamura, The nature of muscarinic receptor binding, in: "Handbook of Psychopharmacology," L. Iversen, S. D. Iversen, and S. H. Snyder, eds., Plenum, New York (1983).
5. H. H. Dale, The action of certain ester and ethers of choline and their relation to muscarine, J. Pharmacol. Exp. Ther. 6:147 (1914).
6. O. Loewi, Uber humorale ubertragbarkeit der herznervenwirkung, Pflugers Arch. Gen. Physiol. 189:239 (1921).
7. A. P. Roszkowski, An unusual type of ganglionic stimulant, J. Pharmacol. Exp. Ther. 132:156 (1961).
8. R. K. Goyal and S. Rattan, Neurohumoral hormonal, and drug receptors for the lower esophageal sphincter, Gastroenterol. 74:598 (1978).
9. B. H. Jaup, R. W. Stockbrugger, and G. Dotevall, Comparison of the action of pirenzepine and 1-hyoscyamine on gastric acid secretion and other muscarinic effects, Scand. J. Gastroenterol. (suppl.) 66:89 (1980).
10. R. Hammer, C. P. Berrie, N. J. M. Birdsall, A. S. C. Burgen, and E. C. Hulme, Pirenzepine distinguishes between different subclasses of muscarinic receptors, Nature 283:90 (1980).
11. R. Hammer and A. Giachetti, Muscarinic receptor subtypes: $M_1$ and $M_2$ biochemical and functional characterization, Life Sci. 31:2991 (1982).
12. M. P. Caulfield, G. A. Higgins, and D. W. Straughan, Central administration of the muscarinic receptor subtype-selective antagonist pirenzepine selectively impairs passive avoidance learning in the mouse, J. Pharm. Pharmacol. 35:131 (1983).
13. M. Watson, H. I. Yamamura, and W. R. Roeske, A unique regulatory profile and regional distribution of [$^3$H]pirenzepine binding in the rat provide evidence for distinct $M_1$ and $M_2$ muscarinic receptor subtypes, Life Sci. 32:3011 (1983).

14. R. B. Barlow, K. J. Berry, P. A. M. Glenton, N. M. Nikolaou, and K. S. Soh, A comparison of affinity constants for muscarinic-sensitive acetylcholine receptors in guinea-pig atrial pacemaker cells at 29° C and 37° C, Brit. J. Pharmacol. 58:613 (1976).
15. D. A. Brown, A. Forward, and S. Marsh, Antagonist discrimination between ganglionic and ileal muscarinic receptors, Brit. J. Pharmacol. 71:362 (1980).
16. M. Watson, W. R. Roeske, and H. I. Yamamura, [$^3$H]Pirenzepine selectively identifies a high affinity population of muscarinic cholinergic receptors in the rat cerebral cortex, Life Sci. 31:2019 (1982).
17. M. Watson, W. R. Roeske, P. C. Johnson, and H. I. Yamamura, [$^3$H]Pirenzepine identifies putative $M_1$, Brain Res., in press.
18. H. I. Yamamura, J. K. Wamsley, P. Deshmukh, and W. R. Roeske, Differential light microscopic autoradiographic localization of muscarinic cholinergic receptors in the brainstem and spinal cord of the rat using [$^3$H]pirenzepine, Eur. J. Pharmacol. 91:1983 (1983).
19. N. J. M. Birdsall, A. S. V. Burgen, C. R. Hiley, and E. C. Hulme, Binding of agonists and antagonists to muscarinic receptors, J. Supramol. Struc. 4:367 (1976).
20. N. J. M. Birdsall, A. S. C. Burgen, and E. C. Hulme, The binding of agonists to brain muscarinic receptors, Mol. Pharmacol. 14:723 (1978).
21. F. J. Ehlert, Y. Dumont, W. R. Roeske, and H. I. Yamamura, Muscarinic receptor binding in rat brain using the agonist [$^3$H]cismethyldioxolane, Life Sci. 26:961 (1980).
22. F. J. Ehlert, H. I. Yamamura, D. J. Triggle, and W. R. Roeske, The influence of guanyl-5'-yl imidodiphosphate and sodium chloride on the binding of the muscarinic agonist, [$^3$H]cismethyldioxolane, Eur. J. Pharmacol. 61:317 (1980).
23. F. J. Ehlert, W. R. Roeske, and H. I. Yamamura, Regulation of muscarinic receptor binding by guanine nucleotides and N-ethyl-maleimide, J. Supramol. Struct. 14:149 (1980).
24. M. M. Hosey, Regulation of antagonist binding to cardiac muscarinic receptors, Biochem. Biophys. Res. Commun. 107:314 (1982).
25. T. W. Vickroy, W. R. Roeske, and H. I.Yamamura, Characterization of a high affinity muscarinic agonist binding site by a rapid filtration technique with [$^3$H]cismethyldioxolane, Fed. Proc. 42:1146 (1983).
26. K. J. Chang, R. C. Deth, and D. J. Triggle, Structural parameters determining cholinergic and anticholinergic activities in a series of 1,3-dioxolanes, J. Med. Chem. 15:243 (1972).

27. T. W. Vickroy, H. I. Yamamura, and W. R. Roeske, Differential regulation of high-affinity agonist binding to muscarinic sites in the rat heart, cerebellum, and cerebral cortex, Biochem. Biophys. Res. Commun., in press.
28. J. M. Stockton, N. J. M. Birdsall, A. S. V. Burgen, and E. C. Hulme, Modification of binding properties of muscarinic receptors by gallamine, Mol. Pharmacol. 23:551 (1983).
29. J. Dunlap and J. H. Brown, Heterogeneity of binding sites on cardiac muscarinic receptors induced by the neuromuscular blocking agents gallamine and pancuronium, Mol. Pharmacol. 24:15 (1983).
30. A. Bartolini, R. Bartolini, and E. F. Domino, Effects of physostigmine on brain acetylcholine content and release, Neuropharmacology 12:15 (1973).
31. F. J. Rathbun and J. T. Hamilton, Effect of gallamine on cholinergic receptor, Can. Anaesth. Soc. J. 17:574 (1970).
32. A. L. Clark and F. Mitchelson, The inhibitory effect of gallamine on muscarinic receptors, Brit. J. Pharmacol. 14:323 (1976).
33. C. P. Berrie, N. J. M. Birdsall, A. S. C. Burgen and E. C. Hulme, Guanine nucleotides modulate muscarinic receptor binding in the heart, Biochem. Biophys. Res. Commun. 87:1000 (1979).
34. L. B. Rosenberger, W. R. Roeske, and H. I. Yamamura, The regulation of muscarinic cholinergic receptors by guanine nucleotides in cardiac tissue, Eur. J. Pharmacol. 56:179 (1979).
35. L. B. Rosenberger, H. I. Yamamura, and W. R. Roeske, Cardiac muscarinic cholinergic receptor binding is regulated by $Na^+$ and guanyl nucleotides, J. Biol. Chem. 255:820 (1980).
36. J.-W. Wei and P. V. Sulakhe, Cardiac muscarinic cholinergic receptor sites: opposing regulation by divalent cations and guanine nucleotides of receptor-agonist interaction, Eur. J. Pharmacol. 62:345 (1980).
37. J.-W. Wei and P. V. Sulakhe, Requirement for sulfhydryl groups in the differential effects of magnesium ion and GTP on agonist binding of muscarinic cholinergic receptor sites in rat atrial membrane fraction, Naunyn-Schmiedeberg's Arch. Pharmacol. 314:51 (1980).
38. R. S. Aronstam, L. G. Abood, and W. Hoss, Influence of sulfhydryl reagents and heavy metals on the functional state of the muscarinic acetylcholine receptor in rat brain, Mol. Pharmacol. 14:575 (1978).
39. S. J. Korn, M. W. Martin, and T. K. Harden, N-ethylmaleimide-induced alteration in the interaction of agonists with muscarinic cholinergic receptors of rat brain, J. Pharmacol. Exp. Ther. 224:118 (1983).

40. M. Waelbroeck, P. Robberecht, P. Chatelain, and J. Christophe, Rat cardiac muscarinic receptors: I. Effects of guanine nucleotides on high- and low-affinity binding sites, Mol. Pharmacol. 21:581 (1982).
41. T. W. Vickroy, M. Watson, H. I. Yamamura, and W. R. Roeske, Agonist binding to multiple muscarinic receptors, Fed. Proc., in press.
42. M. Rodbell, The role of hormone receptors and GTP-regulatory proteins in membrane transduction, Nature 284:17 (1980).
43. E. M. Ross and A. G. Gilman, Biochemical properties of hormone-sensitive adenylate cyclase, Ann. Rev. Biochem. 49:533 (1980).
44. F. Murad, Y.-M. Chi, J. W. Rall, and E. W. Sutherland, Adenyl cyclase III. The effect of catecholamines and choline esters on the formation of adenosine 3'-5' phosphate by preparations from cardiac muscle and liver, J. Biol. Chem. 237:1233 (1962).
45. A. M. Watanabe, M. M. McConnaughey, R. A. Strawbridge, J. W. Fleming, L. R. Jones, and H. R. Besch, Muscarinic cholinergic receptor modulation of β-adrenergic receptor affinity for catecholamines, J. Biol. Chem. 253:4833 (1978).
46. K. H. Jacobs, K. Aktories, and G. Schultz, GTP-dependent inhibition of cardiac adenylate cyclase by mascarinic cholinergic agonists, Naunyn-Schmiedeberg's Arch. Pharmaocl. 310:113 (1979).
47. O. Hazeki and M. Ui, Modification by islet-activating protein of receptor-mediated regulation of cyclic AMP accumulation in isolated rat heart cells, J. Biol. Chem. 256:2856 (1981).
48. W. R. Roeske and H. I. Yamamura, Adrenergic-cholinergic interactions, in: "Adrenoceptors and Catecholamine Action, Part B," G. Kunos, ed., John Wiley and Sons, New York (1983).
49. T. Katada and M. Ui, Direct modification of the membrane adenylate cyclase system by islet-activating protein due to ADP-ribosylation of membrane protein, Proc. Natl. Acad. Sci. USA 79:3129 (1982).

## FACTORS REGULATING LIGAND BINDING TO 5-HYDROXYTRYPTAMINE RECEPTORS IN RAT BRAIN

Tomio Segawa, Yasuyuki Nomura, Hiroaki Nishio,
Jun-ichi Taguchi and Mutsuko Fujita

Department of Pharmacology, Institute of Pharmaceutical Sciences, Hiroshima University School of Medicine
Hiroshima, Japan

### INTRODUCTION

Several factors have been known to regulate neurotransmitter receptor binding to central nervous system (CNS) but so far little is known about the factors which modify 5-hydroxytryptamine (5-HT) binding properties. Recently, in the course of 5-HT binding studies, it was found that ascorbic acid (AA), $Mn^{2+}$, and $HCO_3^-$ could modify ligand binding to central 5-HT receptors in rat.

### MATERIALS AND METHODS

Both sexes of Wistar strain rats weighing 200-250 g were used. Frontal cortex membranes prepared by the method of Peroutka and Snyder (1) were suspended in 50 mM Tris-HCl buffer (pH 7.7 at 25° C) containing 10 μM Nialamide and were incubated at 37° C, either with [$^3$H]5-HT (2 nM) for 10 min or with [$^3$H]spiperone (1 nM) for 15 min. The incubation was terminated by vacuum filtration through Whatman GF/B filters. Radioactivity on the filters was measured by Packard Tri-Carb liquid scintillation spectrometry. Specific [$^3$H]5-HT binding was defined as the difference between total binding and that occurring in the presence of 10 nM 5-HT, while specific [$^3$H]spiperone binding was the difference between total binding and that in the presence of 0.1 mM 5-HT.

Lipid peroxide formation was determined by the procedure of Ohkuma et al. (2).

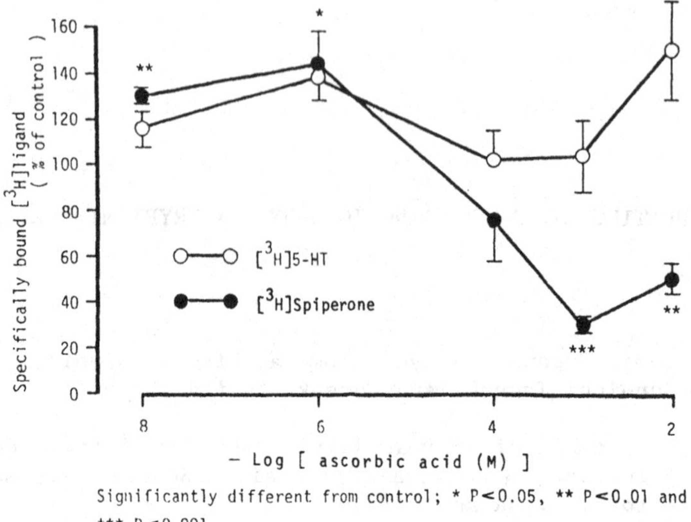


Fig. 1. Effect of ascorbic acid treatment on specific [$^3$H]5-HT and [$^3$H]spiperone binding to rat frontal cortex membranes.

RESULTS AND DISCUSSION

Effect of Ascorbic Acid

When rat frontal cortex membranes were incubated with AA at 37° C for 30 min and after washing the tissues were subjected to binding assays, [$^3$H]spiperone binding was inhibited (Fig. 1). The effect is dose dependent over a concentration range from 1 μM to 1 mM AA, with maximum inhibition occurring at 1 mM. In contrast, specific [$^3$H]5-HT binding was not affected by AA. Scatchard analysis indicated that the number of [$^3$H]spiperone binding sites was decreased significantly by AA treatment.

AA is commonly used as a biological reducing agent. In order to see if chemical reduction is involved in AA-induced decrease of spiperone binding, several reducing agents, dithiothreitol, 2-mercaptoethanol and $NaHSO_3$, were examined on 5-HT and spiperone binding but none of them, at 10 mM, had any significant effect on either 5-HT or spiperone binding. These results ruled out the possibility that chemical reduction is involved in the mechanism of AA-induced decrease of spiperone binding.

Membranes were incubated with 1 mM AA in the presence of EDTA, EGTA, $MnCl_2$, and $LaCl_3$, which are known to inhibit lipid peroxidation. As a reference, the effect of $MgCl_2$, which does not inhibit

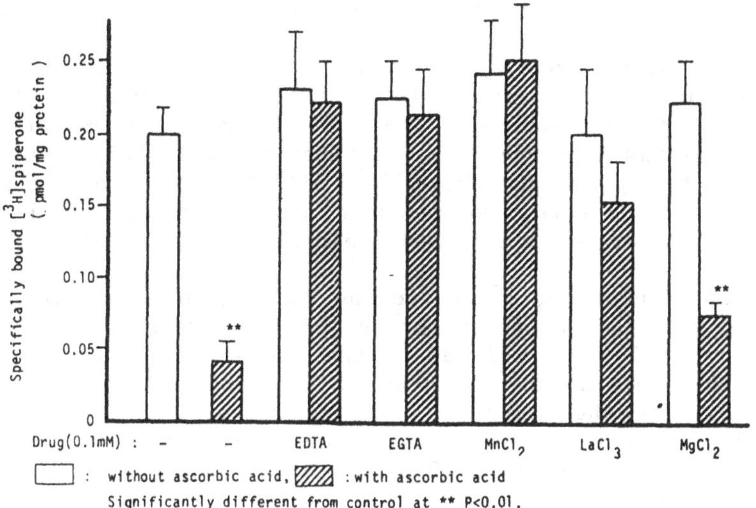


Fig. 2. Effect of reagents and ions on ascorbic acid-induced decrease in specific [$^3$H]spiperone binding to rat frontal cortex membranes.

lipid peroxidation, was also examined. At 0.1 mM, all the reagents and ions known to inhibit lipid peroxidation completely prevented the AA-induced decrease of specific spiperone binding, while $MgCl_2$ did not show protection (Fig. 2). When the membranes were incubated with AA and washed out, EDTA could no longer protect against the AA

Table 1. Effect of chelating agents and metal ions on ascorbic acid-induced increase in lipid peroxide formation in rat frontal cortex membranes

| Reagents (0.1 mM) | Lipid peroxide formation (nmol malonyldialdehyde/mg) |
|---|---|
| Control | 1.3 ± 0.19 |
| Ascorbic acid (1 mM) | 30.0 ± 3.98* |
| + EDTA | 1.6 ± 0.25 |
| + EGTA | 1.4 ± 0.10 |
| + $MnCl_2$ | 1.1 ± 0.09 |
| + $LaCl_3$ | 1.8 ± 0.27 |
| + $MgCl_2$ | 26.2 ± 2.08* |

* P < 0.001 (vs. control)

effect. Therefore, the effect of AA appears to be irreversible.

These chelating agents and metal ions were re-examined with respect to their effect on lipid peroxide formation in frontal cortex membranes. As shown in Table 1, lipid peroxide formation catalysed by AA was significantly inhibited by EDTA, EGTA, $MnCl_2$ and $LaCl_3$, while $MgCl_2$ did not have any effect on AA-induced lipid peroxidation. Therefore, there was good correlation between inhibition of spiperone binding and lipid peroxide formation.

A Scatchard plot for [$^3$H]spiperone binding to normal membranes was compared with that of the membranes previously treated with 1 mM AA. The normal plot was shifted to the left in a parallel manner. Therefore, there was no change in Kd of the binding sites but there was a decrease in the number of binding sites. In contrast, the Scatchard plot was not changed when the membranes were incubated with 1 mM AA in the presence of 0.1 mM EDTA.

When the amount of lipid peroxide formed with varying concentrations of AA was plotted against the amount of spiperone binding, a linear relationship was found to exist with a correlation coefficient of 0.98. Therefore, peroxidation of membrane lipids is inversely proportional to specific [$^3$H]spiperone binding.

Further experiments were performed to see if any lipids could

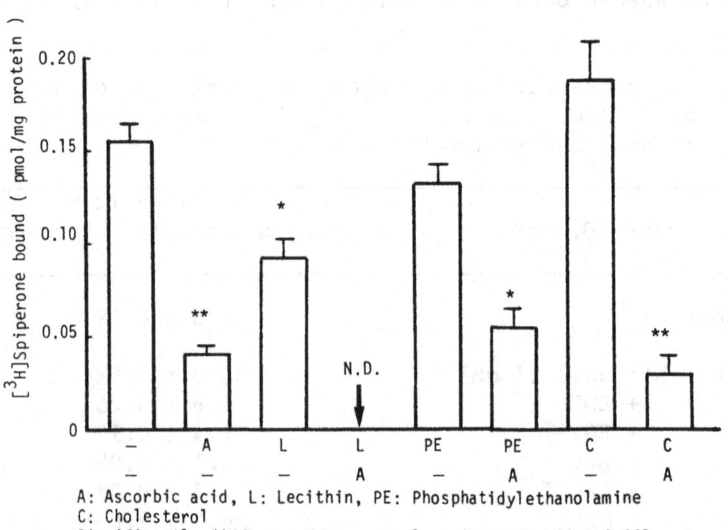


Fig. 3. Effect of lipid treatment on [$^3$H]spiperone binding to rat frontal cortex membranes.

protect the membranes against the AA-induced decrease in spiperone binding. Lipids were incorporated into membranes by the method of Abood and Takeda (4) and the membranes were treated with AA and then subjected to binding assays. The results are shown in Fig. 3. Lecithin, phosphatidylethanolamine and cholesterol could not protect membranes against the AA effect.

## Effects of $Mn^{2+}$ and $HCO_3^-$

As shown in Fig. 4, when binding experiments for [$^3$H]5-HT were performed in 50 mM Tris-HCl buffer medium, addition of $Mn^{2+}$ increased the binding, with maximum enhancement occurring at 5 mM $Mn^{2+}$. The enhancement was much more remarkable in Krebs-Ringer solution, suggesting that ions in Krebs-Ringer solution have a synergistic action on the Mn effect. On the other hand, $Mn^{2+}$ could not enhance [$^3$H]spiperone binding.

In order to determine which ions in Krebs-Ringer solution could have a synergistic action on the Mn effect, the effect of several ions in Krebs-Ringer solution was examined. As shown in Fig. 5, in the presence of sodium bicarbonate, $Mn^{2+}$ increased 5-HT binding remarkably, depending on the concentration of $NaHCO_3$. As similar results could not be obtained with NaCl, the synergistic action of $NaHCO_3$ is due to $HCO_3^-$. Again, $Mn^{2+}$ plus $HCO_3^-$ had no effect on spiperone binding.

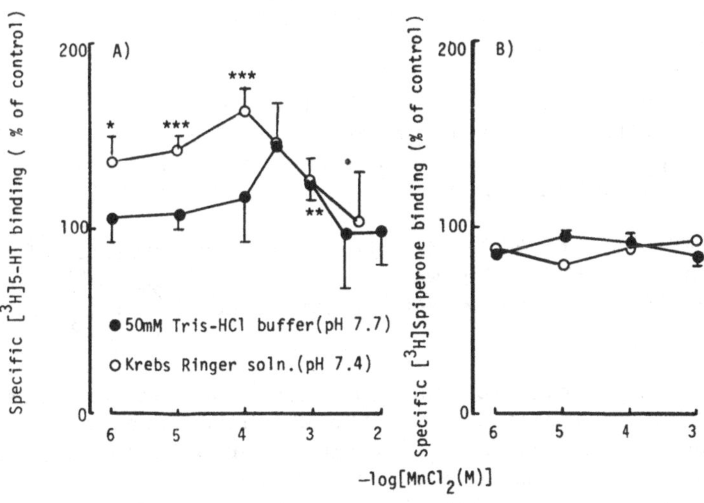


Fig. 4. Effect of $Mn^{2+}$ on specific [$^3$H]5-HT and [$^3$H]spiperone binding to rat frontal cortex membranes.

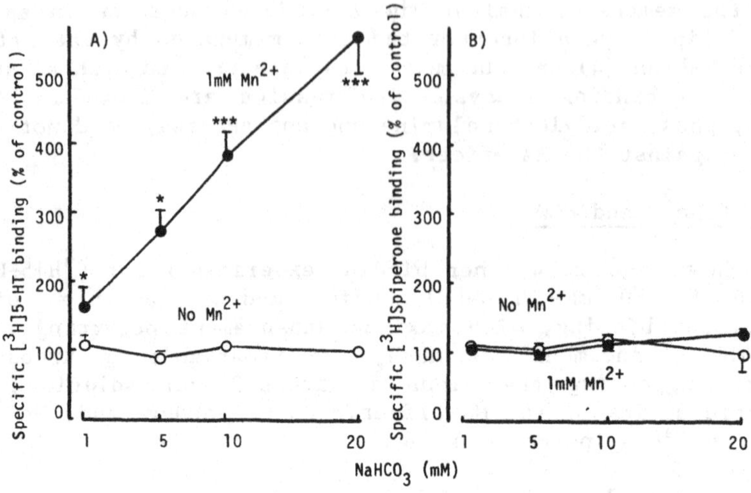


Fig. 5. Effect of $NaHCO_3$ on specific [$^3$H]5-HT and [$^3$H]spiperone binding to rat frontal cortex membranes in the presence of $Mn^{2+}$.

When the concentration of $NaHCO_3$ was fixed at 10 mM and the concentration of $MnCl_2$ was changed, a remarkable synergistic action was obtained at concentrations of $MnCl_2$ from 10 µM to 1 mM. Besides Mn, several other divalent cations, Mg, Cu, Ca and Fe ions were tested to see if they had a similar effect but none of them showed enhancement of 5-HT binding.

Scatchard analysis inidcated that $Mn^{2+}$ either alone or in combination with $HCO_3^-$ shifted the line from the control to the right in a parallel manner. Therefore, only the number of binding sites increased (Fig. 6).

The effect of $Mn^{2+}$ was temperature dependent. Significant enhancement of 5-HT binding was obtained at 37° C, while at 4° C, $Mn^{2+}$ alone did not enhance 5-HT binding and $Mn^{2+}$-$HCO_3^-$ enhanced it slightly.

Subcellular distribution of $Mn^{2+}$-induced enhancement of [$^3$H]5-HT binding indicated that the enhanced binding was localized in the crude mitochondrial $P_2$ fraction. Further subfractionation of $P_2$ revealed that enhanced binding was highest in the synaptosomal $P_2B$ fraction (Fig. 7). Therefore, the enhancement of 5-HT binding only occurred in the synaptic area.

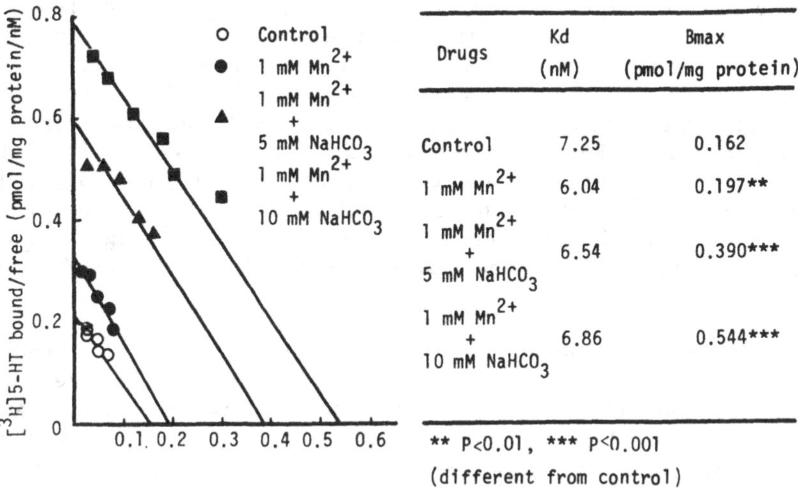


| Drugs | Kd (nM) | Bmax (pmol/mg protein) |
|---|---|---|
| Control | 7.25 | 0.162 |
| 1 mM $Mn^{2+}$ | 6.04 | 0.197** |
| 1 mM $Mn^{2+}$ + 5 mM $NaHCO_3$ | 6.54 | 0.390*** |
| 1 mM $Mn^{2+}$ + 10 mM $NaHCO_3$ | 6.86 | 0.544*** |

** P<0.01, *** P<0.001
(different from control)

Fig. 6. Scatchard analysis of specific [$^3$H]5-HT binding activity in rat frontal cortex membranes.

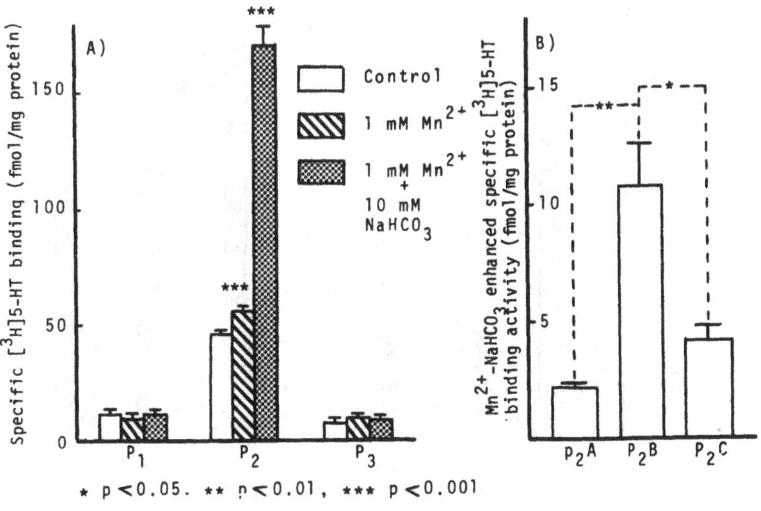


Fig. 7. Subcellular distribution of specific [$^3$H]5-HT binding activity in rat frontal cortex membranes.

In order to see if any membrane components and functional groups might be involved in enhancement of 5-HT binding by $Mn^{2+}$-$HCO_3^-$, the membranes were treated with SH-modifying reagents, N-ethylmaleimide (NEM) or iodoacetic acid (IAA) at 37° C for 10 min; the reagents were then washed out and the membranes were subjected to binding assays. Results are shown in Fig. 8. NEM (0.1 - 1.0 mM) slightly but significantly reduced 5-HT binding while it significantly reduced the $Mn^{2+}$-$HCO_3^-$ enhanced 5-HT binding. IAA had a similar effect but it was less effective. These results suggest that SH radicals in synaptic membranes might be involved in the $Mn^{2+}$-$HCO_3^-$ effect.

The effects of reducing agents on the $Mn^{2+}$-$HCO_3^-$ effect are shown in Fig. 9. The membranes were incubated in the presence of these agents and the effect of $Mn^{2+}$-$HCO_3^-$ was examined. AA and dithiothreitol (DTT) slightly decreased 5-HT binding and 2-mercaptoethanol (2-ME) had no influence on 5-HT binding, but all of them completely inhibited the $Mn^{2+}$-$HCO_3^-$ effect. From these results, together with the results of Fig. 8, it is suggested that S-S bonds rather than SH radicals in membranes might be involved in the $Mn^{2+}$-$HCO_3^-$ effect.

Guanine nucleotides have been shown to decrease receptor affinity for agonists (5-11). Furthermore, $Mn^{2+}$ has been known to reverse the guanine nucleotide effect by stimulating a membrane-bound phosphatase activity that hydrolyzes GTP to GMP and guanosine (12). As shown in Fig. 10, $Mn^{2+}$-$HCO_3^-$ induced enhancement was

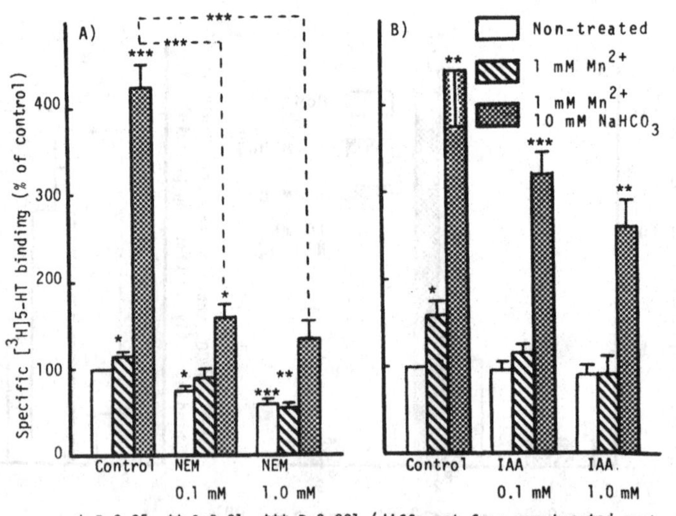


Fig. 8. Effect of pretreatment of N-ethylmaleimide and iodoacetic acid on the enhancement of specific [$^3$H]5-HT binding to rat frontal cortex membranes induced by $Mn^{2+}$-$HCO_3^-$.

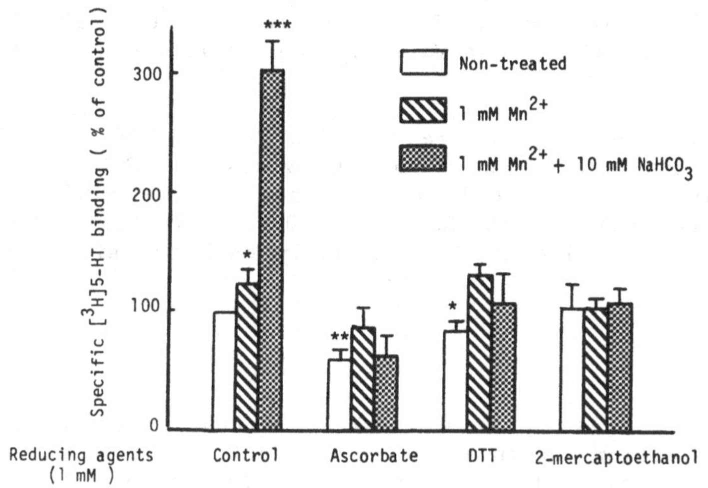


Fig. 9. Effect of reducing agents on the enhancement of specific [$^3$H]5-HT binding to rat frontal cortex membranes induced by $Mn^{2+}$-$HCO_3^-$.

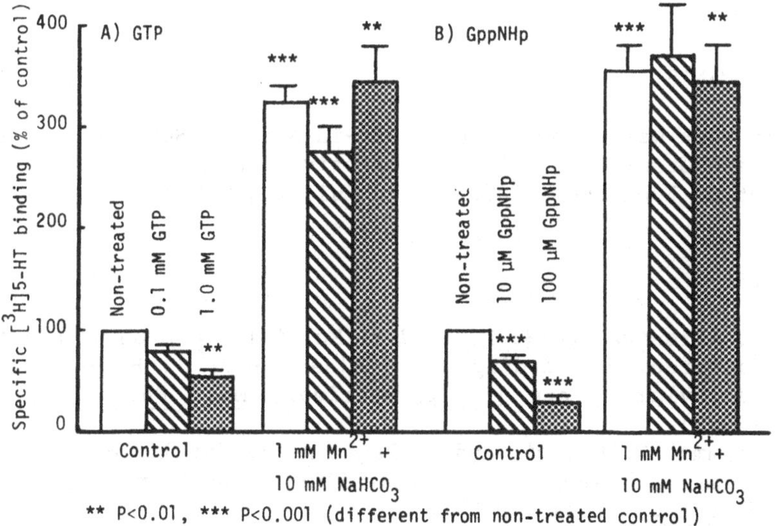


Fig. 10. Effect of pretreatment of GTP and GppNHp on the enhancement of specific [$^3$H]5-HT binding to rat frontal cortex membranes induced by $Mn^{2+}$-$HCO_3^-$.

observed in the presence of GTP and GppNHp (the analogue that is not hydrolyzed by phosphatase). Therefore, it seems unlikely that $Mn^{2+}$ enhances 5-HT binding by stimulating phosphatase activity which in turn hydrolyzes GTP.

The membranes were pretreated with Triton X-100 at 37° C for 10 min and then were subjected to binding assays. Triton X-100 at a concentration of 0.01% had no influence on the binding but it increased the binding significantly at 0.1%. Probably Triton X-100 removed membrane moieties which normally mask the sites for $Mn^{2+}$-$HCO_3^-$.

## CONCLUSION

From these results it was suggested that ascorbate-catalysed lipid peroxidation is involved in the effect of AA on specific [$^3$H]spiperone binding to 5-$HT_2$ receptors. Similar effects have been observed in opioid stereospecific binding in brain homogenates (13) and in dopamine binding in striatal membranes (14). However, as ascorbate-induced generation of peroxide is normally well regulated within the intact cell (13), probably it would not be a regulatory mechanism *in vivo*. Although the mechanisms with which $Mn^{2+}$-$HCO_3^-$ regulates [$^3$H]5-HT binding to frontal cortex membranes remain to be determined, the possibility exists that these endogenous ions play an important role in regulating 5-HT binding to 5-$HT_1$ receptors in CNS.

## ACKNOWLEDGMENTS

This work was supported by Grant-in-Aid for Scientific Research 56480336 and Grant-in-Aid for Co-operative Reseach 00537006 from the Department of Education, Science and Culture, Japan.

## REFERENCES

1. S. J. Peroutka and S. H. Snyder, Multiple serotonin receptors: Differential binding of [$^3$H]serotonin, [$^3$H]lysergic acid diethylamide and [$^3$H]spiroperidol, *Mol. Pharmacol.* 16:687 (1979).
2. H. Ohkuma, N. Ohishi, and K. Yagi, Assay for lipid peroxide in animal tissues by thiobarbituric acid reaction, *Anal. Biochem.* 95:351 (1979).
3. O. H. Lowry, H. J. Rosebrough, A. L. Farr, and R. J. Randall, Protein measurement with the Folin phenol reagent, *J. Biol. Chem.* 193:265 (1951).

4. L. G. Abood and F. Takeda, Enhancement of stereospecific opiate binding to neural membranes by phosphatidyl serine, European J. Pharmacol. 39:71 (1976).
5. M. Rodbell, M. C. Lin, Y. Salomon, C. Londos, J. P. Harwood, B. R. Martin, M. Rendell, and N. Berman, Role of adenine and guanine nucleotides in activity and response of adenylate cyclase system, Adv. Cyclic Nucleotide Res. 5:3 (1975).
6. J. Wolff and G. H. Cook, Activity of thyroid membrane adenylate cyclase by purine nucleotides, J. Biol. Chem. 248:350 (1973).
7. A. J. Blume and C. J. Foster, Neuroblastoma adenylate cyclase: Role of 2-chloradenoisine, prostaglandin E and guanine nucleotides in regulation of activity, J. Biol. Chem. 251:3399 (1976).
8. A. Levitzki, N. Sevilla, and M. L. Steer, The regulatory control of β-adrenergic dependent adenylate cyclase, J. Supramol. Struct. 4:405 (1976).
9. F. Pecker and J. Hanoue, Activity of epinephrine-sensitive adenylate cyclase in rat liver by cytosolic protein nucleotide complex, J. Biol. Chem. 252:2784 (1977).
10. M. Lucas and J. Bockaert, Use of $^3$H-dihydroalprenolol to study β-adrenergic adenylate cyclase coupling in glioma cells: Role of 5'-guanylimidodiphosphate, Mol. Pharmacol. 13:314 (1977).
11. H. Yamamura and M. Rodbell, Hydroxybenzylpindolol and hydroxybenzylpropranolol: Partial β-adrenergic agonists of adenylate cyclase in rat adipocytes, Mol. Pharmacol. 12:693 (1976).
12. S. R. Childers and S. H. Snyder, Differential regulation by guanine nucleotides of opiate agonist and antagonist receptor interaction, J. Neurochem. 34:583 (1980).
13. F. M. Leslie, C. E. Dunlap III, and B. M. Cox, Ascorbate decreases ligand binding to neurotransmitter receptors, J. Neurochem. 34:219 (1980).
14. S. O. Kayaalp and N. H. Neff, Differentiation by ascorbic acid of dopamine agonist and antagonist binding sites in striatum, Life Sci. 26:1837 (1980).

# EVIDENCE FOR $ALPHA_2$ ADRENERGIC RECEPTORS IN BOVINE CEREBRAL ARTERIES

Takashi Taniguchi, Motohatsu Fujiwara[*],
Tetsuya Tsukahara[*], and Hajime Handa[*]

Department of Pharmacology and Department of
Neurosurgery[*], Faculty of Medicine
Kyoto University, Kyoto 606, Japan

## INTRODUCTION

Adrenergic innervation to cerebral arteries is thought to play an important role in controlling cerebral blood flow (1). Histochemical studies have revealed high concentrations of norepinephrine and rich adrenergic innervation from the superior cervical ganglion in the adventitia and outer border of the medial layer of cerebral arteries in various species (2,3). Isolated cerebral arteries are contracted by alpha adrenergic agonists in a dose dependent manner and this contraction is blocked by alpha adrenergic antagonists (4,5). Alpha adrenergic receptors have been classified into $alpha_1$ and $alpha_2$ subtypes (6,7). Sakakibara et al. (8) suggested that adrenergic contraction of the isolated dog basilar artery is mediated by $alpha_2$ receptors. In the present study, we attempted to characterize alpha adrenergic receptors in bovine cerebral arteries using an $alpha_1$ antagonist, [$^3$H]prazosin and an $alpha_2$ antagonist, [$^3$H]yohimbine.

## METHODS

Pia-arachinoid arteries were carefully removed from bovine brain and kept in a freezer (-80° C). THe arteries were minced with scissors and homogenized in 10 volumes of ice-cold 50 mM sodium phosphate buffer (pH 7.4) with a glass homogenizer. The artery homogenates were filtered through two layers of gauze, rehomogenized at setting 10 on a Polytron with a 20 s burst, the homogenates centrifuged at 1,000 X g for 10 min, and the supernatant carefully removed and centrifuged at 100,000 X g for 60 min. The resulting pellet was resuspended at a protein concentration ranging from 1 to 2 mg/ml in buffer. Protein concentration was determined by the method of Lowry et al. (9).

[$^3$H]Prazosin and [$^3$H]yohimbine binding were performed by incubating aliquots of the artery homogenates at 24° C for 30 min in 250 µl of sodium phosphate buffer, containing [$^3$H]prazosin or [$^3$H]yohimbine, in the presence or absence of 10 mM norepinephrine. [$^3$H]Yohimbine bound in the presence of 10 mM norepinephrine was termed "nonspecific" and was subtracted from that obtained in the absence of 10 mM norepinephrine, termed "total binding", to obtain the binding termed "specific binding". The assay was terminated by addition of 3 ml of ice-cold buffer and rapid filtration through Whatman GF/B glass fiber filters under suction. After washing twice with 3 ml of the buffer, the filters were dried in an oven, transferred to counting vials and 8 ml of scintillation fluid was added. Radioactivity was counted in a Packard Tri-Carb scintillation spectrometer (Model 3255).

[$^3$H]Prazosin (specific activity, 17.4 Ci/mmol) and [$^3$H]yohimbine (specific activity, 89.7 Ci/mmol) were purchased from New England Nuclear. Clonidine was a gift from Boehringer, Ingelheim. All other chemicals were of the purest grade commercially available.

## RESULTS

Saturability of [$^3$H]yohimbine binding. Specific binding of increasing concentrations of [$^3$H]yohimbine (4 to 56 nM) was saturable (Fig. 1A). Scatchard analysis indicated a single class of binding sites with an apparent equilibrium dissociation constant ($K_D$) and maximum binding capacity (Bmax) (Fig. 1B) of 21 nM and 720 fmol/mg protein, respectively. The Hill coefficient was calculated from the data in Fig. 1 and was 1.06.

Specificity of [$^3$H]yohimbine binding. The specificity of [$^3$H]yohimbine binding was studied using alpha adrenergic agents (Table 1). Specific alpha$_2$ agents such as yohimbine, tramazoline and clonidine gave $IC_{50}$ values of 55 nM, 270 nM, and 580 nM, respectively. Alpha$_1$ agents such as methoxamine and prazosin had $IC_{50}$ values of 15 µM and 47 µM, respectively. Other alpha agents such as phenylephrine, norepinephrine and epinephrine gave $IC_{50}$ values of over 49 µM.

[$^3$H]Prazosin binding. As determined from the increasing concentration of [$^3$H]prazosin (22 to 223 nM), no specific [$^3$H]prazosin binding was detectable.

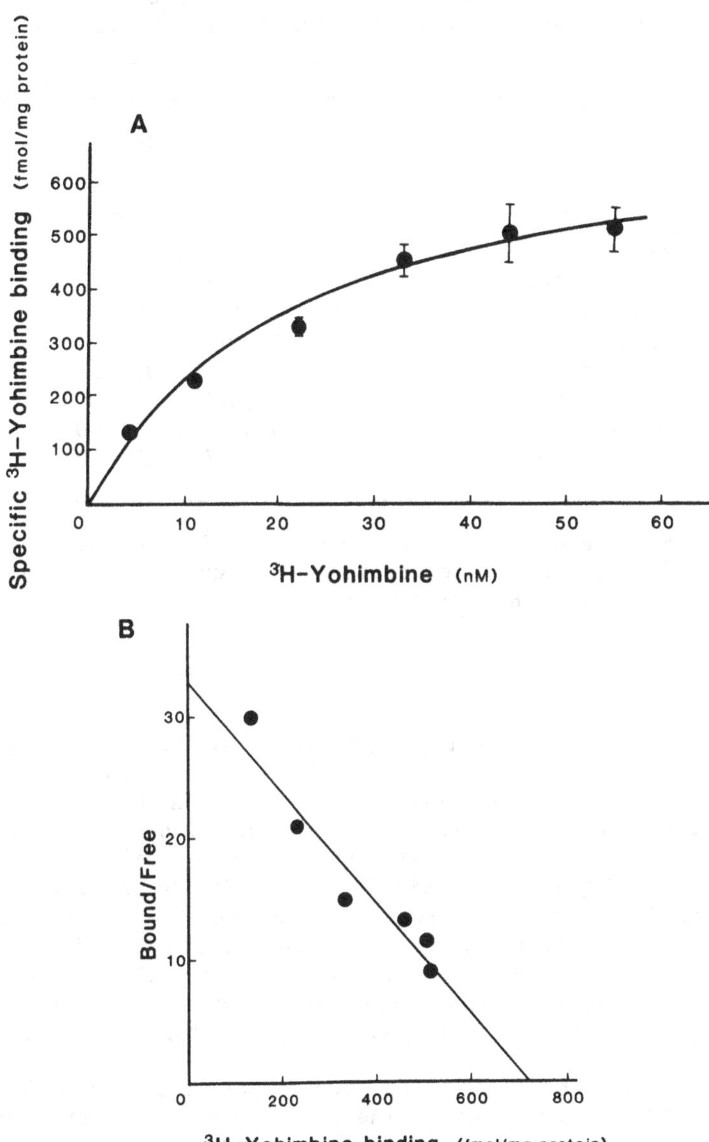


Fig. 1. A, saturation of specific [$^3$H]yohimbine binding to bovine cerebral arteries. Each value is the mean ± S.E. of 4 experiments. B, a Scatchard plot derived from the specific [$^3$H]yohimbine binding data of Fig. 1A. The slope was determined by linear regression analysis (r = 0.96); $K_D$ = 21 nM; Bmax = fmol/mg protein.

Table 1. $IC_{50}$ values for various drugs of [³H]yohimbine binding to bovine cerebral arteries

| Drug | $IC_{50}$ (nM) |
|---|---|
| Yohimbine | 55 ± 2 |
| Tramazoline | 270 ± 20 |
| Clonidine | 580 ± 30 |
| Methoxamine | 15,000 ± 2,400 |
| Prazosin | 47,000 ± 9,700 |
| Phenylephrine | 49,000 ± 1,800 |
| Norepinephrine | 60,000 ± 12,000 |
| Epinephrine | 150,000 ± 27,000 |

Each value is the mean ± S.E. of 3 experiments. $IC_{50}$ is the concentration of drugs that reduces the specific [³H]yohimbine by 50%.

## DISCUSSION

Using [³H]yohimbine and [³H]prazosin, we noted the presence of a large number of $alpha_2$ adrenergic receptor sites in bovine cerebral arteries, whereas specific $alpha_1$ adrenergic binding sites were not apparent. Specific [³H]yohimbine binding was saturable, and of high affinity. Scatchard and Hill plot analyses of the data indicated that [³H]yohimbine binding sites in the cerebral arteries are of a single population. Potent $alpha_2$ agents such as yohimbine, tramazoline and clonidine, at low concentrations, inhibited the binding of [³H]yohimbine. The $alpha_1$ agents such as prazosin and methoxamine, and other alpha adrenergic drugs such as norepinephrine and epinephrine inhibited the binding only at high concentrations.

A large number of binding sites for the alpha adrenergic antagonist [³H]dihydroergocryptine were found in the hog pial membrane (10). Although pial membranes contain a number of tissue elements, Friedman and Davis (10) used the pial membrane as a model of small-sized cerebral blood vessels. They did not classify alpha receptors into subtypes, because [³H]dihydroergocryptine binds both to $alpha_1$ and $alpha_2$ receptors. Harik et al. (11) reported that rat and pig cerebral microvessels lack significant binding sites for [³H]WB-4101, an $alpha_1$ antagonist. They suggested the absence of alpha adrenergic receptor sites in the cerebral vessels, but did not eliminate the possible existence of $alpha_2$ adrenergic receptor sites. Our data suggest the lack of $alpha_1$ adrenergic receptor sites in bovine cerebral arteries and provide evidence for the presence of $alpha_2$ receptors in these same arteries.

Alpha adrenergic receptors have been classified according to their pharmacological nature. Although $alpha_2$ adrenergic receptors were originally proposed to be presynaptic, accumulating evidence suggests the existence of postsynaptic $alpha_2$ adrenergic receptors (12,13,14). Recently Langer and Shepperson (15) presented a scheme showing the function of postsynaptic $alpha_1$ and $alpha_2$ adrenergic receptors on vascular smooth muscles and emphasized that both $alpha_1$ and $alpha_2$ subtypes mediated vasoconstrictions by activating calcium channels. Cerebral arteries may be a typical example where postsynaptic $alpha_2$ adrenergic receptors play a dominant role in vasoconstriction. Sakakibara et al. (8) suggested that contractions of the dog basilar artery are mediated by postsynaptic $alpha_2$ adrenergic receptors. Although we do not eliminate the existence of presynaptic $alpha_2$ adrenergic receptors in the present experiments, our data provide support for vasoconstrictions mediated by postsynaptic $alpha_2$ adrenergic receptors in the cerebral arteries.

## SUMMARY

Specific [$^3$H]yohimbine binding to the homogenates of bovine cerebral arteries was saturable and of high affinity ($K_D$ = 21 nM) with a Bmax of 720 fmol/mg protein. On the other hand, there was no detectable specific [$^3$H]prazosin binding to these tissues. Scatchard and Hill plot analyses of specific [$^3$H]yohimbine binding indicated one class of binding sites. Specific binding of [$^3$H]yohimbine was displaced effectively by $alpha_2$ adrenergic agents and less effectively by $alpha_1$ adrenergic agents. $IC_{50}$ values for adrenergic drugs of [$^3$H]yohimbine binding were as follows: yohimbine, 55 nM; tramazoline, 270 nM; clonidine, 580 nM; methoxamine, 15 μM; prazosin, 47 μM, phenylephrine, 49 μM; norepinephrine, 60 μM; epinephrine, 150 μM. These results suggest that alpha adrenergic receptors in bovine cerebral artery are mostly of the $alpha_2$ subtype.

## ACKNOWLEDGEMENTS

This work was supported by a Grant-in Aid for Specific Project Research (No. 57213016), by a Grant-in-Aid for Cooperative Research (No. 58370008) from the Ministry of Education, Science and Culture, Japan (M.F.) and was supported by a grant for research from the ministry of Health and Welfare, Japan (H.H.).

## REFERENCES

1. L. G. D'Alecy and E. O. Feigel, Sympathetic control of cerebral blood flow in dogs, <u>Circ. Res.</u> 31:267 (1972).

2. N. Ohgushi, Adrenergic fibers to the brain and spinal cord vessels in the dog, Archiv. fur Japanishe Chirurgie 37:294 (1968).
3. B. Hartman and S. Udenfriend, The use of dopamine-β-hydroxylase as a marker for the central noradrenergic nervous system in rat brain, Proc. Natl. Acad. Sci. USA 69:2722 (1972).
4. L. Edvinsson and C. Owman, Pharmacological characterization of adrenergic alpha and beta receptors mediating the vasomotor responses of cerebral arteries in vitro, Circ. Res. 35:835 (1974).
5. S. P. Duckles and J. A. Bevan, Pharmacological characterization of adrenergic receptor of a rabbit cerebral artery in vitro, J. Pharmacol. Exp. Ther. 197:371 (1976).
6. S. Z. Langer, Presynaptic adrenoceptor and regulation of release, in: "The Release of Catecholamine from Adrenergic Neurons," D. M. Panton, ed., Pergamon Press, Oxford, p. 59 (1979).
7. K. Starke and S. Z. Langer, A note on terminology for presynaptic receptors, in: "Presynaptic Receptors," S. Z. Langer, K. Starke and M. L. Dubocovich, ed., Pergamon Press, Oxford, p. 1 (1979).
8. Y. Sakakibara, M. Fujiwara and I. Muramatsu, Pharmacological characterization of the alpha adrenoceptors of the dog basilar artery, Naunyn-Schimiedeberg's Arch. Pharmacol. 319:1 (1982).
9. O. H. Lowry, N. J. Rosebrough, A. L. Farr, and R. J. Randall, Protein measurement with the folin phenol reagent, J. Biol. Chem. 193:265 (1951).
10. A. H. Friedman and J. N. Davis, Identification and characterization of adrenergic receptors and catecholamine-stimulated adenylate cyclase in hog pial membranes, Brain Res. 183:89 (1980).
11. S. I. Harik, V. K. Sharma, J. R. Wetherbee, R. H. Warren, and S. P. Banerjee, Adrenergic receptors of cerebral microvessels, Eur. J. Pharmacol. 61:207 (1977).
12. S. M. Bentley, G. M. Drew, and S. B. Whiting, Evidence for two distinct types of postsynaptic α-adrenoceptor, Br. J. Pharmacol. 61:116P (1977).
13. P. B. M. W. M. Timmermans, H. Y. Kwa, and P. A. van Zwieten, Possible subdivision of postsynaptic α-adrenoceptors mediating pressor responses in the pithed rat, Naunyn-Schmiedeberg's Arch. Pharmacol. 310:189 (1979).
14. S. Z. Langer, R. Massingham, and N. B. Shepperson, Presence of postsynaptic $\alpha_2$-adrenoceptors of predominantly extrasynaptic location in the vascular smooth muscle of the dog hind limb. Clin. Sci. 59:225s (1980).
15. S. Z. Langer and N. B. Shepperson, Recent development in vascular smooth muscle pharmacology: the post-synaptic $\alpha_2$-adrenoceptor, Trends Pharmacol. Sci. 3:440 (1982).

# MODULATION OF ALPHA$_2$-ADRENERGIC RECEPTORS OF RAT VAS DEFERENS BY ADENOSINE RECEPTORS

Yashiro Watanabe and Hiroshi Yoshida

Department of Pharmacology I, Osaka University
School of Medicine, Nakanoshima, Kita-ku
Osaka 530, Japan

It is generally considered that norepinephrine and ATP are co-stored in dense core vesicles in the sympathetic nerve terminals (1,2) and that the two substances are released together on stimulation (3-5). Furthermore, it was reported that the response of vas deferens induced by norepinephrine was augmented by adenosine (6). Adenosine in the synaptic cleft comes from ATP in the nerve terminals and also from the smooth muscles. We examined the effects of adenosine and its derivatives on adrenergic receptors in rat vas deferens and found that adenosine and its derivatives have a novel effect on presynaptic adrenergic $\alpha_2$ receptors through activation of adenosine receptors.

## MATERIALS AND METHODS

### Tissue Preparation and Binding Conditions

Male Wistar rats weighing 200-220 g were decapitated and the vasa deferentia were removed, freed from the serosa coat, blood vessels and connective tissues, and weighed. Then they were homogenized in 40 volumes of 25 mM Tris-HCl buffer (pH 7.4) at 0° C in a polytron for 30 seconds. The homogenate was filtered through 4 layers of nylon cloth. Then the homogenate was centrifuged at 100,000 X g for 20 min and the pellet was suspended in Tris buffer and used as crude membrane fraction. The reaction mixture consisted of 50 mM Tris-HCl (pH 7.4), homogenate or crude membrane fraction from 6.25 mg wet weight of the original tissue, test agents such as adenosine, and [$^3$H]clonidine (2.3-22.5 nM) or [$^3$H]yohimbine (2.0 - 27.0 nM) in a final volume of 1 ml. After incubation at 25° C for 30 min, the reaction mixtures were passed through Whatman GF/F glass filters and washed three times with 5 ml of ice-cold 50 mM Tris-HCl buffer. Then the radioactivity on the filter was measured in a liquid scintillation counter. Specific binding was expressed as the difference between the radioactivity in the presence and absence of 0.1 mM l-norepinephrine.

## Contraction of Vas Deferens by Electrical Stimulation

The isolated vas deferens with a load of 250 mg was placed in a 10 ml organ bath containing Locke's solution and contractile responses were recorded isometrically using a force displacement transducer (SBIT Nihon Koden, Japan) and a polygraph. Electrical stimulations were applied directly through a hooked punctuate electrode inserted into the upper end of the vas deferens. The stimulations were 30 second trains of 1 msec - 1 Hz pulses with a supramaximal voltage (over 30 volts) at 7 min intervals. Clonidine was added 2 min before stimulation and washed out after each stimulation.

## RESULTS

### 1. Increase in Number of [$^3$H]Clonidine Binding Sites of Homogenates of Rat Vas Deferens by Adenosine and Its Derivatives (7)

When 22.5 nM of [$^3$H]clonidine was used, no binding sites were detected in homogenate of untreated rat vas deferens. But after treatment of rats with reserpine (0.5 mg/kg body weight, once a day for 2 days), [$^3$H]clonidine binding sites appeared (1.9 pmoles/g wet weight) in the vas deferens. So, in the following experiments, homogenates of reserpinized rat vas deferens were used.

The [$^3$H]clonidine binding sites had the pharmacological characteristics of $\alpha_2$-adrenergic receptors. In a low concentration range (2.3-22.5 nM), [$^3$H]clonidine labelled a single population of receptors having a Kd value of 3 nM . But at higher concentrations than 22.5 nM, the Scatchard plot of [$^3$H]clonidine binding became curvilinear.

In the presence of adenosine and its derivatives, the amount of binding of [$^3$H]clonidine at 22.5 nM in homogenates in Tris-buffer increased as shown in Fig. 1. Adenosine and 2-chloroadenosine were effective, but AMP and adenine had no effect. Furthermore, it was found that the effect of adenosine was inhibited by theophylline in a dose-dependent manner (Fig. 2). This finding seems to indicate that the increase in clonidine binding caused by adenosine came from activation of adenosine receptors.

As described above, no clonidine binding sites were detected in a homogenate of untreated rat vas deferens. However, even with

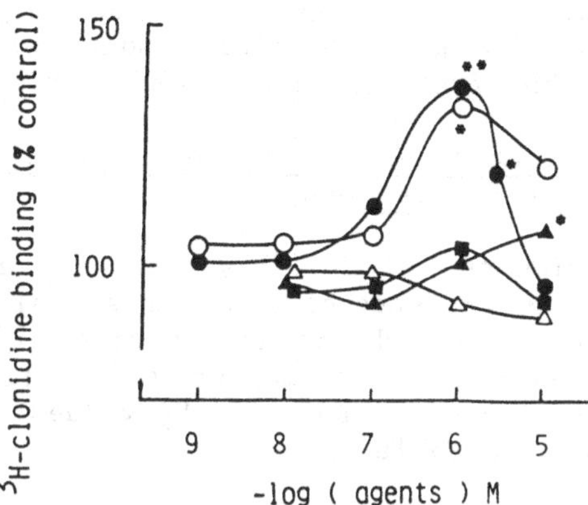


Fig. 1. Effects of increasing concentrations of adenosine (●), 2-chloroadenosine (O), ATP (▲) and adenine (■) on specific binding of [$^3$H]clonidine (22.5 nM). The amount of clonidine binding of homogenate of reserpine-treated rat vas deferens was 2.0 ± 0.3 pmol/g wet weight. This value was expressed as 100%.

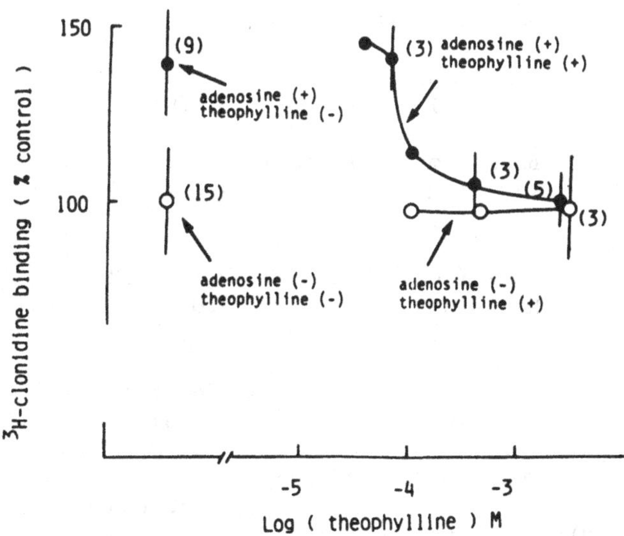


Fig. 2. Effect of theophylline on the increase of [$^3$H]clonidine binding caused by 1 μM adenosine. Presence (●) or absence (O) of 1 μM adenosine.

untreated rat vas deferens, binding sites became detectable in the presence of 2-chloroadenosine with 22.5 nM [$^3$H]clonidine as shown in Table 1. Furthermore the effect of 2-chloroadenosine disappeared in the presence of GTP (10 μM). Similar results were also obtained with the crude membrane fraction instead of the homogenate.

## 2. Effect of 2-Chloroadenosine on [$^3$H]Yohimbine Binding in Homogenates of Rat Vas Deferens

Binding experiments with [$^3$H]yohimbine, an $\alpha_2$-adrenergic antagonist, showed the presence of a monophasic binding site with a Kd of 8.0 in both untreated and reserpinized rat vas deferens. 2-Chloroadenosine increased the number of [$^3$H]clonidine binding sites, but had no influence on Bmax and Kd values of the [$^3$H]yohimbine binding as shown in Table 1.

Table 1. Effects of 2-chloradenosine (2CA), adenosine, theophylline and GTP on [$^3$H]clonidine and [$^3$H]yohimbine binding in vas deferens of untreated rat homogenates

| Treatment | | [$^3$H]Clonidine | [$^3$H]Yohimbine |
|---|---|---|---|
| Control | Bmax (pmol/g tissue) | not detectable | 1.2 ± 0.2 |
| | Kd (nM) | | 8.0 ± 0.8 |
| | (N) | (9) | (6) |
| 2CA ($10^{-5}$M) | Bmax (pmol/g tissue) | 0.8 ± 0.1 | 1.1 ± 0.2 |
| | Kd (nM) | 4.0 ± 0.6 | 8.0 ± 0.9 |
| | (N) | (5) | (3) |
| Adenosine | Bmax (pmol/g tissue) | 0.7 ± 0.2 | -- |
| | Kd (nM) | 3.0 ± 0.40 | |
| | (N) | (5) | |
| GTP ($10^{-5}$M) | | not detectable | |
| | (N) | (3) | |
| 2CA ($10^{-5}$M) and GTP ($10^{-5}$M) | Bmax (pmol/g tissue) | not detectable | 1.5 ± 0.2 |
| | Kd (nM) | | 8.8 ± 1.7 |
| | (N) | (3) | (3) |

Maximal amounts of binding sites (Bmax) and dissociation constants (Kd) were determined by Scatchard analysis. Values are means ± S.E.M.

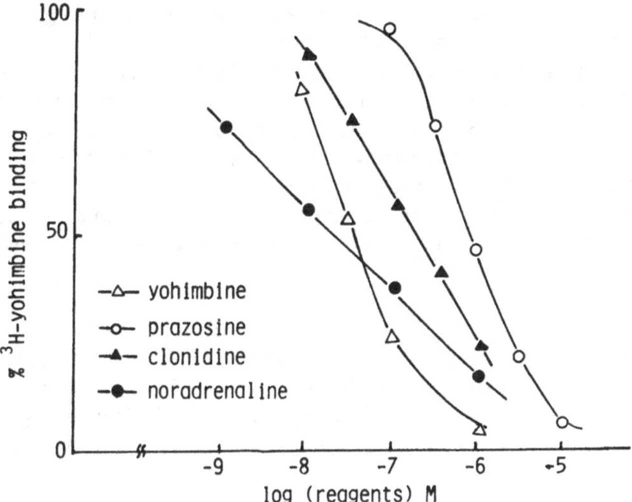


Fig. 3. Inhibition of specific [$^3$H]yohimbine binding to homogenates of vas deferens of control rats. Homogenates were incubated with [$^3$H]yohimbine for 30 min at 25° C in the presence of various concentrations of unlabeled clonidine, norepinephrine, yohimbine and prazosin.

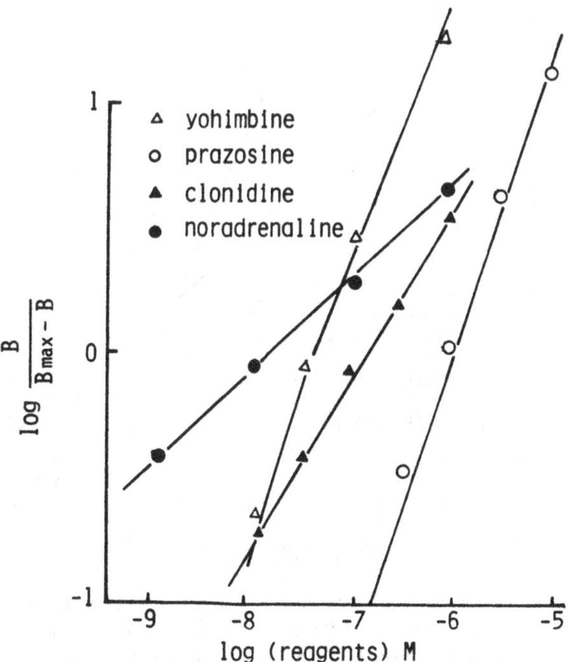


Fig. 4. Psudo-Hill plots of inhibition of specific [$^3$H]yohimbine binding by drugs.

We examined the inhibitory effects of α-adrenergic ligands on the [$^3$H]yohimbine binding. As shown in Fig. 3, yohimbine and prazosin inhibited the binding competitively, the Hill coefficient being about 1.0. However, the inhibition curves for the $\alpha_2$-agonists clonidine and norepinephrine were flat and the Hill coefficients were 0.63 and 0.39 respectively (Fig. 4). This finding suggested the existence of heterologous binding sites having high and low affinities for agonists in $\alpha_2$-adrenergic receptors in rat vas deferens, as already reported in human platelets (8) and NG 108-15 cells (9).

Next we examined the effects of 2-chloroadenosine (10 μM) on [$^3$H]yohimbine binding. 2-Chloroadenosine had no effect on the Kd value or Bmax of [$^3$H]yohimbine binding (Table 1), but enhanced the inhibitory effects of clonidine and norepinephrine on [$^3$H]yohimbine binding, as shown in Fig. 5. It reduced the $IC_{50}$ value of

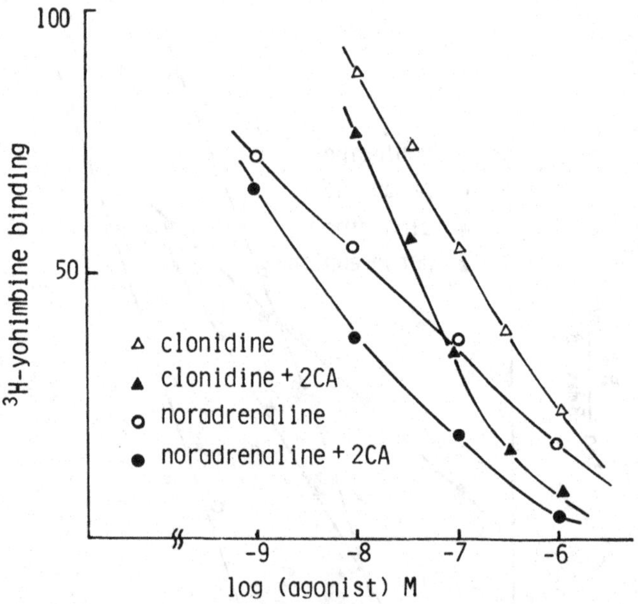


Fig. 5. Effect of 2-chloroadenosine on the inhibitory effects of clonidine and norepinephrine on [$^3$H]yohimbine binding. Homogenates were incubated with 3 nM [$^3$H]yohimbine for 30 min at 25° C in the presence of various concentrations of agonist.

clonidine to 40 ± 20 nM from 110 ± 30 nM(4), though inhibitory effects of antagonists were not influenced by 2-chloroadenosine.

These binding experiments with [$^3$H]clonidine and [$^3$H]yohimbine suggest that activation of adenosine receptors caused a configurational change of the $\alpha_2$-adrenergic receptor and increased the affinity for agonists.

### 3. Effect of 2-Chloroadenosine on Presynaptic $\alpha_2$-Receptors in Intact Rat Vas Deferens

The isolated intact vas deferens was preincubated in an organ bath with 10 μM 2-chloroadenosine in 10 ml of Locke's solution at 37° C for 30 min and then washed 10 times in 30 min at 37° C and homogenized. Even in this case we could observe the effect of 2-chloroadenosine on [$^3$H]clonidine binding, indicating that the site of action of 2-chloroadenosine was on the outside of the plasma membrane and that the effect was rather long lasting. On the other hand, the effect of adenosine derivatives described above could not be observed in denervated rat vas deferens. Accordingly, the effect of adenosine derivatives on the $\alpha_2$-adrenergic receptor seems to be mainly at presynaptic membranes. $\alpha_2$-Adrenergic receptors in presynaptic membranes are known to inhibit transmitter release by activation of the receptor. Therefore, we examined the effect of pretreatment with adenosine derivatives on the inhibitory effect of clonidine on the contractile response of rat vas deferens induced by electrical stimulation. The contraction by electrical stimulation was blocked by tetrodotoxin but not influenced by pretreatment with 10 μM 2-chloroadenosine. However, as shown in Fig. 6, the inhibitory effect of clonidine, which was prevented by yohimbine, on the contractile response by electrical stimulation was strengthened by pretreatment with 2-chloroadenosine with significant decrease in the $IC_{50}$ value of clonidine from 20 nM to 14.9 nM. Moreover, pretreatment with 10 μM 2CA had no influence on the contractile response induced by exogenous norepinephrine or phenylephrine. Accordingly, potentiation of the inhibitory effect of clonidine on the contractile response induced by electrical stimulation seems to be due to increase in the sensitivity of presynaptic adrenergic $\alpha_2$-receptors (Fig. 6).

## DISCUSSION

Our results indicated that the binding of [$^3$H]clonidine increased when adenosine or its derivatives were added to the homogenate or crude membrane fraction of rat vas deferens in Tris buffer. The binding of [$^3$H]yohimbine did not change under these conditions. These findings exclude the possibility that the net amount of $\alpha_2$-adrenergic receptors was increased through protein

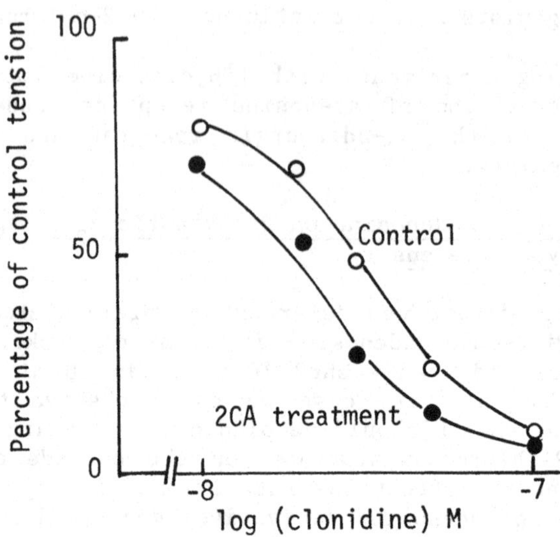


Fig. 6. Typical curves for inhibition by clonidine of contraction induced by electrical stimulation before (O) and after (●) treatment with 2CA (10 μM). Clonidine (10-100 nM) was added 2 min before applying a stimulus for recording dose-response curves.

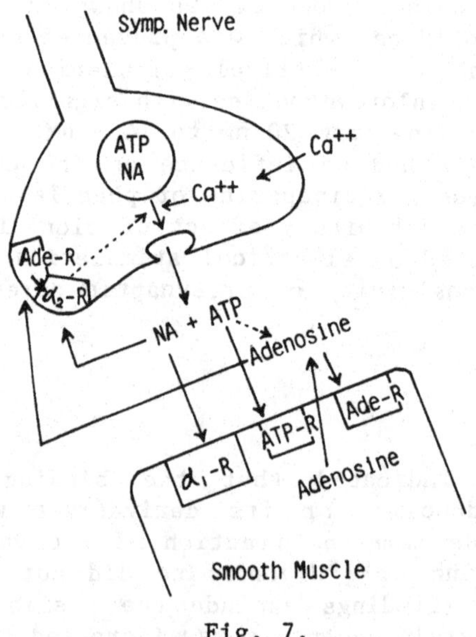


Fig. 7.

synthesis. Thus, the effect seems likely to be due to a change in configuration of $\alpha_2$-adrenergic receptors. Two states of $\alpha_2$-adrenergic receptors with high and low affinities for agonists have been reported in human platelets (8) and NG108-15 cells (9), as well as for adrenergic β-receptors (10,11), muscarinic acetylcholine receptors (12) and others. The high affinity states of these receptors are converted to low affinity states for agonist by GTP and its analogues. These results and our finding that the effect of adenosine derivatives disappeared in GTP suggest that the effect of adenosine derivatives may be to increase the receptor-GTP binding protein complex which is considered to be a high affinity state for agonist.

Activation of adenosine receptors seemed to cause an increase in affinity of the $\alpha_2$-adrenergic receptors for agonist, resulting in potentiation of the inhibitory action of agonists on norepinephrine release from the nerve terminals (Fig.7). However, with denervated rat vas deferens, an increase in clonidine binding induced by adenosine was not observed. Accordingly, $\alpha_2$-adrenergic receptors located on the postsynaptic membrane of the smooth muscle were rather increased by denervation (13) but not modulated by adenosine receptors. The correlation between adenosine receptors and $\alpha_2$-adrenergic receptors seems to be related with their closely adjacent locations on the same membrane.

Acetylcholine (ACh) and VIP (vasoactive intestinal peptide) were reported to be present in postganglionic fibers innervating the submandibular salivary gland and to be released together on stimulation. Furthermore, it was reported that VIP potentiated the ACh induced secretion of saliva and increased the affinity of muscarinic ACh receptors for muscarinic agonists in the salivary gland (14,15). However, in cardiac membranes, no effect of VIP on the muscarinic ACh receptors was observed.

Interactions among co-transmitters at the receptor level seem to have high tissue specificity. The molecular mechanisms of these interactions remain to be clarified, but these direct interactions among co-transmitter receptors seem to have functional significance.

## SUMMARY

The effects of adenosine and its derivatives on binding of $\alpha_2$-adrenergic receptor ligands with homogenates of rat vas deferens were examined.

Adenosine and 2-chloradenosine increased the amount of [$^3$H]clonidine binding. ATP had a slight effect, but AMP and adenine had none. The effect of adenosine was inhibited by theophylline. On the other hand, [$^3$H]yohimbine binding was not influenced by 2-chloradenosine. The inhibitory effect of clonidine on [$^3$H]yohimbine binding was strengthened by 2-chloradenosine, indicating an increase in affinity of $\alpha_2$-adrenergic receptors for clonidine.

Similar results were obtained with intact rat vas deferens pretreated with 2-chloradenosine. Then, it was found that presynaptic inhibition of noradrenaline release by clonidine was increased by pretreatment with 2-chloradenosine as judged by the contractile response of vas deferens induced by electrical stimulation.

From these results it was concluded that adenosine and its derivatives cause a configurational change in presynaptic $\alpha_2$-adrenergic receptors through activation of adenosine receptors. The induced configuration of $\alpha_2$-receptors with high affinity for agonists seemed to be that of the active functional state, because the inhibitory effect of clonidine on norepinephrine release was potentiated by pretreatment with 2-chloradenosine.

ACKNOWLEDGEMENTS

This work was partly supported by Grant-in-Aid for scientific research from Ministry of Education, Science and Culture of Japan. The authors also wish to acknowledge the secretarial assistance provided by Mrs. M. Nakamura.

REFERENCES

1. U. S. von Euler, T. Lishajko, and L. Stjarner, Catecholamine and adenosine triphosphate in isolated adrenergic nerve granules, Acta Physiol. Scand. 59:495 (1963).
2. A. D. Smith, Subcellular localization of noradrenaline in sympathetic neurons, Pharmacol. Reviews 24:435 (1972).
3. W. W. Douglas and A. M. Poisner, Evidence that the secreting adrenal chromaffin cell releases catecholamines directly from ATP rich granules, J. Physiol. 183:236 (1966).
4. P. Stevens, R. L. Robison, K. Van Dyke, and R. E. Stitzel, Studies on the synthesis and release of ATP-$^3$H in the isolated cat adrenal gland, J. Pharmacol. Exp. Therap. 181:463 (1972).
5. C. Su, Neurogenic release of purine compounds in blood vessels, J. Pharmacol. Exp. Therap. 195:159 (1975).
6. M. I. Holck and B. H. Marks, Purine nucleotide and nucleoside interactions on normal and subsensitive alpha adrenoreceptor responses in guinea-pig vas deferens, J. Pharmacol. Exp. Therap. 205:104 (1978).

7. Y. Watanabe, R. T. Lai, and H. Yoshida, Increase of $^3$H-clonidine binding sites induced by adenosine receptor agonists in rat vas deferens in vitro, Europ. J. Pharmacol. 86:265 (1983).
8. B. S. Tsai and R. T. Lefkowitz, Agonist specific effects of guanine nucleotides on alpha-adrenergic receptors in human platelets, Mol. Pharmacol. 16:61 (1979).
9. D. J. Kahn, J. C. Mitrius, and D. C. U'Prichard, Alpha-2 adrenergic receptors in neuroblastoma x glioma hybrid cells characterization with agonist and antagonist radioligands and relationship to adenylate cyclase, Mol. Pharmacol. 21:17 (1982).
10. B. B. Hoffman, D. M. Kilpatrick and R. J. Lefkowitz, Heterogeneity of radioligand binding to α-adrenergic receptors. Analysis of guanine nucleotide regulation of agonist binding in relation to receptor subtypes, J. Biol. Chem. 255:4645 (1980).
11. M. Rodbell, The role of hormone receptors and GTP-regulatory proteins in membrane transduction, Nature 284:17 (1980).
12. S. Uchida, K. Matsumoto, K. Takeyasu, H. Higuchi, and H. Yoshida, Molecular mechanism of the effects of guanine nucleotide and sulfhydryl reagent on muscarinic receptors in smooth muscles studied by radiation inactivation, Life Sci. 31:201 (1982).
13. Y. Watanabe, R. T. Lai, H. Maeda, and H. Yoshida, Reserpine and sympathetic denervation cause an increase of postsynaptic $\alpha_2$-adrenoreceptors in rat vas deferens, Europ. J. Pharmacol. 80:105 (1982).
14. J. M. Lundberg, A. Anggard, J. Fahrenkrug, T. Hokfelt and V. Mutt, Vasoactive intestinal polypeptide in cholinergic neurons of exocrine glands: Functional significance of coexisting transmitters for vasodilation and secretion, Proc. Natl. Acad. Sci. USA 77:1651 (1980).
15. J. M. Lundberg, B. Hedlung, and T. Bartfai, Vasoactive intestinal polypeptide enhances muscarinic ligand binding in cat submandibular salivary gland, Nature 295:147 (1982).

# NEUROMODULATORY ROLES OF ADENOSINE RECEPTORS COUPLING TO THE CALCIUM CHANNEL AND ADENYLATE CYCLASE

Yoichiro Kuroda

Department of Neurochemistry
Tokyo Metropolitan Institute for Neurosciences
Fuchu-shi, Tokyo 183, Japan

## INTRODUCTION

It has been shown that ATP is stored together with acetylcholine or catecholamines in synaptic vesicles and released from nerve terminals by electrical stimulation. The released ATP is degraded to adenosine extracellularly. Adenosine derivatives also might be released from excited postsynaptic neurons. Therefore, during and after stimulation, adenosine derivatives are accumulated in the synaptic cleft. Physiological functions of ATP and adenosine have been reported. In the peripheral nervous system, adenosine reduces the quantum content and frequency of miniature end-plate potentials in the rat phrenic nerve-diaphragm preparation (1). In the central nervous system, we found that extracellular addition of adenosine derivatives caused two effects on the post-synaptic potentials (PSP) evoked by electrical stimulation of lateral olfactory tract (LOT) and recorded from the surface of olfactory cortex slices: one is the direct inhibition of PSP (2,3) and the other is indirect facilitation via cyclic AMP which appears after the removal of adenosine (4). By iontophoretic studies, it has been shown that adenosine derivatives have a depressant action on the firing of neurons in several regions of the rat brain (5). These effects are suggested to be presynaptic and mediated by two different types of adenosine receptor on the presynaptic membrane, which regulate transmitter release by changing the intracellular concentration of $Ca^{2+}$ or cyclic AMP in the nerve terminals. The physiological significance of adenosine derivatives on neurotransmission in the mammalian central nervous system is discussed especially in relation to facilitation as a possible mechanism of post-tetanic potentiation (PTP) and heterosynaptic facilitation (HSF) which are considered to be basic electrophysiological models of learning and memory (6).

## ACCUMULATION OF ADENOSINE DERIVATIVES IN THE SYNAPTIC CLEFT

### Release of Adenosine Derivatives

By electrical or high $K^+$ stimulation, [$^{14}$C] adenosine derivatives were released from cortical synaptosomes whose ATP pool was prelabeled by [$^{14}$C] adenosine (Fig. 1) (7). The release was $Ca^{2+}$ dependent and therefore similar to that of other putative neurotransmitters. About 50% of the released [$^{14}$C] derivatives was found to be [$^{14}$C] adenosine which is probably formed after degradation of the released ATP by ATPase and 5'-nucleotidase on the surface of the synaptic membrane. Only 8% of the [$^{14}$C] material remained as [$^{14}$C] adenine nucleotides.

Release of ATP from synaptosomes by high $K^+$ stimulation was directly monitered by the ATP-luciferin-luciferase system (8). Adenosine derivatives also may be released from postsynaptic neurons which are excited. Contribution of pre- and post-synaptic release of adenosine derivatives to the local concentration in the synaptic

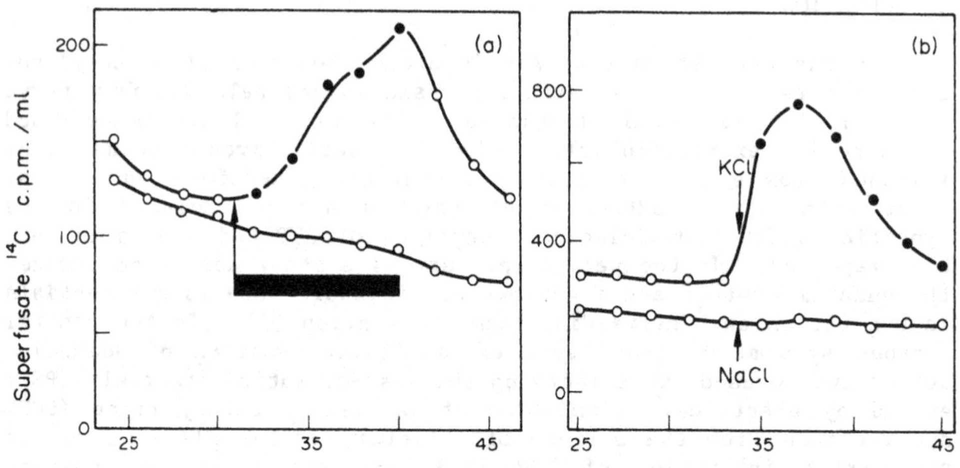


Fig. 1. Release of [$^{14}$C] adenosine derivatives from synaptosome beds by electrical stimulation and by 50 mM-KCl. Synaptosome beds were incubated for 40 min in 5 ml of glucose-bicarbonate medium containing 0.2 µCi/ml of [8-$^{14}$C] adenosine (3.85 µM). The beds were transfered into fresh medium and superfused at approx. 3.5 ml/min. Superfusates were collected at 2 min intervals and counted. (a) Electrical pulses were applied for 8 min (black bar), starting 32 min after beginning superfusion, at 10 V, 0.4 msec duration and 50 Hz (•); Unstimulated control (o). (b) At the time indicated by arrows, conc. KCl (•) or NaCl (o) was injected to give 50 mM final concentration.

cleft has not yet been determined quantitatively.

### Physiological Conditions for the Accumulation of Adenosine

When the synaptosomes were incubated with Krebs-Ringer medium, they took up micromolar concentrations of [$^{14}$C] adenosine actively from the medium. After the uptake, 88% of the [$^{14}$C] adenosine was phosphorylated to [$^{14}$C] adenine nucleotides in the synaptosomes (7). This suggests that a coupling system of uptake and phosphorylation specific for adenosine exists in nerve terminals. Adenosine may be inactivated by deamination to inosine or by diffusion. Because of these homeostatic mechanisms, adenosine derivatives would not be expected to accumulate in the synaptic cleft *in vivo* under moderate stimulation. However, adenosine can accumulate when the release of adenosine derivatives overcomes the maximum capacity for the inactivation, for example, under tetanic stimulation. In fact, when LOT was tetanically stimulated, adenosine derivatives were released from olfactory cortex slices *in vitro* (9), which were used in the following electrophysiological observations.

## MODULATION OF TRANSMITTER RELEASE BY ADENOSINE

### Direct Inhibition by Adenosine Derivatives

Olfactory cortex slices are one of the *in vitro* mammalian brain preparations suitable for examining the effects on neuronal activities of substances administered in known concentrations (10,11). Application of adenosine (5 μM to 1 mM) to the incubation medium immediately decreased the amplitude of PSP evoked by the electrical stimulation of LOT, whereas the amplitude of presynaptic fiber potentials was not influenced (Fig. 2) (2).

The inhibitory effect continued in the presence of adenosine for as long as 20 min; however, such effects disappeared rapidly after the removal of the extracellular adenosine. Moreover, after extensive rapid washings, the PSP often rebounded.

ATP, ADP and 5'-AMP showed a similar inhibitory action to that of adenosine. Adenine, inosine, hypoxanthine, guanine, guanosine, GTP, cytidine, uridine and thymidine (100 μM to 1 mM) showed no significant effects on the PSP (3). Theophylline (2 mM) completely inhibited the effects of 100 μM adenosine or ATP. Dipyridamole, known as an inhibitor of adenosine uptake, potentiated the inhibition of PSP at submaximum doses of adenosine (3). Therefore, the action of adenosine derivatives is caused extracellularly through interaction with a specific adenosine receptor on the cell surface. The adenosine receptor for inhibition of the PSP (ADO-1) appeared to be different from the one (ADO-2) which is coupled to the adenylate cyclase system in the slices (12) (see the following section).

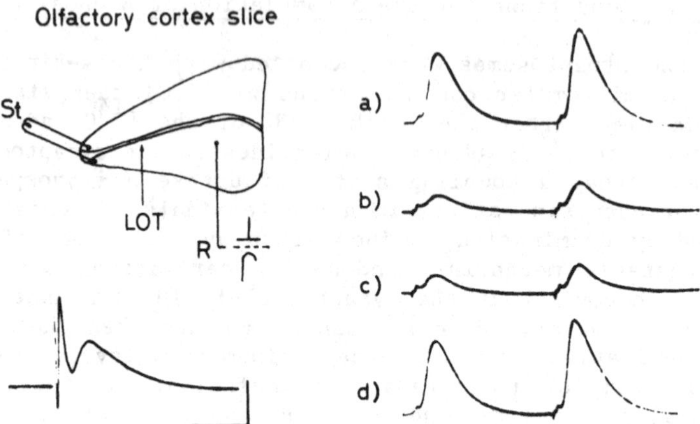


Fig. 2. Inhibition and potentiation of PSP by adenosine in olfactory cortex slices *in vitro*. Olfactory cortex slices of guinea pig were incubated in Krebs-Ringer-glucose medium gassed with (95% $O_2$ and 5% $CO_2$ ). The frontal end of the LOT was stimulated with rectangular pulses of 10 V in strength and 0.1 msec in duration at 2 Hz. When the recording electrode was located on the LOT, presynaptic fiber potentials were recorded; these were followed by PSP (lower left; calibration: 20 msec, 2 mV). a) When the electrode was situated on the olfactory tubercle area (as shown upper left), PSPs were recorded as control; b) when 100 μM adenosine was added to the medium, the amplitude of PSP was decreased to about 40% of the original within 1 min; c) it was not changed by a further 10 min incubation with adenosine; d) by the removal of adenosine from the medium it rebounded over the original level within 2 min.

Is the site of action of adenosine presynaptic or postsynaptic? In other words, is the adenosine receptor located on nerve terminals or on the postsynaptic neurons? An excitatory action of glutamate, a putative transmitter in the terminals of LOT, on olfactory cortex neurons was observed by unit recording. The postsynaptic action of glutamate was not changed by the addition of adenosine derivatives at the same concentration as that showing inhibition of the PSP in identical slices. Intracellular recording of neurons in the cortex slices has not yet been carried out because of practical difficulty.

Instead, intracellular recording from a periamygdaloid cortex neuron, which has similar properties to the prepyriform cortex, has been done. Adenosine (1 μM to 10 mM) had no effect on membrane resistance and membrane potentials (13). Recently, it has been found that adenosine appears to increase the magnitude of frequency facilitation observed in the synapses of LOT (14). These results indicate that adenosine does not act postsynaptically but inhibits transmitter release presynaptically, resulting in the decrease of PSP amplitude. It has been suggested that adenosine decreases the release of excitatory transmitter at the neuromuscular junction (1).

## Adenosine Receptor Coupling to Calcium Channels

It is generally accepted that the release of transmitter is highly dependent on extracellular $Ca^{2+}$ and is caused by the increase of intracellular $Ca^{2+}$ concentration due to the depolarization-dependent influx of $Ca^{2+}$ into nerve terminals. Correlations between the amplitude of PSP and the concentration of extracellular $Ca^{2+}$ observed in the olfactory cortex slices (15) suggest that the amount of released transmitter depends on the extent of the $Ca^{2+}$ influx into the presynaptic terminals, which may in turn be affected by extracellular $Ca^{2+}$ in the medium. When the $Ca^{2+}$ concentration in

Table 1. Effects of adenosine and cyclic AMP analogs on high $K^+$-dependent $^{45}Ca$ influx into synaptosomes.

| Agents | Concentration (μM) | $^{45}Ca$ influx (%) |
|---|---|---|
| None | | [100] |
| 2-Chloroadenosine | 50 | 63 |
| 2-Chloroadenosine<br>+Theophylline | 50<br>880 | 111 |
| Dibutyryl cyclic AMP | 100 | 186 |
| Dibutyryl cyclic GMP | 100 | 80 |

A synaptosome preparation from guinea pig cerebral cortex was preincubated with adenosine deaminase (0.5 U/ml) in Ringer solution. Aliquots of the suspension were added to $^{45}CaCl_2$ (5 μCi) and test agents. High $K^+$ (30 mM) treatments were carried out for 15 sec and the synaptosomes were collected and counted for radioactivity.

the medium was increased to 5.6 mM from the normal level (1.2 mM), the inhibition of PSP by adenosine (100 μM) was significantly prevented. Similar elevation of extracellular $Mg^{2+}$ caused a small effect, which was not significant (16). Thus, a possible mechanism can be proposed in which adenosine inhibits $Ca^{2+}$ influx, so preventing the release of transmitter.

We found that adenosine inhibited the depolarization-dependent influx of $Ca^{2+}$ into nerve terminals. High $K^+$-dependent $^{45}Ca$ accumulation in cortical synaptosomal preparation was decreased by the addition of 2-chloroadenosine, an analog of adenosine; this inhibition was blocked by theophylline, an antagonist of adenosine receptors, in the medium (Table 1).

More recently, we obtained evidence that adenosine directly inhibited voltage-dependent Ca channels. $Ca^{2+}$ currents observed in whole cell clamped neuroblastoma cells were decreased by the addition of adenosine analogs (Kuroda et al. in preparation). Therefore, the adenosine receptor for the inhibition of PSP (ADO-1) appeared to couple to voltage-dependent Ca channels located in the nerve endings.

## Facilitation by Adenosine-Induced Increase of Cyclic AMP

When checking the reversibility of PSP inhibition by adenosine, the amplitude of PSP was often increased to a level higher than the original amplitude after the removal of adenosine (4). This effect was investigated more carefully; it appeared that the potentiation of PSP was more significant after washing two or three times with fresh medium in a short time (less than 30 sec). Control experiments without adenosine treatment showed no increase of PSP amplitude. The increase was 20 to 30% and continued for quite a while (several minutes) after the removal of adenosine (17). The results suggest that adenosine induces some biochemical change in the synapses of LOT, which lasts even after the disappearance of extracellular adenosine.

It has been reported that adenosine is one of the agents which increase the level of cyclic AMP in brain slices (18). In olfactory cortex slices, adenosine derivatives increase the level of cyclic AMP (3). As one of the criteria to check the possible involvement of cyclic AMP, analogs of cyclic AMP were applied to the slices in order to test whether they mimic the action of adenosine derivatives. Dibutyryl cyclic AMP and 8-bromo cyclic AMP did increase the amplitude of the PSP by 20 to 30%, similar to the treatment with adenosine (12).

Similar potentiation by cyclic AMP analogs has been observed in *Aplysia* ganglia (19), and in certain peripheral systems (20,21). Such potentiations appeared to be linked to a facilitation of transmitter release. In olfactory cortex slices, it is difficult to obtain direct evidence for a facilitation of transmitter release mediated by adenosine-induced increase of cyclic AMP. However, if this mechanism does occur, cyclic AMP should be increased presynaptically.

### Adenosine Receptor Coupling to Adenylate Cyclase

The endogenous level of cyclic AMP in incubated synaptosomes from olfactory cortex slices was investigated. It apeared that the synaptosomal suspension already contained exogenous adenosine. Preincubation with theophylline or with adenosine deaminase decreased both the exogenous level of adenosine and the intrasynaptosomal level of cyclic AMP. The level of cyclic AMP was restored by the addition of adenosine agonists, especially 2-chloroadenosine (Fig. 3) (22). This increase was antagonized by deoxyadenosine and was not inhibited by dipyridamole. These results indicate that the adenosine derivatives in the synaptic cleft regulate the level of cyclic AMP in nerve terminals through an adenosine receptor (ADO-2) coupling to a cyclic AMP generation system, presumably adenylate cyclase. Membrane fragments containing adenylate cyclase, which inevitably contaminate synaptosome preparations, cannot contribute to the change of cyclic AMP levels caused by adenosine, since there is no substrate ATP in the medium.

The presynaptic increase of cyclic AMP in response to adenosine, potentiation of PSP by adenosine treatment, and its mimicry by cyclic AMP analogs suggest that adenosine accumulated in the synaptic cleft can increase the intracellular concentration of cyclic AMP, which then facilitates the release of transmitter from the nerve terminals. One of the possible mechanisms is a facilitation of $Ca^{2+}$ influx into nerve terminals. In fact, $^{45}Ca$ accumulation in synaptosomes dependent on high $K^+$ was increased by the addition of dibutyryl cyclic AMP. Dibutyryl cyclic GMP at the same concentration did not increase but decreased the $Ca^{2+}$ accumulation, indicating that the effect of dibutyryl cyclic AMP is due to the cyclic AMP moiety, not the dibutyryl residue (Table 1).

## SUBTYPES OF ADENOSINE RECEPTOR

### Two Adenosine Receptors in Nerve Terminals

Evidence for the existence of two adenosine receptors (ADO-1 and ADO-2) at nerve terminals in mammalian brain (Fig. 4) is as follows:

1. When time courses of PSP inhibition and cyclic AMP formation in olfactory cortex slices are examined carefully from 5 to 25 sec after the adddition of adenosine, the inhibition of PSP had

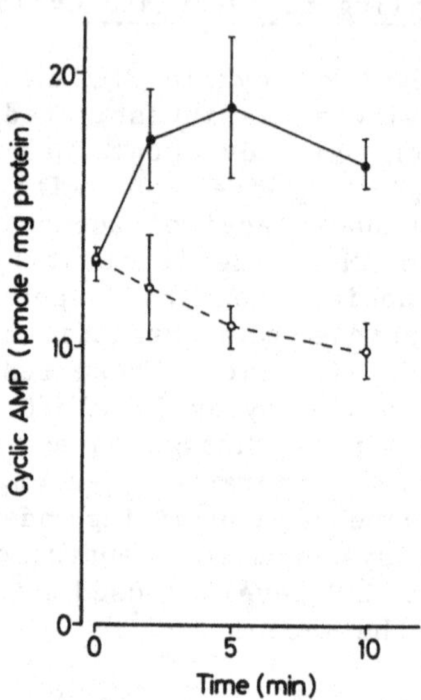


Fig. 3. Time course of the stimulation of cyclic AMP levels by 2-chloroadenosine in adenosine deaminase-treated synaptosomes. Synaptosomes from guinea pig cerebral cortex were preincubated with 0.5 U/ml adenosine deaminase. The suspension was incubated without (o) or with (•) 2-chloroadenosine (100 μM).

already occurred to nearly full extent, at a time when the level of cyclic AMP had not shown any significant increase (12). The increase of cyclic AMP in nerve terminals induced by adenosine also showed a similar slow time course.

2. Analogs of cyclic AMP were applied to the slices in order

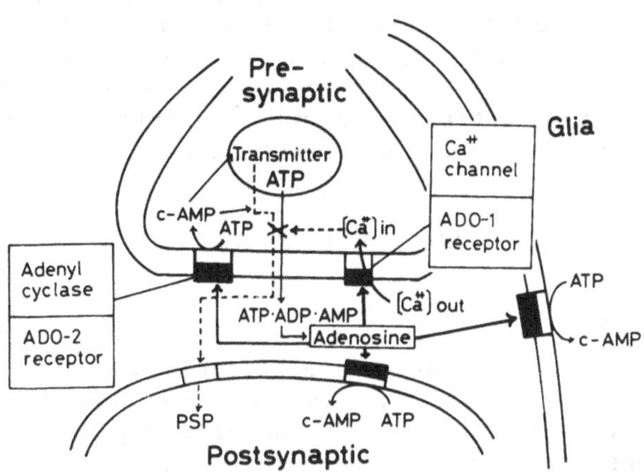


Fig. 4. "Samsara" model of dual control of transmitter release by two types of adenosine receptor. Adenosine derivatives in the synaptic cleft act on the ADO-1 receptor coupling to $Ca^{2+}$ channels on the presynaptic membrane resulting in a decrease of $Ca^{2+}$ influx, which reduces the release of transmitter. Another presynaptic receptor (ADO-2) coupling to adenylate cyclase can cause facilitation of transmitter release mediated by the increase of cyclic AMP in presynaptic terminals. (Modified from ref. 12)

to test whether they mimic the action of adenosine derivatives. Dibutyryl cyclic AMP and 8-bromo cyclic AMP did not decrease but increased the amplitude of PSP (12). Thus, facilitation of PSP can be mediated by adenosine receptors which increase the level of cyclic AMP, while inhibition must be mediated by another type of receptor.

3. Dose-response curves for the two actions are different. For the inhibition of PSP, the $EC_{50}$ was about 1 μM, whereas it was about 10 μM for the increase of cyclic AMP.

4. $N^6$-cycloheptyladenosine, NAD, and β,γ-methylene ATP caused the PSP inhibition but did not increase the level of cyclic AMP significantly in olfactory cortex slices.

Van Calker et al. (23) also reported two types of adenosine receptor that inhibit ($A_1$) or enhance ($A_2$) the formation of cyclic AMP in cultured glial cells. The ADO-2 receptor in terminals of LOT appears to be similar to the $A_2$ receptor in cultured glial cells. Whether the ADO-1 receptor corresponds to the inhibitory $A_1$ receptor is unknown. More precise pharmacological studies are necessary using specific antagonist and agonists to distinguish such receptors in different tissues and with different functions.

## SYNAPTIC PLASTICITY AND ADENOSINE RECEPTORS

### Synaptic Depression

It is well known that end-plate potentials are depressed during high frequency stimulation, at which time also the release of adenosine derivatives overcomes the maximum capacity for inactivation and adenosine derivatives accumulate in the synaptic cleft. This phenomenon, synaptic depression, has been thought to be due to the depletion of transmitter storage in the terminals. It is interesting to note, however, that this depression could also be explained by the inhibition of transmitter release by adenosine derivatives which accumulated in the synaptic cleft. Extracellular adenosine might also accumulate due to abnormal metabolism, reduced reuptake, and/or release of adenosine derivatives. Spreading depression in brain might be caused by such accumulation of adenosine derivatives, which then diffuse and spread out to the neighboring region of cerebral cortex and depress the neuronal activity.

### Post-tetanic Potentiation and Heterosynaptic Facilitation

In olfactory cortex slices, PTP was observed after the tetanic stimulation of LOT (10). A 20 to 40% potentiation was sustained for 0.5 to 3 min. The magnitude and duration are similar to those for the potentiation of PSP by adenosine treatment; the magnitude is also similar to that given by the addition of cyclic AMP analogs. Furthermore, tetanic stimulation of LOT increases the release of adenosine derivatives from the olfactory cortex slices (9). These data support the idea that PTP at terminals of LOT is mediated by adenosine-induced increase of cyclic AMP. At such synapses, adenosine is proposed to be the first messenger that increases the level of cyclic AMP, the second messenger. For HSF, adenosine can diffuse to the neighboring synapses, which are common in the central nervous system, and increase the level of cyclic AMP in the unstimulated nerve terminals. Thus, stimulation of one pathway can cause a facilitation at a different pathway (Fig. 5).

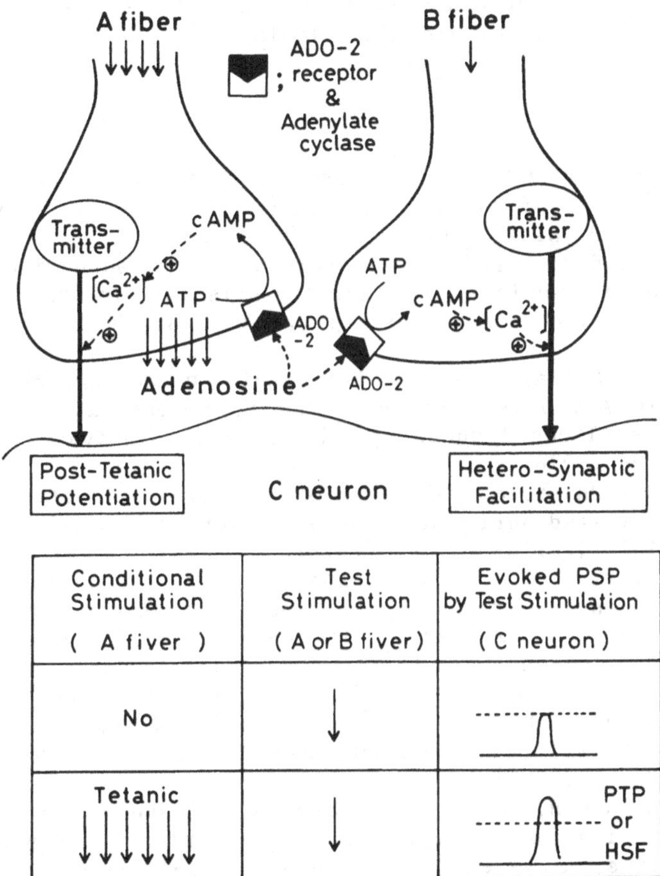


| Conditional Stimulation ( A fiver ) | Test Stimulation ( A or B fiver) | Evoked PSP by Test Stimulation ( C neuron ) |
|---|---|---|
| No | | |
| Tetanic | | PTP or HSF |

Fig. 5. Possible mechanism of heterosynaptic facilitation. Tetanic stimulation of fiber A releases adenosine, which diffuses to the neighboring terminals of fiber B. Adenosine acts on the ADO-2 receptor in the fiber B terminals and causes the increase of cyclic AMP. Test stimulation of fiber B evokes a stimulated release of transmitter from the terminals, depending on the conditional stimulation of fiber A, and HSF is observed in the C neuron.

At other synapses, other neurotransmitters or modulators can accumulate in the synaptic cleft and increase the presynaptic cyclic AMP, thereby mediating PTP or HSF. It is interesting to note that PTP and HSF are considered as possible electrophysiological mechanisms involved in the processes of learning and memory (24). In fact, a serotonin-induced increase of cyclic AMP mediates HSF which has been described as learning in *Aplysia* (25).

## REFERENCES

1. B. L. Ginsborg and G. D. S. Hirst, The effect of adenosine on the release of the transmitter from the phrenic nerve of the rat, *J. Physiol.* (London), 224:629 (1972).
2. Y. Okada and Y. Kuroda, Inhibitory action of adenosine and adenosine analogs on neurotransmission in the olfactory cortex slice of guinea pig, *Eur. J. Pharmacol.* 61:137 (1980).
3. Y. Kuroda, M. Saito, and K. Kobayashi, Concomitant changes in cyclic AMP level and postsynaptic potentials of olfactory cortex slices induced by adenosine derivatives, *Brain Res.* 109:196 (1976).
4. Y. Kuroda, Physiological roles of adenosine derivatives which are released during neurotransmission in mammalian brain, *J. Physiol.* (Paris) 74:463 (1978).
5. J. W. Phillis, G. K. Kostopoulos and J. J. Limacher, A potent depressant action of adenine derivatives on cerebral cortical neurons, *Eur. J. Pharmacol.* 30:125 (1975).
6. Y. Kuroda, Neuronal plasticity and adenosine derivatives in mammalian brain, *in*: "Physiology and Pharmacology of Adenosine Derivatives," J. W. Daly, Y. Kuroda, J. W. Phillis, H. Shimizu, and M. Ui, eds., Raven Press, New York, (1983).
7. Y. Kuroda and H. McIlwain, Uptake and release of $^{14}C$ adenine derivatives at beds of mammalian cortical synaptosomes in superfusion system, *J. Neurochem.* 22:691 (1974).
8. T. D. White, Release of ATP from a synaptosomal preparation by elevated extracellular $K^+$ and by veratridine, *J. Neurochem.* 30:329 (1978).
9. I. H. Heller and H. McIlwain, Release of $^{14}C$ adenine derivatives from isolated subsystems of the guinea pig brain: Actions of electrical stimulation and of papaverine, *Brain Res.* 53:105 (1973).
10. C. Yamamoto and H. McIlwain, Electrical activities in thin sections from the mammalian brain maintained in chemically defined media in vitro, *J. Neurochem.* 13:1333 (1966).

11. Y. Kuroda, Brain slices: Assay systems for the neurotoxicity of environmental pollutants and drugs on mammalian central nervous system, in: "Mechanisms of Toxicity and Hazard Evaluation," H. Holmstedt, R. Lauwerys, M. Mercier, and M. Roberfroid, eds., Elsevier/North-Holland Biomedical Press, Amsterdam, p. 59 (1980).
12. Y. Kuroda and K. Kobayashi, Physiological role of presynaptic "adenosine receptors" in mammalian brain, Jpn. J. Pharmacol. 29:45p (1979).
13. C. N. Scholfield, Depression of evoked potentials in brain slices by adenosine compounds, Brit. J. Pharmacol. 63:239 (1978).
14. Y. Kuroda and K. Kobayashi, Feed-back regulation of synaptic transmission by adenosine derivatives in mammalian brain, Neurosci. Lett. [Suppl.] 2:2 (1979).
15. C. D. Richards and R. Sercombe, Calcium, magnesium and the electrical activity of guinea pig olfactory cortex in vitro, J. Physiol. (London) 211:571 (1970).
16. Y. Kuroda, M. Saito, and K. Kobayashi, High concentration of calcium prevents the inhibition of postsynaptic potentials and the accumulation of cyclic AMP induced by adenosine in brain slices, Proc. Jpn. Acad. 52:86 (1976).
17. Y. Kuroda and K. Kobayashi, Post-tetanic potentiation can be mediated by adenosine-induced increase of cyclic AMP in the presynaptic terminal, Proc. Intern. Union Physiol. Soc. 14:534 (1980).
18. A. Sattin and T. W. Rall, The effect of adenosine and adenine nucleotides on the cyclic adenosine 3',5'-phosphate content of guinea pig cerebral cortex slices, Mol. Pharmacol. 6:13 (1970).
19. T. Shimahara and L. Tauc, Cyclic AMP induced by serotonin modulates the activity of an identified synapse in Aplysia by facilitating the active permeability to calcium, Brain Res. 127:168 (1977).
20. G. F. Wooten, N. B Thoa, I. J. Kopin and J. Axelrod, Enhanced release of dopamine-β-hydroxylase and norepinephrine from sympathetic nerves by dibutyryl cyclic adenosine monophosphate and theophylline, Mol. Pharmacol. 9:178 (1973).
21. L. Cubeddu, E. Barnes, and N. Weiner, Release of norepinephrine and dopamine hydroxylase by nerve stimulation. IV. An evaluation of a role for cyclic AMP, J. Pharmacol. Exp. Ther. 193:105 (1975).
22. K. Kobayashi, Y. Kuroda, and M. Yoshioka, Change of cyclic AMP level in synaptosomes from cerebral cortex: Increase by adenosine derivatives, J. Neurochem. 36:86 (1981).
23. D. Van Calker, M. Muller, and B. Hamprecht, Adenosine regulates via two different types of receptors: The accumulation of cyclic AMP in cultured brain cells, J. Neurochem. 33:999 (1979).

24. J. C. Eccles, "The Understanding of the Brain," 2nd edition, McGraw-Hill, New York (1977).
25. M. Klein and E. Kandel, Presynaptic modulation of voltage-dependent $Ca^{2+}$-current: Mechanism for behavioral sensitization in Aplysia californica, Proc. Natl. Acad. Sci. USA 75:3512 (1978).

# INTERACTION OF ENKEPHALIN WITH OPIOID RECEPTORS OF BRAIN MEMBRANES: REGULATION BY ESSENTIAL TRANSITION METALS, GUANINE NUCLEOTIDES AND TEMPERATURE

Norio Ogawa, Sachiko Mizuno and Akitane Mori

Institute for Neurobiology
Okayama University Medical School
Okayama, Japan

## INTRODUCTION

Guanine nucleotides [guanosine-5'-triphosphate: GTP, and guanylyl-5'-imidophosphate: Gpp(NH)p] play an important role in regulating β-noradrenergic (1), α-noradrenergic (2), dopamine (3) and other neurotransmitter receptors. Guanine nucleotides also decrease the receptor binding of tritiated opiate agonists but not that of the antagonists (1,4-6).

Guanine nucleotides and divalent cations may control the physiological responses to neurotransmitters and hormones by affecting the sensitivity of the receptor to the agonist and the coupling of the receptor and adenylate cyclase (7).

In the present study, we describe the interaction among guanine nucleotides, transition metals and enkephalin (ENK) receptors, and the important role of incubation temperature in the ENK actions of *in vitro* receptor binding assays.

## MATERIALS AND METHODS

### Materials

[$^3$H]D-Ala$^2$-Met$^5$-enkephalinamide ([$^3$H]ENK; specific activity 51 Ci/mmol) and [$^3$H]naloxone (specific activity 46 Ci/mmol) were respectively purchased from Amersham International (Buckinghamshire, England) and from New England Nuclear Co. (Boston, MA). D-Ala$^2$-Met$^5$-enkephalinamide (ENK) was purchased from Peptide Institute Co. (Osaka). Naloxone was a gift from Endo Laboratories, Inc. (NY).

Guanine nucleotides were purchased from Sigma Chemical Co. (St. Louis, MO). All other reagents and chemicals were reagent grade.

Receptor Preparation

For receptor preparation, crude synaptic membrane was prepared by the methods described previously (8,9). Briefly, adult male Sprague-Dawley rats weighing 200-250 g were decapitated and the whole brains were rapidly removed and homogenized with 10 volumes of ice-cold 0.32 M sucrose in two 10 sec bursts in a Polytron PT-10 homogenizer. The homogenate was centrifuged at at 900 x g for 10 min, and the resulting supernatant was centrifuged at 11,500 x g for 20 min. The pellet was suspended in 10 volumes of Tris-HCl buffer (50 mM, pH 7.6) and centrifuged at 11,500 x g for 20 min. The pellet was resuspended in 10 volumes of the same buffer and kept at -70° C. Before the receptor binding assay, the crude synaptic membrane was homogenized with excess Tris-HCl buffer and centrifuged at 11,500 x g for 20 min. The pellet was resuspended in Tris-HCl buffer at a final concentration of 1.2 mg/ml.

Receptor Binding Assay

The ENK or naloxone receptor binding assay was carried out essentially by a method previously described by us (8,10,11). Briefly, crude synaptic membrane (600 μg protein), pretreated at 25° C for 10 min or in ice for 30 min (referred henceforth as 0° C for convenience) with or without the test nucleotides, was mixed with increasing concentrations of [$^3$H]ENK (2-35 nM) or [$^3$H]naloxone (2-35 nM) in the presence or absence of unlabeled 1 μM ENK or 1 μM naloxone in a final volume of 1 ml Tris-HCl buffer. All experiments with ENK contained 50 μg/ml of bacitracin to inhibit proteolytic breakdown (12).

The tubes were incubated at 25° C for 30 min or on ice (0° C) for 2 hr, and filtered rapidly through Whatman GF/C glass fiber filters with two 3 ml washes of ice-cold Tris-HCl buffer. The radioactivity on the filters was counted in a liquid scintillation spectrometer. Specific binding was the difference between the radioactivity bound to the the receptor in the presence of 1 μM ENK or 1 μM naloxone and in their absence.

Since it has been shown that $Na^+$ ions (100 mM) decrease the specific binding of opiate agonists and increase that of antagonists (13), all receptor binding assays were carried out in the presence and absence of 100 mM NaCl.

Scatchard plots of the data were analyzed according to Marquardt (14) using a computer (NEC system PC-8800).

Effects of $Cu^{2+}$ and $Zn^{2+}$ on ENK Receptor Binding

Cupric chloride ($CuCl_2$) or zinc chloride ($ZnCl_2$) at a concentration of 50 μM was added to the incubation mixture of receptor binding assays to observe the change in specific binding. The sample with only Tris-HCl buffer (50 mM Tris-HCl buffer, pH 7.6) was used for the control.

Thermostability of Receptor Molecules

Crude synaptic membranes were treated for 30 min at 37° C or for 3 min at 56° C. The treated crude synaptic membranes were then centrifuged at 50,000 x g for 20 min, and the pellet was resuspended in the original volume of Tris-HCl buffer. The binding activity of [$^3$H]ENK was determined in this resuspension. A receptor preparation that was not exposed to high temperature but incubated in ice for 30 min and centrifuged for the same period was used as the control.

RESULTS

Effects of Guanine Nucleotides on Binding with Varying Sodium and Incubation Environments

In the presence of 100 mM NaCl, Gpp(NH)p produced a remarkable reduction in the receptor binding of [$^3$H]ENK in a concentration-dependent manner at both 25° C and 0° C incubation (Fig. 1-A, B). This Gpp(NH)p with sodium induced lowering of ENK receptor binding reflected a decrease in the number of binding sites at 25° C, although it reflected a decrease in binding affinity at 0° C (Fig. 2). In the absence of sodium at 25° C, high concentrations of Gpp(NH)p did not reduce ENK binding (Fig. 1-A). However, without sodium at 0° C, Gpp(NH)p produced a significant increase in ENK binding in a dose-dependent fashion (Fig. 1-B). This Gpp(NH)p-induced increase in ENK receptor binding reflected an increase in binding affinity without alternating binding capacity (Fig, 2). On the other hand, Gpp(NH)p had no effect on the receptor binding of [$^3$H]naloxone in similar or varying assay environments (Fig. 1-C, D).

Similar results were obtained when GTP was used as a substitute for Gpp(NH)p (data not shown).

In further experiments the effect of other nucleotides were examined upon [$^3$H]ENK and [$^3$H]naloxone binding (Table 1). In the presence of sodium, Gpp(NH)p produced a significant decrease in the specific binding of [$^3$H]ENK. A similar decrease in ENK binding was observed when GTP and GDP were tested, while GMP, ATP, ADP and AMP were less or not effective at 100 μM concentrations. These nucleotide effects on ENK bindings were essentially the same when incubations were carried out at 25° C or 0° C.

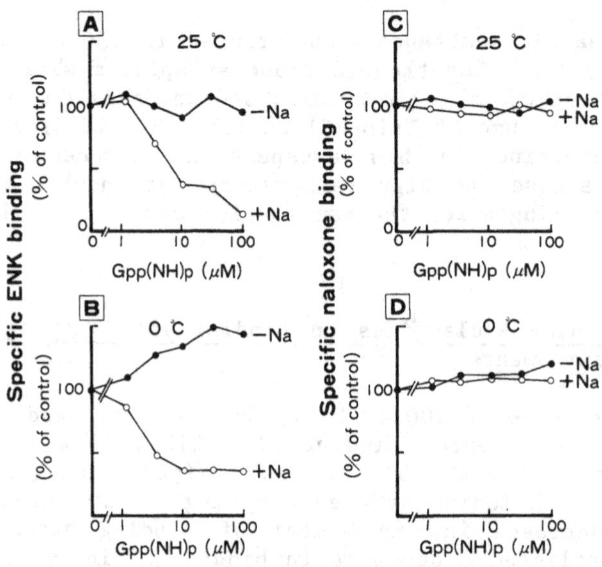


Fig. 1. Effect of different concentrations of Gpp(NH)p on [$^3$H]ENK and [$^3$H]naloxone binding. (A), [$^3$H]ENK binding at 25°C incubation; (B), [$^3$H]ENK binding at 0°C incubation; (C), [$^3$H]naloxone binding at 25°C incubation; (D), [$^3$H]naloxone binding at 0°C incubation. Rat synaptic membranes were incubated with [$^3$H]-labeled ligands and Gpp(NH)p at 1 - 100 μM in the presnece (+Na) or absence (-Na) of 100 mM NaCl. Dark circle represent % [$^3$H]-labeled ligands bound in the absence of NaCl and Gpp(NH)p; open circles represent % bound in the absence of Gpp(NH)p and in the presence of 100 mM NaCl.

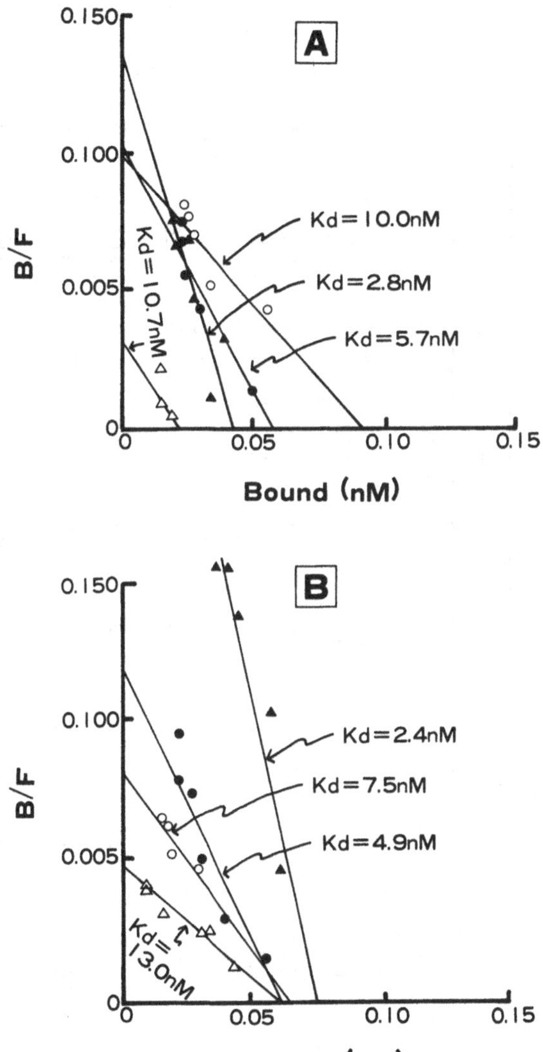


Fig. 2. Scatchard plots of saturation binding data of the effects of NaCl(100 mM) and Gpp(NH)p (50 µM) on [$^3$H]ENK binding. [A], 25°C incubation; [B], 0°C incubation. Experimental details are described in the Materials and Methods section. ●, No addition; ○, 100 mM NaCl; ▲, 50 µM Gpp(NH)p; △, 100 mM NaCl + 50 µM Gpp(NH)p.

Table 1. Effect of Nucleotides on [$^3$H]ENK and [$^3$H]Naloxone Binding to Brain Synaptic Membranes of Rats

| Nucleotide | μM | Specific Binding ( % of control ) | | | | | | | |
|---|---|---|---|---|---|---|---|---|---|
| | | ENK binding | | | | Naloxone binding | | | |
| | | Treated temperature 25°C | | Treated temperature 0°C | | Treated temperature 25°C | | Treated temperature 0°C | |
| | | -Na | +Na | -Na | +Na | -Na | +Na | -Na | +Na |
| None | 0 | 100 | 100 | 100 | 100 | 100 | 100 | 100 | 100 |
| Gpp(NH)p | 50 | 94 | 24 | 121 | 31 | 95 | 107 | 119 | 107 |
| | 100 | 94 | 14 | 144 | 37 | 102 | 93 | 120 | 107 |
| GTP | 50 | 90 | 21 | 130 | 30 | 86 | 89 | 105 | 94 |
| | 100 | 86 | 13 | 173 | 18 | 101 | 112 | 118 | 109 |
| GDP | 50 | 87 | 30 | 122 | 33 | 93 | 116 | 119 | 95 |
| | 100 | 79 | 23 | 154 | 38 | 96 | 115 | 120 | 90 |
| GMP | 100 | 118 | 90 | 109 | 67 | 97 | 110 | 105 | 110 |
| ATP | 100 | 119 | 78 | 132 | 74 | 99 | 100 | 119 | 106 |
| ADP | 100 | 125 | 87 | 122 | 82 | 106 | 99 | 108 | 104 |
| AMP | 100 | 123 | 95 | 117 | 92 | 94 | 106 | 109 | 102 |

Receptor binding assays were carried out in the presence (+Na) or absence (-Na) of 100 mM NaCl.

In the absence of sodium, however, Gpp(NH)p, GTP and GDP produced a pronounced increase in ENK receptor binding at 0° C, although no increase was observed in the binding of [$^3$H]ENK at 25° C. On the other hand, GMP, ATP, ADP and AMP had negligible effects on [$^3$H]ENK binding (Table 1).

Guanine nucleotides and adenine nucleotides failed to change [$^3$H]naloxone binding in all tested assay environments (Table 1).

## Interactions of Gpp(NH)p and Metal Ions with [$^3$H]ENK Receptor Binding

Although it has been known that manganese selectively enhances agonist binding to opioid receptors, the interaction of metal ions and guanine nucleotides has not been fully characterized in brain membranes (4).

In the present study the Gpp(NH)p-metal ion interaction in both the absence and presence of sodium were evaluated (Fig. 3). First, an attempt was made to determine how the metal ion itself interacted in affecting [$^3$H]ENK binding. $Cu^{2+}$ and $Zn^{2+}$ were found to largely inhibit the specific binding of [$^3$H]ENK in both the absence and presence of sodium. $Cu^{2+}$ was more potent than $Zn^{2+}$ in decreasing ENK binding. In the presence of sodium however, $Zn^{2+}$ had no effect on ENK binding at 25° C. Effects of $Cu^{2+}$ and $Zn^{2+}$ on [$^3$H]naloxone binding were similar to those on ENK binding. The question of whether metal ions altered the number of binding sites or affected the affinity of the binding was explored by determining the effect of increasing [$^3$H]ENK concentrations on specific binding in the presence or absence of 50 µM of metal ions (Table 2). The main effect of $Cu^{2+}$ on ENK binding was to reduce the number of binding sites, although the $Zn^{2+}$ effect was mainly due to a decrease in binding affinity (Table 2).

The effects of metal ions and Gpp(NH)p appeared to be largely additive, but the effect of $Zn^{2+}$ and Gpp(NH)p appeared to be offset in the presence of sodium at 0° C (Fig. 3). The Scatchard plots of data of the combined effect of Gpp(NH)p and $Cu^{2+}$ were similar to those of $Cu^{2+}$ alone (Table 2). The combined effect of Gpp(NH)p and $Zn^{2+}$ was also similar to that of $Zn^{2+}$ alone, except in the case of 0° C incubation without sodium (Table 2). $Zn^{2+}$ failed to have any effect on the Gpp(NH)p-induced increase in ENK receptor binding at 0° C without sodium (Fig. 3, Table 2). These results indicated that the mechanisms of action of $Cu^{2+}$ and $Zn^{2+}$ differed from each other.

## Thermostability of ENK and Gpp(NH)p Binding Sites

The thermostability of [$^3$H]ENK binding sites and Gpp(NH)p effective sites (binding protein) was examined (Fig. 4). ENK receptor binding and the effects of Gpp(NH)p were not changed in

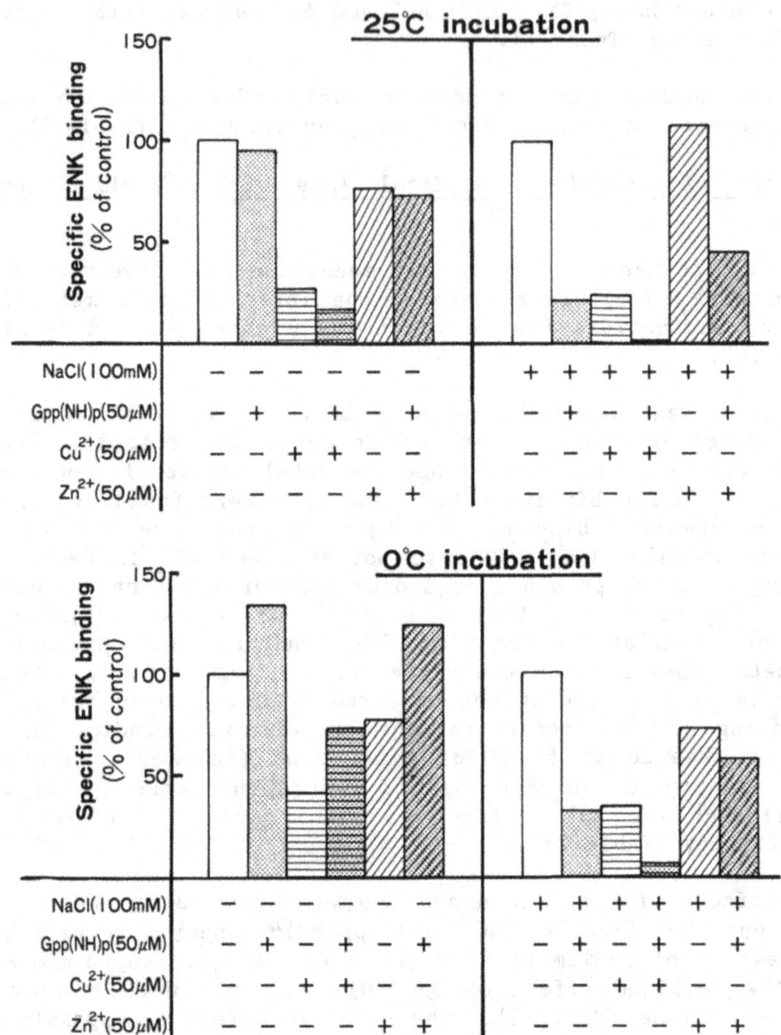


Fig. 3. Interactions of Gpp(NH)p and metal ions during ENK receptor binding. $CuCl_2$ or $ZnCl_2$ at 50 µM was added to the incubation mixture in receptor binding assays with or without 50 µM Gpp(NH)p and/or 100 mM NaCl. Incubations were carried out at 25°C (upper panel) or at 0°C (lower panel).

Table 2. Effects of $Cu^{2+}$, $Zn^{2+}$ and Gpp(NH)p on $[^3H]$ENK Binding to Brain Synaptic Membranes of Rats

| | | 25°C incubation | | 0°C incubation | |
|---|---|---|---|---|---|
| | | affinity | site | affinity | site |
| Gpp(NH)p | −Na | * | * | ↑↑ | * |
| | +Na | * | ↓↓ | ↓ | * |
| $Cu^{2+}$ | −Na | * | ↓↓ | * | ↓↓ |
| | +Na | * | ↓↓ | * | ↓↓ |
| $Cu^{2+}$ + Gpp(NH)p | −Na | * | ↓↓ | ↓ | ↓ |
| | +Na | * | ↓↓ | * | ↓↓ |
| $Zn^{2+}$ | −Na | ↓ | ↑ | ↓ | * |
| | +Na | ↓ | ↑ | ↓ | * |
| $Zn^{2+}$ + Gpp(NH)p | −Na | ↓ | ↑ | ↓ | ↑ |
| | +Na | ↓ | * | ↓ | ↓ |

Saturation experiments were performed as described in the Materials and Methods section, and Scatchard plots of the data were analyzed according to Marquardt (14) using a computer. ↑↑, Remarkable increase; ↑, increase; ↓↓, remarkable decrase; ↓, decrease; *, little or no change.

synaptic membranes treated at 37° C for 30 min in varying assay environments. Without sodium, heating at 56° C for 3 min had no effect on either receptor binding or the Gpp(NH)p effect. On the other hand, prior exposure of the synaptic membrane to heating at 56° C for 3 min with sodium reduced ENK binding with the active Gpp(NH)p binding protein at 25° C incubation but without it at 0° C. These results indicate that thermal treatment destroyed the nucleotide binding sites but not the ENK binding sites. And this suggests that there may be two different binding sites for Gpp(NH)p, because the Gpp(NH)p effects varied at different binding assay temperatures.

## DISCUSSION

It is well known that guanine nucleotides selectively decrease the binding of [$^3$H]opiate-agonists to receptor sites (5,6,15,16). This effect is specific, since GMP is much less potent than GDP and GTP, and adenine nucleotides are much less potent than guanine nucleotides (15,16). The main finding of the present study is that guanine nucleotides strongly increased ENK binding without sodium at 0° C in a dose-dependent manner (Fig. 1). Furthermore, we confirmed that guanine nucleotides reduced the receptor binding activity for [$^3$H]ENK with sodium at 25° C and 0° C.

Guanine nucleotides (50 µM Gpp(NH)p or GTP) potentiated ENK binding by 50-60% over the control at 0° C, incubated without sodium. This novel nucleotide-induced increase of ENK binding reflects an increase in the affinity of binding sites (Fig. 2-B). This nucleotide-induced ENK binding is specific, since GMP and adenine nucleotides are much less potent than Gpp(NH)p or GTP (Table 1). The dramatic increase in ENK binding by guanine nucleotides in the absence of sodium at 0° C indicates that sodium and temperature play important roles in the action of guanine nucleotides in regulating opioid receptor function.

Cupric ions remarkably reduced the specific binding of ENK, although zinc ions had little effect on ENK binding. Simon and Groth (17) have demonstrated an essential SH-group at the opioid receptor, and Marzullo and Friedhoff (18) have pointed out that the ability of transition cations to inhibit opioid receptor binding corresponds to their affinity for protein SH-groups. Furthermore, it was shown that cupric ions could modify opioid receptor functions through a redox reaction (19). We have already pointed out that $Cu^{2+}$ increased lipid peroxide content of synaptic membranes, although $Zn^{2+}$ had no effect on lipid peroxide levels (20). These findings indicate that there may be oxidation of the opioid receptors by transition metal ions. The additive effect of $Cu^{2+}$ and Gpp(NH)p on reducing the ENK receptor binding suggests that different coupling mechanisms are involved in $Cu^{2+}$ and Gpp(NH)p action.

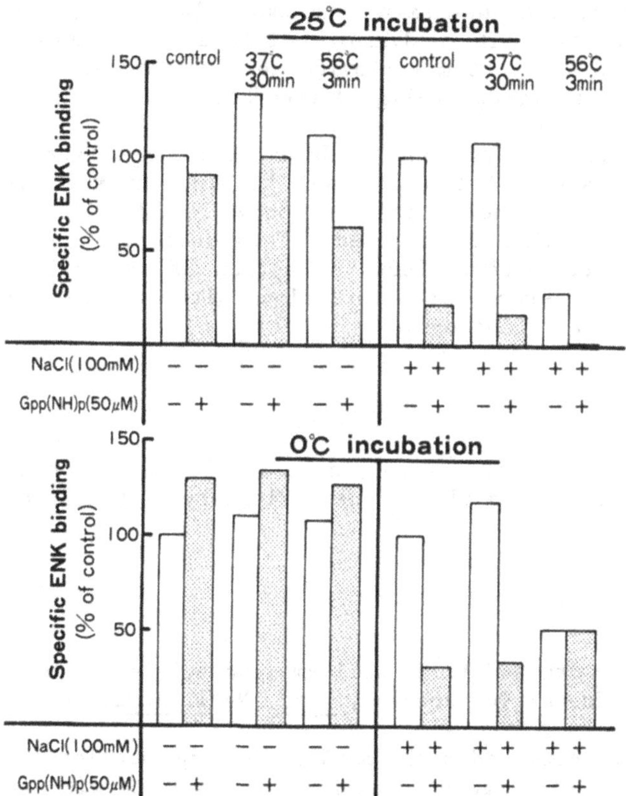


Fig. 4. Thermostability of ENK binding sites and Gpp(NH)p binding sites. Synaptic membranes were treated for 30 min at 37°C or 3 min at 56°C, and receptor binding assays were carried out as described in the Materials and Methods section.

When binding assays were carried out at 0° C, the prior exposure of synaptic membranes to heat (56° C, 3 min) reduced ENK binding and eliminated the nucleotide-induced decrease in ENK binding (Fig. 4). In contrast, at 25° C incubation this heated receptor preparation showed reduced ENK binding while retaining an active Gpp(NH)p binding protein. These results may indicate that Gpp(NH)p affects different sites at 25° C and 0° C. Although we have proposed two possible effective sites of nucleotides, it is not possible at present to determine the exact location of the guanine nucleotide effects. Thermodynamic differences between receptor binding at 25° C and 0° C may provide new insights into the molecular consequences of receptor-ligand interactions.

In any case, the present study reveals no effect of guanine nucleotides on antagonist (naloxone) binding in the presence of sodium, and supports the notion that only ENK binding to receptors is affected by guanine nucleotides. This selective specificity of guanine nucleotides may reflect the action of nucleotides in coupling agonist (ENK)-receptor interactions with adenylate cyclase or other physiological effectors.

## ACKNOWLEDEGMENTS

This work was supported in part by a grant-in-aid for scientific research from the Japanese Ministry of Education, Science and Culture.

## REFERENCES

1. R. J. Lefkowitz and L. T. Williams, Catecholamine binding to the β-adrenergic receptor, Proc. Natl. Acad. Sci. USA 74:515 (1977).
2. D. C. U'Prichard and S. H. Snyder, Guanine nucleotide influences on [$^3$H]ligand binding to α-noradrenergic receptors in calf brain, J. Biol. Chem. 253:3444 (1978).
3. I. Creese, T. Prosser, and S. H. Snyder, Dopamine receptor binding: Specificity, localization and regulation by ions and guanine nucleotides, Life Sci. 23:495 (1978).
4. A. J. Blume, Interactions of ligands with opiate receptors of brain membranes: Regulation by ions and nucleotides, Proc. Natl. Acad. Sci. USA 75:1713 (1978).
5. S. R. Childers and S. H. Snyder, Differential regulation by guanine nucleotides of opiate agonist and antagonist receptor interactions, J. Neurochem. 34:583 (1980).
6. K. -J. Chang, E. Hazum, A. Killian, and P. Cuatrecasas, Interactions of ligands with morphine and enkephalin receptors are differentially affected by guanine nucleotide, Mol. Pharmacol. 20:1 (1981).

7. M. Rodbell, M. G. Lin, Y. Salomon, C. Londos, J. P. Harwood, B. R. Martin, M. Rendell, and N. Berman, Role of adenine and guanine nucleotides in the activity and response of adenylate cyclase system, Adv. Cyclic Nucleotide Res. 5:3 (1975).
8. N. Ogawa, Y. Yamawaki, H. Kuroda, and T. Ofuji, Effects of bromocriptine on receptor binding of methionine-enkephalin, Neuroscience Lett. 23:215 (1981).
9. N. Ogawa, Y. Yamawaki, H. Kuroda, I. Nukina, Z. Ota, M. Fujino, and N. Yanaihara, Characteristics of thyrotropin releasing hormone (TRH) receptor in rat brain, Peptides 3:669 (1982).
10. N. Ogawa, Y. Yamawaki, H. Kuroda, I. Nukina, and T. Ofuji, Differentiation of agonist conformation and antagonist conformation in multiple opioid receptors, Neuroscience Lett. 27:205 (1981).
11. N. Ogawa, S. Mizuno, A. Mori, and H. Kuroda, Chronic dihydroergotoxine administration sets on receptors for enkephalin and thyrotropin releasing hormone in the aged-rat brain, Peptides, in press.
12. R. Simantov, S. R. Childers, and S. H. Snyder, The opiate receptor binding interactions of [$^3$H]methionie enkephalin, an opioid peptide, Mol. Pharmacol. 14:69 (1978).
13. C. B. Pert and S. H. Snyder, Opiate receptor binding of agonists and antagonists affected differentially by sodium, Mol. Pharmacol. 10:868 (1974).
14. D. W. Marquardt, An algorithm for least-squares estimation of nonlinear parameters, J. Soc. Indust. Appl. Math. 11:431 (1963).
15. G. W. Pasternak and S. H. Snyder, Identification of novel high-affinity opiate receptor binding in rat brain, Nature 253:563 (1975).
16. S. R. Childers and S. H. Snyder, Guanine nucleotides differentiate agonist and antagonist interactions with opiate receptors, Life Sci. 23:759 (1978).
17. E. J. Simon and J. Groth, Kinetics of opiate receptor interaction by sulphydryl reagents: evidence for a conformational change in the presence of sodium ions, Proc. Natl. Acad. Sci. USA 72:2404 (1975).
18. G. Marzullo and A. J. Friedhoff, An inhibitor of opiate receptor binding from human erythrocytes identified as glutathione-copper complex, Life Sci. 21:1559 (1977).
19. G. Marzullo and B. Hine, Opiate receptor function may be modulated through an oxidation-reduction mechanism, Science 208:1171 (1980).
20. S. Mizuno, N. Ogawa, and A. Mori, Differential effects of some transition metal cations on the binding of β-carboline-3-carboxylate and diazepam, Neurochem. Res. 8:873 (1983).

# $Ca^{2+}$-ACTIVATED, FATTY ACID-DEPENDENT GUANYLATE CYCLASE IN SYNAPTIC PLASMA MEMBRANES AND ITS REQUIREMENT FOR $Ca^{2+}$ AND Mg-GTP IN THE ACTIVATION

Takeo Asakawa, Masako Takano, Keiichi Enomoto and Kazuko Hayama

Department of Pharmacology, Saga Medical School Nabeshima Nabeshima-machi, Saga 840-01, Japan

## INTRODUCTION

Cyclic GMP in an intact cell system was reported to increase by various biologically active compounds, including muscarinic cholinergic agonists and excitatory amino acids, and supposed to act as intracellular messenger of the various neurotransmitters and hormones. Guanylate cyclase, a cyclic GMP-generating enzyme, in a cell free system can not be stimulated by agents increasing cyclic GMP contents. This suggests that the cyclase is activated by a substance generated or a process taking place within cells upon hormone stimulation. Thus cellular components such as $Ca^{2+}$, fatty acids and their metabolites that are intracellularly liberated on the stimulation have been supposed to be possible physiological activators of guanylate cyclase in intact cells.

Unsaturated fatty acids were originally shown to activate guanylate cyclase in plasma membranes of fat cells (1, 2) and Balb 3T3 fibroblasts (3) and in particulate fraction from rat brains and livers (2, 4). Much work has so far been reported on the activation. The mechanism of the activation of membrane-bound guanylate cyclase appears to involves a direct lipid-protein interaction rather than peroxide or free radical formation, but still remains to be further clarified. Soluble guanylate cyclase in platelets and some tissues were also demonstrated (5-10) to be activated by unsaturated fatty acids, while the fatty acids inhibited the soluble cyclase in brain and liver (2). Nitric oxide and other agents which generate nitroxy free radicals greatly activated soluble guanylate cyclase but not membrane-bound gaunylate cyclase. Differing from NO-generating agents, cholinergic agents and stimulants such as histamine, fatty acids and the lymphocyte mitogen phytohemagglutinin and concanavalin A were shown to require extracellular $Ca^{2+}$ for their elevating effects on cyclic GMP levels in intact cell systems. And $Ca^{2+}$ was demonstrated to partly stimulate membrane-bound guanylate cyclase activity in some tissues (11-14). The interaction of hormone and neurotransmitters with their receptors occurs on plasma membranes. Membrane-bound guanylate cyclase, rather than soluble guanylate cyclase, is close to these receptors and is supposed to play a functional role for expression of the action of hormones and neurotransmitters.

In this paper, the activation of gunaylte cyclase is shown to require Mg-GTP in addition to $Ca^{2+}$ and unsaturated fatty acid. We describe the possible mechanism of the activation and some other properties of $Ca^{2+}$-activated, fatty acid-dependent guanylate cyclase in synaptic plasma membranes.

## MATERIALS AND METHODS

Synaptic plasma membraanes of rat cerebral cortex were prepared according to the method of Whittaker et al (15) with the modification by Yoshida and co-workers (16, 4). The final sample of synaptic plasma membranes was stored in liquid nitrogen.

Guanylate cyclase assay was carried out as follows (1, 4). The standard reaction mixture, unless otherwise indicated, contained 40 mM Tris-HCl buffer, pH 8.0, 2 mM [8-$^3$H]GTP (8 mCi/mmol), 4.4 mM $MgCl_2$, 1 mM cyclic GMP, 16 mM phosphocreatine, 7 units of creatine phosphokinase and enzyme in a total volume of 0.25 ml. After 30 min of incubation at 37°C, the reaction was terminated by the addition of 1.75 ml of TCA (final concentration, 5 % w/v). The radioactive cyclic GMP formed was isolated by sequential chromatography on alumina and Dowex 1 x 8. The conditions for chromatography on alumina and Dowex 1 x 8 were reported in the previous paper (17). With aliquots of the final cyclic GMP fractions from Dowex 1 x 8, the absorbance at 256 nm was determined to estimate the recovery of cyclic GMP through the entire procedure.

The pre-incubation of guanylate cyclase for the activation was carried out in the medium containing 40 mM Tris-HCl, pH 8.0, 1.0 mM Mg-GTP, 2.4 mM $MgCl_2$, 250 μM $CaCl_2$, 16 mM phosphocreatine, 28 units/ml of creatine phosphokinase and 600 nmol/ml of linoleic acid. After 8 min incubation at 37°, synaptic plasma membranes were washed by centrifugation and assayed for the enzyme activity.

## RESULTS AND DISCUSSION

Guanylate cyclase in synaptic plasma membranes was assayed

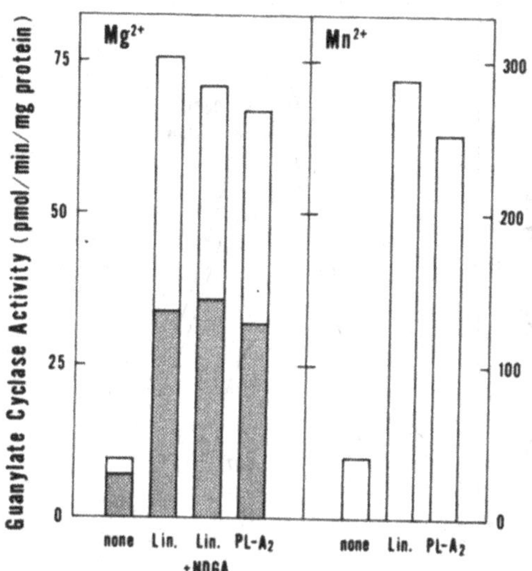


Fig. 1. Activation by linoleic acid and $Ca^{2+}$ of guanylate cyclase in synaptic plasma membranes. Guanylate cyclase was assayed with $Mg^{2+}$ (left) and with $Mn^{2+}$ (right) as a divalent cation in the absence (None) and the presence of linoleic acid (Lin) or linoleic acid plus NDGA (Lin + NDGA). The membranes incubated with phospholipase $A_2$ ($PL-A_2$) was gathered by centrifugation and assayed for the cyclase activity. $CaCl_2$ at 80 µM (open bars) or 1 mM EGTA (hatched bars) was present in the reaction mixtures for the $Mg^{2+}$-guanylate cyclase activity.

with $Mg^{2+}$ as a divalent cation and stimulated as much as 10-fold by the presence of both a low concentration of $Ca^{2+}$ and unsaturated fatty acid (Fig. 1). Linoleic acid was used in this study as unsaturated fatty acid and other unsaturated fatty acids, arachidonic acid and oleic acid, have almost the same efficacy.

Unsaturated fatty acids in intact cells are actually liberated from phospholipids in membrane matrix by phospholipase $A_2$ probably present in plasma membranes. The pre-incubation of synaptic plasma membranes with phospholipase $A_2$ was found to result in a great enhancement of guanylate cyclase activity (Fig. 1). Accordingly, unsaturated fatty acid enzymatically liberated as well as that exogenously added can probably be an activator of $Ca^{2+}$-activated, $Mg^{2+}$-associated guanylate cyclase. On the other hand, gunaylate cyclase activity assayed with $Mn^{2+}$ as a divalent cation, as shown in Fig. 1, was also stimulated to a great extent by unsaturated fatty acid and phospholipase $A_2$ reaction.

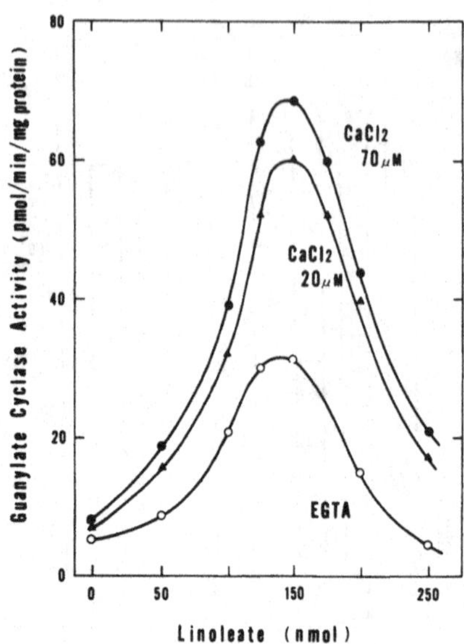


Fig. 2. Effect of increasing concentrations of linoleic acid on $Ca^{2+}$-activated, fatty acid-dependent guanylate cyclase in synaptic plasma membranes. The enzyme was assayed in the presence of linoleic acid as indicated with 20 μM and 70 μM $CaCl_2$ or with 1 mM EGTA.

The activation of $Mg^{2+}$-associated guanylate cyclase was dependent on the concentration of unsaturated fatty acid and exhibited a biphasic curve (Fig. 2). The peak level of activation depended on a $Ca^{2+}$-concentrations present in the assay medium. The concentration of the fatty acid for the peak activation, however, was not influenced by a $Ca^{2+}$-concentration. The fatty acid-dependent activity of guanylate cyclase exhibited a biphasic response to $Ca^{2+}$. The addition of 80 μM $CaCl_2$ to the standard assay medium produced about 2.2-2.4-fold stimulation. At higher than 100 μM $CaCl_2$, the cyclase was remarkably inhibited dependently on an increase in a $Ca^{2+}$-concentration. Using an EGTA-buffering system, we determined the free concentration of $Ca^{2+}$ for the stimulation of the fatty acid-dependent guanylate cyclase (Fig. 3). Free $Ca^{2+}$ required for half the maximal stimulation was estimated to be 0.15 μM and for the maximal stimulation it was 0.6 μM. The ratio of the activity associated with EGTA and with 0.6 μM free $Ca^{2+}$ depended on the concentration of Mg-GTP present in the assay medium. The peak level of stimulation (3.8-fold) was caused by 0.3 to 0.4 mM Mg-GTP. This concentration for the maximal sitmulation is similar to a physiological GTP concentration found in intact cells.

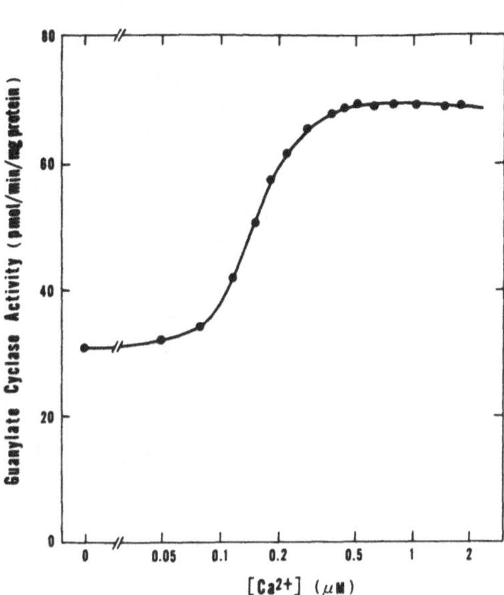


Fig. 3.
Effect of free concentrations of $Ca^{2+}$ on $Ca^{2+}$-activated, fatty acid-dependent guanylate cyclase in synaptic plasma membranes. Guanylate cyclase was assayed in the presence of linoleic acid in the standard reaction mixtures including 2 mM EGTA-buffering systems with various ratios of EGTA/$Ca^{2+}$.

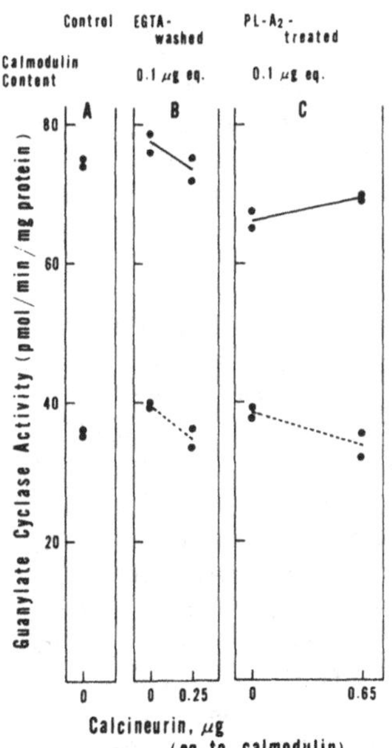


Fig. 4.
Effect of calcineurin (calmodulin-binding protein) on $Ca^{2+}$-activated, fatty acid-dependent guanylate cyclase in synaptic plasma membranes. The enzyme activity was assayed in the presence of calcineurin with (A, B) and without (C) linoleic acid under the presence of 80 μm $CaCl_2$ (●—●) or 1 mM EGTA (●--●). A; the membranes untreated, B; the membranes washed extensively with 1 mM EGTA, C; the membranes incubated with phospholipase $A_2$ followed by washing. Calmodulin content was determined by using calmodulin-deficient phosphodiesterase purified from bovine heart (20).

Calmodulin is found in various tissues of many species and shown to be the multifunctional $Ca^{2+}$-binding protein which confers a $Ca^{2+}$-sensitivity upon a variety of $Ca^{2+}$-dependent enzymes and cellular responses including brain adenylate cyclase, Tetrahymena guanylate cyclase (13) and erythrocyte ATPase. In view of these

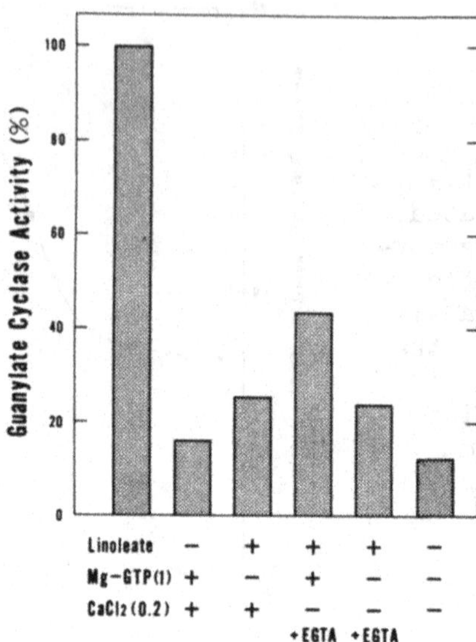


Fig. 5. Requirement for Mg-GTP and $Ca^{2+}$ in the activation by linoleic acid of $Ca^{2+}$-activated, fatty acid-dependent guanylate cyclase. Synaptic plasma membranes were pre-incubated with test compounds at 37° and washed by centrifugation. The pellets were assayed for the cyclase activity.

findings it was deemed important to examine whether the stimulatory effect of $Ca^{2+}$ on the fatty acid-dependent, $Mg^{2+}$-guanylate cyclase is mediated by calmodulin. The synaptic plasma membranes were extensively washed with 1 mM EGTA to remove calmodulin as reported for membranes from other sources. After the washing about 40 % of total calmodulin (0.51 µg/mg protein) still remained and the $Ca^{2+}$-stimulated activity of the cyclase did not diminish to any extent. The addition of calmodulin purified from pig brains by the method of Yagi and co-workers (18) or calmodulin-binding protein from bovine brains (19) to the synaptic plasma membranes washed with EGTA did not produce a significant effect on the $Ca^{2+}$-stimulated cyclase activity (Fig. 4). These resulsts may exclude a possibility that calmodulin is involved in the $Ca^{2+}$-stimulation of the fatty acid dependent $Mg^{2+}$-guanylate cyclase.

The time course of the $Ca^{2+}$-activated, fatty acid-dependent activity of the cyclase was studied. The enhancement of the activity with the presence of linoleic acid and $Ca^{2+}$ was induced after about 5 min-lag. But with the presence of EGTA, the time course exhibited a straight line. Thus this was tempting to

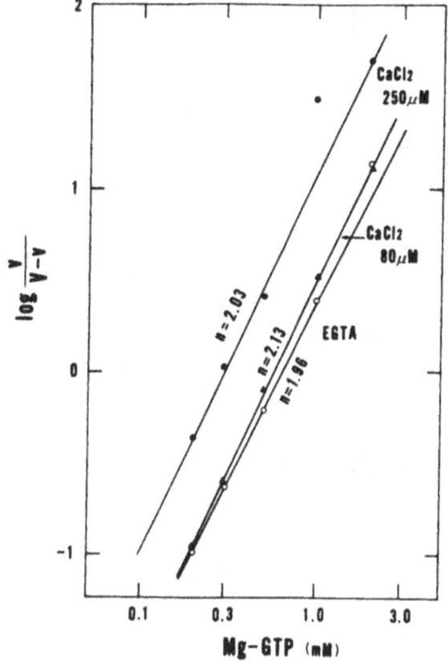


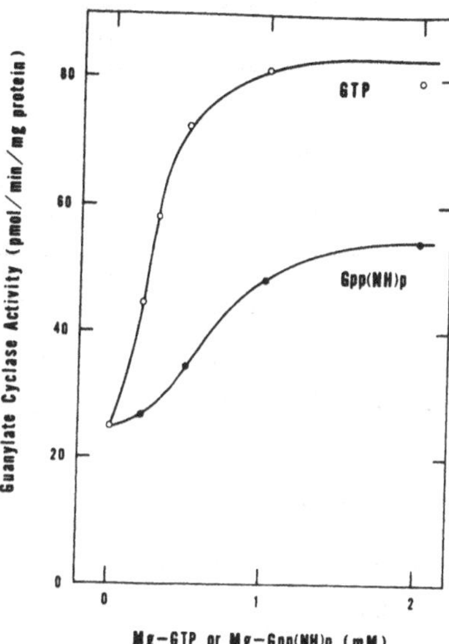


Fig. 6. (left) Hill plots of the activating effects of Mg-GTP on $Ca^{2+}$-activated, fatty acid-dependent guanylate cyclase. The pre-incubation for the activation was carried out as described under "MATERIALS AND METHODS" except that $Ca^{2+}$-concentrations were changed to 80 μM ( ▲ ) and 250 μM ( ● ), or in the place of $CaCl_2$, 1 mM EGTA ( ○ ) was added. After activation, the enzyme was assayed in the standard reaction mixtures with 1 mM EGTA in the place of $CaCl_2$.

Fig. 7. (right) Activating effect of GTP and Gpp(NH)p on $Ca^{2+}$-activated, fatty acid-dependent guanylate cyclase. The pre-incubation for the activation was carried out as described under "MATERIALS AND METHODS".

suppose that the cyclase was probably during the lag time changed to an activated form. To study this view, synaptic plasma membranes were pre-incubated with linoleic acid and test substances to determine a requirement for the activation. After pre-incubation the pellets obtained by washings with centrifugation were assayed for guanylate cyclase activity. The activation of the cyclase was shown to require for a co-existing $Ca^{2+}$ and $Mg^{2+}$-GTP besides unsaturated fatty acid in the maximal activation (Fig. 5). But the requirement for $Ca^{2+}$ was not so absolute and the omission of $Ca^{2+}$ from the complete system for the activation produced about 40 % of the activation.

Table 1. Kinetic constants of $Mg^{2+}$-GTP for the activation and for the enzyme reaction of $Ca^{2+}$-activated, fatty acid-dependent guanylate cyclase.

| Kinetic Constants Mg-GTP | Km | | Vmax | Hill analysis n | |
|---|---|---|---|---|---|
| for Activation | (mM) | | | | |
| + $CaCl_2$ 250 μM | 0.30 | | 118.1 | 2.03 | |
| + $CaCl_2$ 80 μM | 0.58 | | 87.3 | 2.13 | |
| + EGTA | 0.63 | | 57.7 | 1.96 | |
| for Cyclase Reaction | | | | | |
| + $CaCl_2$ 80 μM | 0.97 | 0.42 | 111.5 (51.3) | 2.00 | 2.00 |
| + EGTA | 0.99 | 0.59 | 75.1 (36.2) | 1.97 | 1.98 |

The effect of increasing concentration of Mg-GTP was studied under the presence of linoleic acid and $Ca^{2+}$. The activation was increased with the increase in the concentration of Mg-GTP and reached the plateau level at more than 1 mM. The maximal level of the activation is dependent on a concentration of $Ca^{2+}$ present in the activating medium. Hill plots of Mg-GTP for activation with co-existing 250 μM and 80 μM $CaCl_2$ and 1 mM EGTA gave Hill coefficients of about 2.03, 2.13 and 1.96 respectively, indicating a positive cooperative binding of Mg-GTP to the guanylate cyclase (Fig. 6). Therefore, the concentration of Mg-GTP was tentatively expressed as $S^2$ and Hofstee plots were employed to analyze the data. The plots exhibited a straight line for each with 250 μM and 80 μM of $CaCl_2$ and 1 mM EGTA. Apparent K' value of Mg-GTP were calculated to be 0.30, 0.58 and 0.63 mM for activation in the presence of 250 μM and 80 μM $CaCl_2$ and 1 mM EGTA respectively (Table 1). From these values, the affinity of Mg-GTP to the cyclase tends to increase with the presence of $Ca^{2+}$. Mg-GTP, on the other hand, is a substrate of the guanylate cyclase reaction. Km values of Mg-GTP were obtained by Lineweaver-Burk analysis. In this experiment, guanylate cyclase in synaptic plasma membranes was pre-activated by incubation with $Ca^{2+}$, Mg-GTP and linoleic acid, and washed by centrifugation. The activity in the pellets was assayed with various concentrations of Mg-GTP in the presence and absence of $Ca^{2+}$. Each activity was revealed to have 2 values of Km. High Km values (0.97-0.99 mM) were nearly the same for both activities with $Ca^{2+}$ and with EGTA. A low Km value (0.42 mM) of the activity with $Ca^{2+}$ was a little smaller than that (0.59 mM) with EGTA (Table 1). These values of Km were quite different from the K' value of Mg-GTP for the activation. These results suggest that the binding site of Mg-GTP for the activation differs from that for the cyclase reaction. The activating effect of GTP is rather specific. GMP, guanosine, cyclic GMP and ATP had little effects. However, Gpp(NH)p, guanylylimidodiphosphate, was partly effective and 2.0 mM of Gpp(NH)p had about 50 % effect of GTP (Fig. 7).

The effect of $Ca^{2+}$ on the activation of $Ca^{2+}$-activated, fatty acid-dependent guanylate cyclase was concentration-dependent and exhibited a saturation curve with an increase in $Ca^{2+}$-concentration. About 200 µM of $CaCl_2$ produced the maximal activation and half the maximal activation was induced with 60 µM $CaCl_2$. In the absence of Mg-GTP little activation was observed even with an increase in the $Ca^{2+}$-concentration. It was a problem whether the pre-activated cyclase by the fatty acid under the presence of $Ca^{2+}$ and Mg-GTP still required free $Ca^{2+}$ for the activity. To examine this problem, $CaCl_2$ was added to the assay mixtures for the pre-activated enzyme. The addition of $CaCl_2$ did not produce a significant enhancement of the activity and at a slightly higher concentration markedly reduced the activity, probably due to the presence of enough amounts of free $Ca^{2+}$, contaminated from the preceding activating incubation. However, EGTA reduced the activity by about 25 to 30 %. Thus the cyclase

Table 2. Effect of inhibitors of fatty acid oxidation on the activation of $Ca^{2+}$-activated, fatty acid-dependent guanylate cyclase.

| Addition | Guanylate cyclase activity (pmol/min/mg protein) | % |
|---|---|---|
| in preincubation | | |
| None | 84.4 | 100 |
| Nordihydroguaiaretic acid (NDGA) (20 µM) | 89.8 | 106.4 |
| Indomethacin (10 µM) | 88.0 | 104.3 |
| Aspirin (0.5 mM) | 80.8 | 95.7 |
| Chlorpromazine (0.3 mM) | 1.3 | 1.5 |

The pre-incubation for the activation was carried out as described under "MATERIALS AND METHODS".

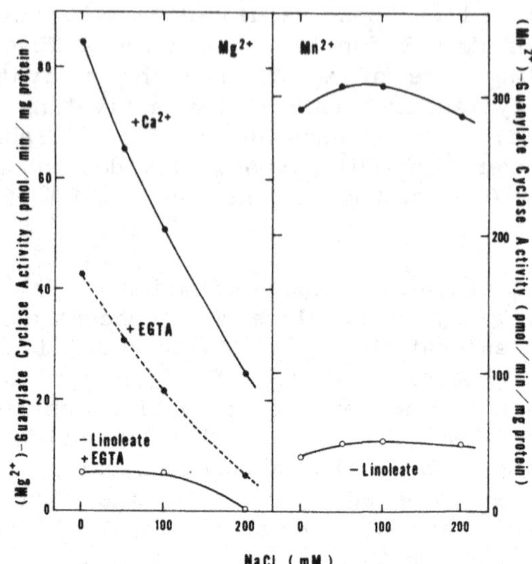


Fig. 8. Effect of NaCl on $Mg^{2+}$- and $Mn^{2+}$-guanylate cyclase activity. Enzyme was assayed with $Mg^{2+}$ (left) and $Mn^{2+}$ (right) in the presence ( -●- ) and absence ( -○- ) of linoleic acid. $CaCl_2$ at 80 µM or 1 mM EGTA was present in the reaction mixtures for the $Mg^{2+}$-activity.

pre-activated seems to still need $Ca^{2+}$ for the maximal activation. The concentration of $Ca^{2+}$ for the maximal activity was calculated to be appox. 0.6 µM, by using a EGTA-buffering system.

Fatty acid peroxide was reported to activate soluble guanylate cyclase in platelets (5-7). It was possible that the activation of the membrane-bound guanylate cyclase by unsaturated fatty acid in this paper involved a transformation of the fatty acid to a hydroperoxide derivatives during pre-incubation through the action of a lipoxygenase or a cyclooxygenase present in the membranes. To test this view, the pre-incubation for the activation was carried out in the presence of nordihydroguaiaretic acid (NDGA), a lipoxygenase inhibitor, and aspirin and indomethacin, cyclo-oxygenase inhibitors. As shown in Table 2, these inhibitory drugs did not suppressed the ability of linoleic acid with $Ca^{2+}$ and Mg-GTP to activate the cyclase in synaptic plasma membranes. Thus the fatty acid itself is probably an activating moiety of the guanylate cyclase in synaptic plasma membranes. Chlorpromazine (0.3 mM) was found to completely suppressed the activation. This drug has been shown to interact with membrane-phospholipids, and thereby seems to inhibit the activation of the guanylate cyclase in synaptic plasma membranes.

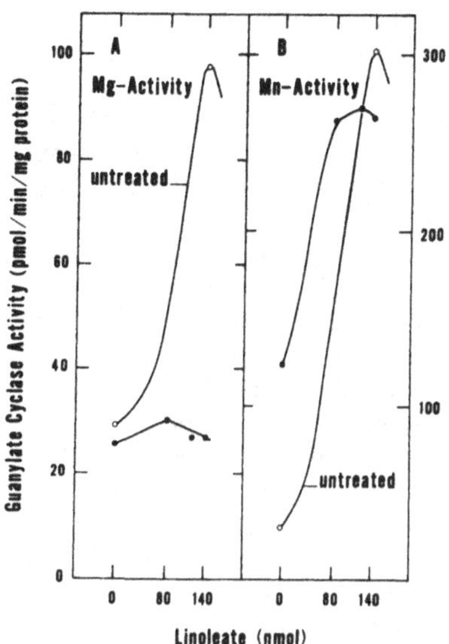


Fig. 9. Solubilization of guanylate cyclase by alkylglucoside. Synaptic plasma membranes were incubated with 30 mM n-octyl-β-D-glucopyranoside. The supernatant obtained by centrifugation at 100,000 x g was assayed for guanylate cyclase activity with $Mg^{2+}$ plus $Ca^{2+}$ (left) or $Mn^{2+}$ (right) in the presence of linoleic acid indicated.

In the case of $Mn^{2+}$-associated guanylate cyclase, linoleic acid produced as much as 8-fold activation and in contrast with the $Mg^{2+}$-associated activity, GTP and $Ca^{2+}$ were not required for the activation. From this result the presence of $Mn^{2+}$, not of $Mg^{2+}$, in the pre-incubation seems to cancel the requirement for GTP and $Ca^{2+}$ in the activation by linoleic acid. Monovalent cations such as $Na^+$, $K^+$ and $NH_4^+$ were found to have an inhibitory effect on the Mg-activity. The fatty acid-dependent activities assayed with $Ca^{2+}$ and with EGTA were remarkably inhibited by NaCl, dependently on the concentration (Fig. 8). The activity without linoleic acid was not suppressed. In contrast with the $Mg^{2+}$-cyclase, $Mn^{2+}$-associated guanylate cyclase was not diminished with the presence of NaCl (Fig. 8). These findings seem to suggest that the activation of the Mg-cyclase by the fatty acid includes a Na-sensitive process which is not necessary for the activation of the $Mn^{2+}$-cyclase.

On this view, a solubilizing effect of alkylglucopyranoside was examined on the $Mg^{2+}$- and $Mn^{2+}$-associated cyclase of synaptic plasma membranes. After the incubation of the membranes with alkylglucopyranoside about 50 % of the Mn-associated activity was found in the supernatant from 100,000 x g centrifugation and the

$Mn^{2+}$-activity was greatly stimulated by linoleic acid (Fig. 9). While $Mg^{2+}$-associated guanylate cyclase activity was detected only by less than 20 % in the supernatant and the stimulation by linoleic acid was little (Fig. 9). When we take into consideration the result that fatty acid-dependent $Mn^{2+}$-guanylate cyclase was solubilized with relatively high recovery, compared with that of the $Mg^{2+}$-cyclase, there is a possibility that the $Mn^{2+}$-cyclase and the $Mg^{2+}$-cyclase are different enzyme molecule. The alternative possibility is that the $Mg^{2+}$-cyclase has an additional protein component which is activated by unsaturated fatty acid dependently on Mg-GTP and $Ca^{2+}$. If we assume the latter possibility in the case, a question arises whether the GTP-binding regulatory proteins of adenylate cyclase system are related to the activation of the $Mg^{2+}$-guanylate cyclase. The treatment of the synaptic plasma membranes with cholera toxin (21) which was reported to ADP-ribosylate the stimulatory regulatory protein (Ns), increased adenylate cyclase activity approx. 10-fold but the $Mg^{2+}$-guanylate cyclase activity was reduced only about 20 %. The ADP-ribosylation of the inhibitory regulatory protein (Ni) with IAP (islet activating protein) (22) did not affect the activation of the $Mg^{2+}$-cyclase. Thus the regulatory protein system of adenylate cyclase does not concern the activation of the $Ca^{2+}$-activated, fatty acid-dependent $Mg^{2+}$-guanylate cyclase.

## ACKNOWLEDGEMENTS

We thank Dr. Michio Ui in Hokkaido University for the generous gift of IAP.

## REFERENCES

1. T. Asakawa, C. Johnson, J. Ruiz, I. Scheinbaum, T. R. Russell, and R. J. Ho, Activation by "feedback regulator" and some properties of guanylate cyclase of plasma membrane of rat epididymal fat cells, Biochem. Biophys. Res. Comm. 72:1335 (1976).
2. T. Asakawa, I. Scheinbaum, and R. J. Ho, Stimulation of guanylate cyclase activity by several fatty acids, Biochem. Biophys. Res. Comm. 73:141 (1976).
3. D. Wallach, and I. Pastan, Stimulation of guanylate cyclase of fibroblasts by free fatty acids, J. Biol. Chem. 251:5802 (1976).
4. T. Asakawa, M. Takenoshita, S. Uchida, and S. Tanaka, Activation of guanylate cyclase in synaptic plasma membranes of cerebral cortex by free fatty acids, J. Neurochem. 30:161 (1978).
5. D. B. Glass, W. Frey, D. W. Carr, and N. D. Goldberg, Stimulation of human platelet guanylate cyclase by fatty acid, J.

Biol. Chem. 252:1279 (1977).
6. H. Hidaka, and T. Asano, Stimulation of human platelet guanylate cyclase by unsaturated fatty acid peroxides, Proc. Natl. Acad. Sci. USA 74:3657 (1977).
7. G. Graff, J. H. Stephenson, D. B. Glass, M. K. Haddox, and N. D. Goldberg, Activation of soluble splenic cell guanylate cyclase by prostaglandin endoperoxides and fatty acid hydroperoxides, J. Biol. Chem. 253:7662 (1978).
8. C. J. Struck, and H. Glossmann, Soluble bovine adrenal cortex guanylate cyclase: Effect of sodium nitroprusside, nitrosamines, and hydrophobic ligands on activity, substrate specificity and cation requirement, Naunyn-Schmiedeberg's Arch. Pharmacol. 304:51 (1978).
9. J. M. Braughler, C. K. Mittal, and F. Murad, Purification of soluble guanylate cyclase from rat liver, Proc. Natl. Acad. Sci. USA 76:219 (1979).
10. D. Leiber, and S. Harbon, The relationship between the carbachol stimulatory effect on cyclic GMP content and activation by fatty acid hydroperoxides of a soluble guanylate cyclase in the guinea pig myometrium, Mol. Pharmacol. 21:654 (1982).
11. D. Wallach, and I. Pastan, Stimulation of membranous guanylate cyclase by concentrations of calcium that are in the physiological range, Biochem. Biophys. Res. Comm. 72:859 (1976).
12. J. Levilliers, F. Lecot. and J. Pairault, Modulation by substrate and cations of guanylate cyclase activity in detergent-dispersed plasma membranes from rat adipocytes, Biochem. Biophys. Res. Comm. 84:727 (1978).
13. S. Nagao, Y. Suzuki, Y. Watanabe, and Y. Nozawa, Activation by a calcium-binding protein of guanylate cyclase in Tetrahymena pyriformis, Biochem. Biophys. Res. Comm. 90:261 (1979).
14. N. Narayanan, and P. V. Sulakhe, Magnesium- and manganese-supported guanyalte cyclase in guinea pig heart: Subcellular distribution and some properties of the microsomal enzyme, Int. J. Biochem. 13:1133 (1981).
15. V. P. Whittaker, I. A. Michaelson, and R. J. A. Kirkland, The separation of synaptic vesicles from nerve-ending particles ("synaptosomes"), Biochem. J. 90:293 (1964).
16. K. Saito, S. Uchida, and H. Yoshida, Calcium binding of isolated synaptic membranes from rat cerebral cortex, Jap. J. Pharmacol. 22:787 (1972).
17. T. Asakawa, T. R. Russell, and R. J. Ho, Purification and succinylation of cyclic GMP from large volume samples and radioimmunoassay of succinyl cyclic GMP, Biochem. Biophys. Res. Comm. 68:682 (1976).
18. M. Yazawa, M. Sakuma, and K. Yagi, Calmodulins from muscles of marine invertebrates, scallop and sea anemone, J. Biochem. 87:1313 (1980).

19. R. W. Wallace, T. J. Lynch, E. A. Tallant, and W. Y. Cheung, Purification and characterization of an inhibitor protein of brain adenylate cyclase and cyclic nucleotide phosphodiesterase, J. Biol. Chem. 254:377 (1978).
20. H. C. Ho, T. S. Teo, R. Desai, and J. H. Wang, Catalytic and regulatory properties of two forms of bovine heart cyclic nucleotide phosphodiesterase, Biochim. Biophys. Acta 429:461 (1976).
21. K. Enomoto, and T. Asakawa, Evidence for the presence of a GTP-depenent regulatory component of adenylate cyclase in myelin from rat brain, J. Neurochem. 40:434 (1983).
22. T. Murayama, and M. Ui, Loss of the inhibitory function of the guanine nucleotide regulatory component of adenylate cyclase due to its ADP ribosylation by islet-activating protein, pertussis toxin, in adipocyte membranes, J. Biol. Chem. 258:3319 (1983).

# MODULATION OF NEURORECEPTOR FUNCTIONS BY LIPOMODULIN, A PHOSPHOLIPASE INHIBITORY PROTEIN

Fusao Hirata, Yoshitada Notsu, Keiichi Matsuda and Toshio Hattori

Laboratory of Cell Biology
National Institute of Mental Health
Bethesda, Maryland 20205

Glucocorticoids are hormones from the adrenal cortex. Secretion of these hormones is controlled by ACTH from the pituitary gland. The regulation of glucocorticoid secretion by ACTH is influenced by a variety of factors, including stress and depression. Therapeutically, glucocorticoids are often used to treat patients with chronic inflammation and immunological diseases, such as systemic lupus erythematosus and rheumatoid arthritis. Anti-inflammatory action of glucocorticoids is now proposed to be associated with the induction of synthesis of phospholipase inhibitory proteins, macrocotin in macrophages (1), renocortin in kidney cells (2) and lipomodulin in neutrophils (3). All these proteins are immunologically and biologically related (4). We isolated this protein from neutrophils treated with glucocorticoids and examined its effects on receptor functions and on neural development. We found that lipomodulin can regulate functions of certain types of receptors and differentiate neuronal cells to the adrenergic phenotype.

## INDUCTION OF LIPOMODULIN IN INTACT CELLS

Stimulation of many, if not all, receptors results in release of arachidonic acid which, in turn, is converted to prostaglandins, leukotrienes and other hydroxyl derivatives. This release of arachidonic acid is probably due to the activation of phospholipase(s) in the cells. When rabbit peritoneal neutrophils are stimulated with synthetic chemoattractants such as fMet-Leu-Phe, they release arachidonic acid from phospholipids in their membranes. The treatment of these cells with glucocorticoids causes marked decrease in their release of arachidonic acid as well as in their capability of chemotaxis (5). When the membranes of the glucocorticoid-treated neutrophils are solubilized by detergents, and then submitted to Sephadex G-200 column chromatography, the fractions with 40,000 m.w. are able to inhibit porcine pancreatic phospholipase $A_2$ _in vitro_. Recently, we have purified this protein to near homogeneity by the conventional methods of purification and named it as lipomodulin (6). Related proteins have been partially purified from glucocorticoid treated macrophages (macrocortin) and kidney cells (renocortin) (1,2). The molecular weights of these proteins are 40,000, 25,000 and 15,000, respectively. Since all these peptides can crossreact with anti-lipomudulin antibody and can be detected during the purification (7,8), we believe that these peptides are all related. Although the induction of synthesis of these proteins by glucocorticoids is rapid in macrophages, it generally takes 10 to 16 hrs for other tissues or cells to attain the maximal level. The induction can be blocked by actinomycin D and cycloheximide, inhibitors of protein synthesis, and the order of potency of various glucocorticoids in the induction of lipomodulin synthesis appears to parallel that of their potencies in binding to the cytosolic receptors and in their anti-inflammatory activities. These results support the earlier hypothesis proposing that the physiological responses of glucocorticoids require the synthesis of protein(s) following the transfer of the glucocorticoid receptor complex to nuclei (1,2,3).

## PROPERTIES OF LIPOMODULIN

Lipomodulin, purified from rabbit neutrophils, has a m.w. of 40,000 as judged by sodium dodecylsulfate (SDS) gel electrophoresis. In the culture media conditioned by glucocorticoid-treated neutrophils, lymphocytes and macrophages, the smaller peptides with m.w. 25,000 and 15,000 can also be detected, when lipmodulin is immuno-precipitated by anti-lipomodulin antibody or is purified by the biochemical procedures. The recovery of m.w. 40,000 species becomes poor when protease inhibitors such as aprotenin are absent from the buffers during the purification. These results suggest that the smaller peptides are fragments of the m.w. 40,000 species.

Lipomodulin can inhibit a variety of phospholipases including phospholipase $A_2$ and phosphatidylinositol phospholipase C. Judging by the amount required for half maximal inhibition, lipomodulin appears to be more specific for phospholipase $A_2$ (6). Lipomodulin changes the _Vmax_ of porcine pancreatic phospholipase $A_2$ but not its _Km_ for phospholipid substrates. Since a stoichiometric amount of lipomodulin is required to inhibit phospholipase $A_2$ maximally, a one-to-one complex is suggested to be formed between phospholipase $A_2$ and lipomodulin. Since detergents such as sodium deoxycholate and high concentrations of $Ca^{2+}$ reverse the inhibition of phospholipase $A_2$ by lipomodulin, it is likely that lipomodulin binds to the hydrophobic region of the phospholipase $A_2$ molecules where $Ca^{2+}$ binding sites are located.

## PHOSPHORYLATION-DEPHOSPHORYLATION OF LIPOMODULIN

When purified lipomodulin is treated with cyclic AMP-dependent kinase, a time dependent inactivation is observed. The degree of inactivation of lipomodulin parallels the amount of phosphate incorporated, suggesting that phosphorylated lipomodulin is inactive. When this phosphorylated lipomodulin is dephosphorylated by alkaline phosphatase, the capacity of lipomodulin to inhibit phospholipase $A_2$ recovers. These findings indicate that the reversible phosphorylation-dephosphorylation process regulates the activity of lipomodulin in intact cells.

To confirm this interpretation, the neutrophils incubated with $^{32}P$ are stimulated with fMet-Leu-Phe, and the phosphorylated lipomodulin is immunoprecipitated by anti-lipomodulin antibody. The phosphorylated lipomodulin thus obtained from intact neutrophils is comigrated on SDS-gel electrophoresis with lipomodulin purified from $^{35}S$-methionine labeled neutrophils or labeled with ATP-[$\gamma$-$^{32}P$] by cyclic AMP-dependent kinase. Preliminary experiments with several activators and inhibitors of protein kinases suggest that the kinases involved in the *in vivo* phosphorylation of lipomodulin might be protein kinase C and tyrosine phosphorylating kinase. The kinetic analysis of the time course shows that there is a close association between the rate of arachidonic acid release and the amount of phosphorylated lipomodulin. Since the maximal phosphorylation of lipomodulin is attained at 45 sec after the stimulation with fMet-Leu-Phe and the amount of phosphorylated lipomodulin gradually returns to the control level, the dephosphorylation appears to take place *in vivo* as well (6). The phosphorylation-dephosphorylation of lipomodulin is also detected in bradykinin-stimulated fibroblasts, and $\alpha$-adrenergic or opiate receptor stimulated neuroblastoma cells (unpublished data). These results suggest that increased phospholipid turnover mediated through a variety of receptors occurs as a consequence of phosphorylation of lipomodulin.

## EFFECT OF LIPOMODULIN ON CELLULAR RESPONSES TO VARIOUS STIMULI

It is generally known that glucocorticoids have no obvious primary effect on the responses of various cells, whereas they inhibit or enhance the physiological responses induced by other hormones or neurotransmitters. Stimulation of various receptors results in the release of arachidonic acid from phospholipids in membranes. Glucocorticoids can generally block this release of arachidonic acid, thereby influencing receptor functions. For instance, glucocorticoid treatment causes the inhibition of arachidonic acid release from bradykinin-stimulated fibroblasts and chemotactic peptide-stimulated neutrophils. Thus, the corticoid treatment blocks their ultimate responses, cyclic AMP formation or chemotaxis, respectively. Similarly, glucocorticoids can inhibit α-adrenergic receptor or opiate receptor mediated phospholipid turnover in neuroblastoma cells. All these effects of glucocorticoids can be mimicked by purified lipomodulin. Since lipomodulin can reduce the arachidonic acid release mediated through these receptors by inhibiting phospholipase(s) in these cells, one can assume that some receptor-receptor interaction such as inhibition of prostaglandin receptors by α-adrenoreceptors (as measured by cyclic AMP formation) might be a result of changes of phospholipids occurring in the vicinities of these receptors, in addition to the changes in the direct interactions among the components of these receptor systems such as nucleotide stimulatory and inhibitory factors. Alternatively, the interactions among these components of receptors can be influenced by changes in phospholipid compositions and structures. Lipomodulin can mimic a variety of the actions of glucocorticoids such as immunosuppression inhibition of IgE and IgG synthesis (8), anti-inflammation (1,2,3) and anti-leukemic activity (9). Some of these actions have been suggested to be due to the inhibition of phospholipase $A_2$ but the underlying mechanism remains to be established.

## EFFECT OF LIPOMODULIN ON NEURAL DIFFERENTIATION

Glucocorticoids are reported to induce β-adrenoreceptors in lung tissues. These steroids also increase the number of β-adrenoreceptors in $C_6$ astrocytoma cells. Part of this glucocorticoid effect is due to the inhibition of desensitization of β-adrenoreceptors. During the treatment with glucocorticoids, the morphologies of these cells appreciably change. Since lipomodulin promotes the cellular differentiation of U 937 cells, a human histiocytic lymphoma cell line, as measured by morphological changes and antibody dependent cellular cytotoxicity (9), the effect of lipomodulin on neuronal differentiation was examined, using NH15CA2 cells, a hybrid-hybrid cell line which produces both catecholamines and acetylcholine. When this cell line is treated with lipomodulin, the cells have an outgrowth of less branched and shorter neurites. The cells express the adrenergic phenotype as measured by the activity of tyrosine hydroxylase and by the uptake of norepinephrine (10). In contrast, NH15CA2 cells become cholinergic when these cells are cultured with proteases, enzymes that cleave lipomodulin. The amount of lipomodulin on the cell surface measured by the binding of antilipomodulin antibody inversely paralleled the extent of cholinergic expression as measured by choline acetyltransferase. Since glucocorticoids are reported to change the neural differentiation by acting on neuronal target cells with which cells the synapses are formed, lipomodulin is suggested to be one of the factors which is secreted from these target cells and determines the

primary choice of neurotransmitter to be produced in neuronal cells. Thus, by releasing lipomodulin, glucocorticoids can affect the receptor functions of other cells besides those of the cells on which this hormone primarily acts. Certain types of cells such as leukocytes, fibroblasts and glial cells are now demonstrated to release lipomodulin when treated with glucocorticoids. Thus, lipomodulin can act not only on cells from which this peptide is secreted, but also on other cells.

## CONCLUSIONS

Some receptors require directly or indirectly increased turnover of phospholipids for signal transduction. In addition, alterations in the phospholipid composition induced by the turnover of phospholipids appear to be more or less involved in receptor-receptor interaction. Glucocorticoids can modulate a variety of receptor functions by inducing the synthesis of lipomodulin. Lipomodulin inhibits phospholipases by forming a one-to-one complex. Since lipomodulin, synthesis of which is induced by glucocorticoids, can be released from certain types of cells after stimulation of receptors, some receptor functions in other cells can also be affected by inhibiting phospholipases with the released lipomodulin.

## REFERENCES


1. G. J. Blackwell, R. Carnuccio, M. DiRosa, R. J. Flower, J. Ivanyi, C. S. J. Langham, L. Parente, P. Persico, and J. Wood, Suppression of arachidonate oxidation by glucocorticoid induced antiphospholipase peptides, Adv. Prostaglandin, Thromboxane, Leukotriene Res. 11:65 (1983).
2. B. Rothhut, J. F. Cloix, and F. Russo-Marie, Dexamethasone induces the synthesis of renocortin, two antiphospholipase proteins in rat renomeduallary interstitial cells in culture, Adv. Prostaglandin, Thromboxane, Leukotriene Res. 12:51 (1983).
3. F. Hirata, Lipomodulin: a possible mediator of the action of glucocorticoids, Adv. Prostaglandin, Thromboxane, Leukotriene Res. 11:73 (1983).
4. F. Hirata, Y. Notsu, M. Iwata, L. Parente, M. DiRosa, and R. J. Flower, Identification of several species of phospholipase inhibitory protein(s) by radio-immunoassay for lipomodulin, Biochem. Biophys. Res. Commun. 109:223 (1982).

5. F. Hirata, E. Schiffman, K. Venkatasuburamanian, D. Salomon, and J. Axelrod, A phospholipase $A_2$ inhibitory protein in rabbit neutrophils induced by glucocorticoids, Proc. Natl. Acad. Sci. USA 77:2533 (1980).
6. F. Hirata, The regulation of lipomodulin, a phospholipase inhibitory protein in rabbit neutrophils by phosphorylation, J. Biol. Chem. 256:7730 (1981).
7. T. Hattori, F. Hirata, T. Hoffmann, A. Hizuta, and R. Heberman, Inhibition of human natural killer activity and antibody dependent cellular cytotoxicity by lipomodulin, a phospholipase inhibitory protein, J. Immunol. 131:662 (1983).
8. F. Hirata, Role of lipomodulin, a phospholipase inhibitory protein, in immunoregulation, Adv. Inflammation Res. 11:77 (1983).
9. T. Hattori, T. Hoffmann and F. Hirata, Differentiation of a histiocytic lymphoma cell line by lipomodulin, a phospholipase inhibitory protein, Biochem. Biophys. Res. Commun. 111:551 (1983).
10. T. Hattori, E. Mazy, B. Hamprecht, Y. Notsu and F. Hirata, Adrenergic phenotype in NH15CA2 neuroblastoma cells: induction by lipomodulin, submitted.

# ONTOGENETIC DEVELOPMENT OF THE SPECIFIC [$^3$H]NITRENDIPINE BINDING SITES IN THE RAT WHOLE BRAIN

H. Matsubayashi, S. Kito, E. Itoga, K. Mizuno
and R. Miyoshi

Third Department of Internal Medicine
Hiroshima University School of Medicine
1-2-3 Kasumi, Minami-ku, Horoshima 734, Japan

## INTRODUCTION

Ca ion channel antagonists have been attracting much attention as therapeutic drugs for cardiovascular disorders such as angina pectoris and hypertension, especially from the viewpoint of pharmacological mechanism. Recently, through Ehlert and Itoga's experiments, the Ca ion channel antagonist was confirmed to bind specifically, not only to vascular smooth muscle but also to gastrointestinal smooth muscle and even to the neuron itself (1). This broke new ground in the study of the pharmacological action of Ca ion channel antagonists.

Ca ion is known not only to activate muscle contractile elements but also to be related to impulse conduction of the central and peripheral nerves and to regulate release of neurotransmitters from nerve terminals. It is considered that Ca ion plays important roles in intramembrane transduction and intracellular transmission as one of the second messenger systems. This calcium messenger system interacts with the cyclic AMP system in complicated manners sometimes called synarchic regulation (2). Ca antagonists inhibit the various physiological activities of Ca ion described above. In general, Ca ion antagonists include di- and trivalent cations (for example, Mn ion), local anesthetics (procaine), antibiotics (neomycin C), tricyclics (trifluoperazine), antiepileptics (dilantin) and Ca ion channel blockers (verapamil) (3).

In this study we tried to investigate how nitrendipine behaves as a Ca ion channel antagonist from the viewpoint of binding studies with use of various subcellular fractions. In addition, studies

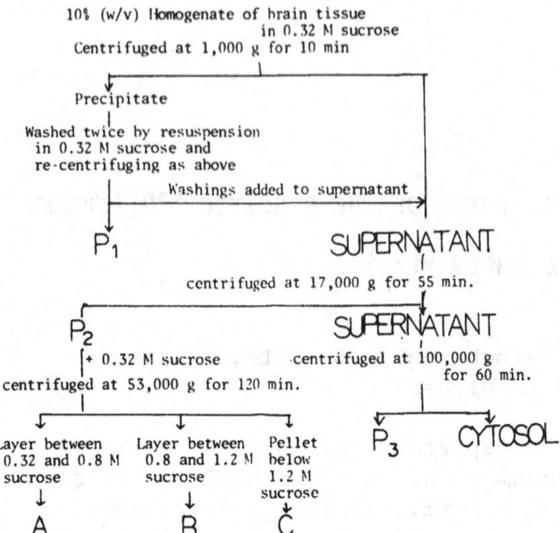


Fig. 1. Method of preparing subcellular fractions of rat brain tissue (4,5).

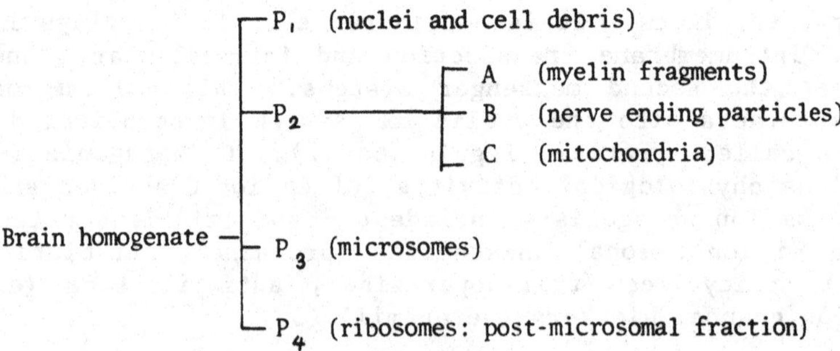


Fig. 2. Nomenclature adopted in describing subcellular fractions (4).

were done on the ontogenic development of such nitrendipine binding sites in the brain, measured biochemically as well as autoradiographically.

## MATERIALS AND METHODS

### Drugs

Tritiated nitrendipine ($[^3H]$NTD) was used as a radioactive ligand (specific activity 70 Ci/mmol), and nifedipine as a non-radioactive ligand.

### Preparation of Subcellular Fractions

Wistar strain male rats (200 g) were sacrificed by decapitation. The whole brains were rapidly removed and fractionated into $P_1$, $P_2$, and $P_3$ according to Whittaker's method (Fig. 1 and Fig. 2) (4,5). The $P_2$ fraction was further subfractionated into A, B and C fractions by the sucrose gradient method. According to Schoemaker and Itoga's method (6), we prepared tissue aliquots, adding 50 mM Tris/HCl buffer (pH 7.4) to each fraction, and performed saturation studies using $[^3H]$NTD as a radioactive ligand and nifedipine as a displacer under the conditions of 25° C, 60 min incubation. Protein was measured by Lowry's method using bovine serum albumin as a standard (7).

### Investigation of the Ontogenetic Development

Receptor assay. Adult females were caged with potent males in pairs over night and the next day was designated as day one of pregnancy. Twenty days after mating, fetuses were decapitated. Similarly, whole brains of animals at 1 day, 3 days, 7 days, 14 days, 28 days, 56 days and 90 days after birth were removed, homogenized, and a crude membrane fraction prepared (Fig. 3) (5). We performed saturation binding studies as above as a function of developmental stage.

Autoradiography. We performed autoradiography of rat brains at each stage described. The animals were sacrificed by decapitation or infusion with physiological saline. Cryostat sections of 10 μm thickness of brain tissue were preincubated in 0.05 M Tris/HCl buffer for 10 min. Incubation with 0.6 nM $[^3H]$nitrendipine in the same buffer for 2 hr at 0° C followed preincubation, and then the sections were apposed to a sheet of tritium-sensitive film.

Autoradiograms were exposed for 2 to 8 weeks. The exposed pictures were computer-analyzed by IBAS II (Zeiss). For the purpose of comparative studies among various ontogenic stages, the exposing period, developing time and other conditions were fixed constant.

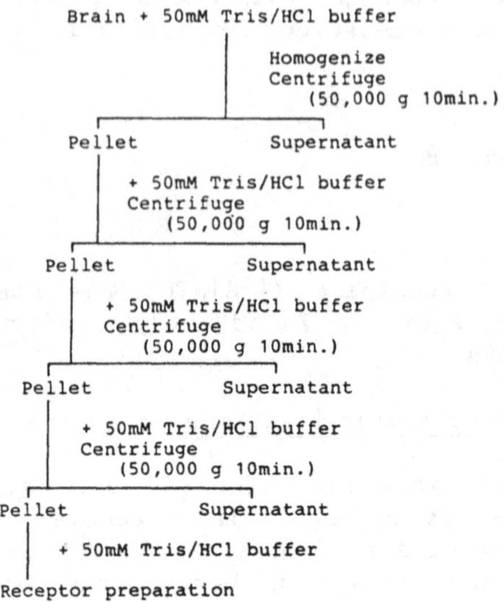


Fig. 3. Method of preparing whole homogenate of rat brain tissue (5).

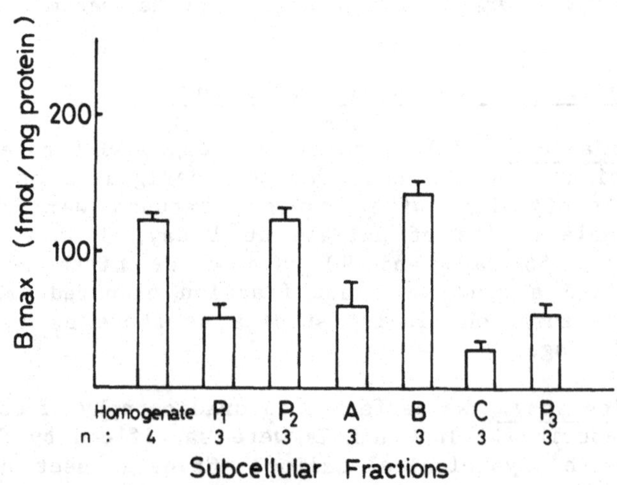


Fig. 4. Bmax values of specific [$^3$H]nitrendipine binding sites in various subcellular fractions. Homogenate: crude membrane fraction at 56 day postnatal stage; $P_1$: nuclei and cell debris; $P_2$: mitochondria - synaptosomal fraction; A: myelin fragments; B: nerve ending particles; C: mitochondria; $P_3$: microsomes; n: number of repeated saturation studies all performed in triplicate.

Table 1. Bmax values of various subcellular fractions

| fraction | n | Bmax (fmol/mg protein) |
|---|---|---|
| Homogenate | 4 | 121.5 ± 3.7 |
| $P_1$ | 3 | 51.4 ± 9.7 ** |
| $P_2$ | 3 | 122.1 ± 12.1 |
| A | 3 | 60.1 ± 19.9 ** |
| B | 3 | 141.9 ± 11.7 |
| C | 3 | 27.5 ± 7.7 ** |
| $P_3$ | 3 | 56.3 ± 9.3 ** |

The values shown are means ± s.d. of 3-4 independent determinations. **: significant difference from $P_2$ fraction, when examined by Student's *t* test, $p < 0.01$.

## RESULTS

### Subcellular Fractions

In our study, specific binding of [$^3$H]nitrendipine was high with a Bmax of 122.1 ± 12.1 (s.d.) fmol/mg protein in the $P_2$ fraction, especially in $P_2$-B which had a Bmax of 141.9 ± 11.7 fmol/mg protein (Table 1). These values were 2.5 to 3 times higher than those of other fractions (Fig. 4). There were no significant differences in [$^3$H]nitrendipine binding among whole homogenate, $P_2$ and $P_2$-B fractions when examined by Student's *t* test (level of significance was less than 0.05). This means that experiments with use of the whole homogenate can be regarded as showing the state of [$^3$H]nitrendipine binding to the synaptosomal fraction.

### Ontogenetic Development

Receptor assay. Results of Scatchard analysis of [$^3$H]nitrendipine binding to the whole homogenate of the rat brain in parallel with the course of ontogenetic development are shown in Fig. 5. In that figure, we did not show Scatchard analysis of 56 and 90 day old rats because they were almost identical to that of 28 day old rats.

Bmax values increased almost linearly from the late fetal stage to 28 days after birth, when they reached the adult level. After 28 days the values showed no recognizable change (Fig. 6). As for developmental change of the Kd value, there was a significant difference between the Kd value at 20 days of gestation and that at 7 days after birth (Fig. 7). There was no notable change of Kd value after the 7 day postnatal stage. In brief, the Kd value

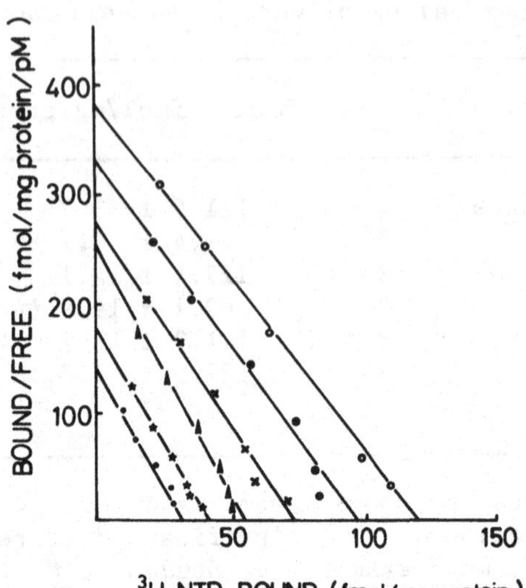


Fig. 5. Scatchard plots of specific [$^{3}$H]nitrendipine binding in rat brains at various pre- and postnatal stages. ✱ : 20 days of gestation; ★: 1-day-old animal; ▲: 3-day-old animal; X: 7-day-old animal; ●: 14-day-old animal; O: 28-day-old animal. The Scatchard plots of 56-day-old and 90-day-old rats were not shown here. They were almost the same as that of a 28-day-old rat.

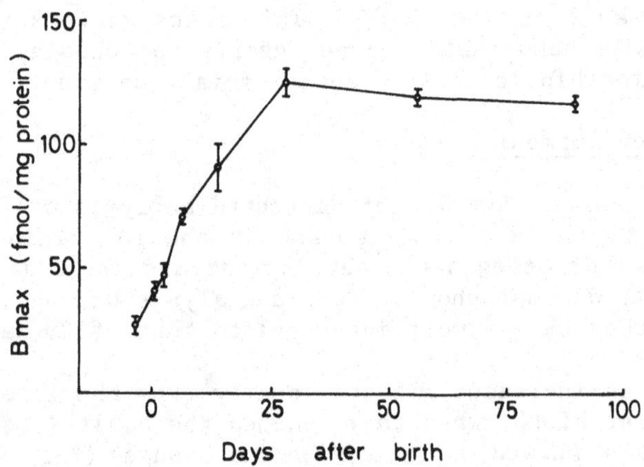


Fig. 6. Developmental change of Bmax value in [$^{3}$H]nitrendipine binding experiments. Each point represents the mean ± s.d. for 3-5 individual determinations.

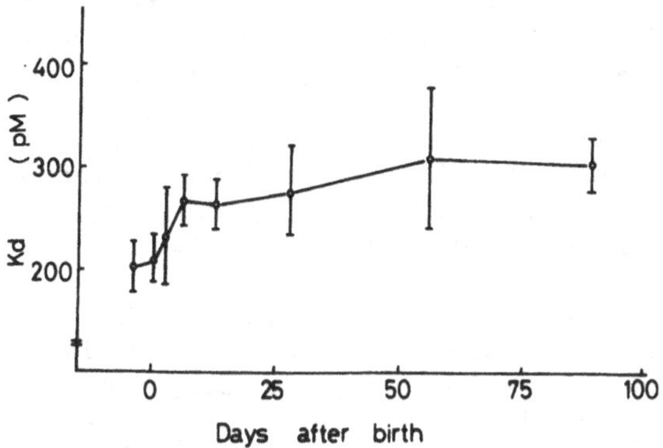


Fig. 7. Developmental change of Kd value in [$^3$H]nitrendipine binding experiments. Each point represents the mean ± s.d. for 3-5 individual determinations.

Table 2. Developmental change of Bmax and Kd value in [$^3$H]nitrendipine binding experiments.

| | [ n ] | Kd [pM] | B max [fmol/mg protein] |
|---|---|---|---|
| 20 days of gestation | [ 3 ] | 203.4 ± 19.4 | 29.5 ± 3.1 |
| 1-Day-old | [ 4 ] | 212.4 ± 19.0 | 40.8 ± 4.5 * |
| 3-Day-old | [ 4 ] | 233.7 ± 47.7 | 47.8 ± 5.7 ** |
| 7-Day-old | [ 4 ] | 266.0 ± 28.8 * | 70.5 ± 3.3 ** |
| 14-Day-old | [ 3 ] | 262.4 ± 28.6 * | 91.4 ± 10.2 ** |
| 28-Day-old | [ 5 ] | 277.7 ± 44.3 ** | 125.4 ± 6.4 ** |
| 56-Day-old | [ 4 ] | 307.8 ± 67.5 | 121.5 ± 3.7 ** |
| 90-Day-old | [ 3 ] | 303.3 ± 25.1 ** | 118.4 ± 3.4 ** |

The values shown are means ± s.d. of 3-5 independent determinations. *: significant difference from 20 days of gestation, when examined by Student's _t_ test, $p < 0.05$ **: significant difference from 20 days of gestation, $p < 0.01$.

showed an increase throughout the prenatal to early postnatal periods and by 7 days after birth, the Kd value moved from high (Kd = 200 pM) to low affinity (Kd = 300 pM) (Table 2).

Autoradiography. An autoradiogram of an adult rat brain section through the interpeduncular nucleus showed higher densities of [$^3$H]nitrendipine binding sites in the hippocampus, olfactory bulb, superior colliculus and interpeduncular nucleus (Fig. 8 - f).

The distribution of specific nitrendipine binding sites in frontal sections through the mammillary body was autoradiographically examined, and we compared the results among brains of various developmental stages (Fig 8, a-f). In the brain autoradiograms of a fetus of 20 days of gestation and of 1 day and 3 day old rats, there was no differentiation of receptor distribution according to the brain structures. At the stage of 14 days, the vague contour of the hippocampus came out autoradiographically. In an autoradiogram of a 28-day-old rat, the distribution of [$^3$H]NTD binding sites was more differentiated between the gray and white matters.

It was noteworthy that nitrendipine binding sites were prominent in the hippocampus. The receptor density in the hippocampus increased with developmental stage and reached the adult level by 28 days after birth. As previously mentioned, the olfactory bulb, superior colliculus and interpeduncular nucleus also included the receptors in high densities.

## DISCUSSION

Nitrendipine, classified as a Ca ion channel blocker derived from dihydropyridine, was proved to inhibit Ca influx into the cell electrophysiologically. In 1982, it was found that the Ca ion channel antagonist bound to the neuron itself by Ehlert and Itoga (1).

The present study confirms that nitrendipine bound to the plasma membrane of the synaptosome. There were many fewer binding sites on the mitochondrial and microsomal membranes than on the synaptosomal membrane. It was concluded that nitrendipine was an ion channel blocker which acted on the plasma membrane. This fact is also supported by the hitherto accumulated data on Ca ion channel antagonists. For instance, D 600, a member of this drug group, is ineffective in skinned smooth muscle and myocardial preparations, and intracellular D 600 is ineffective in inhibiting Ca currents in barnacle muscle. Furthermore, very high concentrations of D 600 or verapamil are needed to block Ca movements in intracellular organelles such as the sarcoplasmic reticulum and mitochondria (3). Conversely, the fact that there are few binding sites for the ion

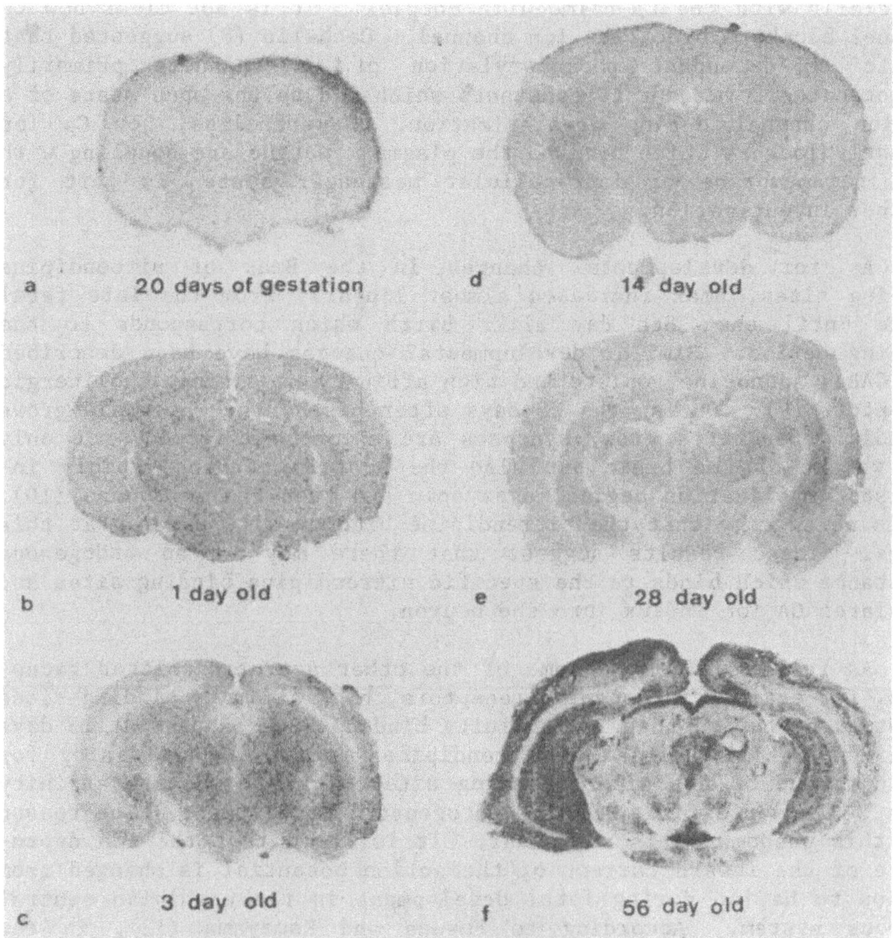


Fig. 8. Developmental change of autoradiographically observed [$^3$H]nitrendipine binding sites at the frontal section through the mammillary body. a: 20 days of gestation, b: 1 day old, c: 3 day old, d: 14 day old, e: 28 day old, f: 56 day old.

channel blocker on the mitochrondrial and microsomal membranes seems to indicate that there are other mechanisms of Ca ion transport on these membranes instead of the plasma membrane-type ion channel. On the other hand, it is an established fact that trifluoperazine and some antipsychotic agents, which have sometimes been designated as Ca antagonists in a wide sense, appear to exert their antagonism through binding to calmodulin (3). According to Ehlert and Itoga (1), displacement of nitrendipine binding by haloperidol showed an $IC_{50}$ of as high as 1,787 nM. All these results show that nitrendipine does not interfere with the intracellular messenger system, especially with the Ca-calmodulin complex. It is not clear how Ca channel blockers block the ion channel. Cachelin (8) suggested that cyclic AMP-dependent phosphorylation of Ca channels primarily promoted the forward rate constants which led to the open state of a Ca ion channel during depolarization. Nevertheless, how Ca ion channel blockers which bind to the plasma membrane are coupling with the intramembrane or intracellular messenger system is left for further investigation.

As for developmental changes in the Bmax of nitrendipine binding sites, Bmax increased almost linearly from the late fetal stage until the 28th day after birth which corresponds to the weaning period. Similar developmental changes have been described for GABA, adenosine, opiate and high affinity muscarinic cholinergic receptors (9). During the 28 days after birth, the rat brain grows rapidly. Dendrites grow, synapses are vigorously formed, not only the weight of the brain but also the protein content rapidly increases, myelination begins, eyes open and then the rat weans (10). It is noteworthy that the nitrendipine binding site matures at this stage. These results suggest that there may be an endogenous substance which binds to the specific nitrendipine binding sites and regulates CA ion influx into the neuron.

As for Kd values in some of the other neurotransmitter receptors, for example in dopamine receptors, high affinity binding sites exist at birth and then low affinity binding sites appear at 15 days after birth (11). In the nitrendipine binding, the affinity for brain tissue of the single binding site changes from high affinity to a 50% lower affinity during ontogenetic development. The reason for this phenomenon is not clear. It is known that the ion dependence of the inward current of the action potential is changed from Ca ion to Na ion during fetal development in the amphibian central nervous system. According to Fukuda and Kameyama (12), in the examination of action potentials produced in the regenerating neurites of guinea pig dorsal root ganglion cells, it appears that the new membrane initially contains voltage dependent Ca ion channels, and that these later disappear. Meiri (13) and others described that following transection of the giant axons of the adult cockroach which are normally generating sodium ion dependent action potentials, recordings from the proximal stump close to the lesion indicated that the membrane was initially inexcitable, later produced Ca ion action potentials, and finally generated normal Na ion dependent spikes. Therefore, the Ca ion channel is considered to be a very early developing structure ontogenetically. The fact that there are rich densities of nitrendipine binding sites in the hippocampus, interpeduncular nucleus, olfactory bulb and superior colliculus, which are phylogenetically early structures, indicates the Ca ion channel plays important roles in the growth or development of a living creature.

At present, it is unclear how the Ca ion channel blocker binding site is coupling with the Ca ion channel. Nevertheless, it is worth studying Ca ion channel blocker binding sites in order to understand the Ca ion channel and Ca ion itself which is one of the second messenger systems.

## SUMMARY AND CONCLUSIONS

1) [$^3$H]Nitrendipine binding sites are localized much more in the synaptosomal membrane than in the mitochondrial and microsomal membranes. 2) To use a whole homogenate (crude membrane fraction) for the purpose of observing nitrendipine binding sites in the synaptosomal membrane instead of $P_2$-B fraction has its reason. 3) The Bmax of specific nitrendipine binding in the rat brain increases linearly until it reaches to the adult level after birth. 4) High affinity binding at early developmental stages shifts to low affinity after the 7 day postnatal stage. 5) Autoradiographically, nitrendipine binding sites are rich in the interpeduncular nucleus, olfactory bulb, hippocampus and superior colliculus in an adult rat. 6) In autoradiography, the density in the hippocampus increases gradually until it reaches the adult level at the 28 day postnatal stage.

## ACKNOWLEDGEMENTS

The authors express their sincere gratitude to Prof. Henry I. Yamamura, University of Arizona Health Sciences Center, for his advice.

## REFERENCES

1. F. J. Ehlert, W. R. Roeske, E. Itoga, and H. I. Yamamura, The binding of [$^3$H]nitrendipine to receptors for calcium channel antagonists in the heart, cerebral cortex and ileum of rats, Life Sci. 30:2191 (1982).
2. H. Rasmussen, "Calcium and cAMP as Synarchic Messengers," A Wiley-Interscience Publication, New York (1981).
3. G. B. Weiss, "New Perspectives on Calcium Antagonists," American Psychological Society, Bethesda (1981).
4. E. G. Gray and V. P. Whittaker, The isolation of nerve endings from brain: an electron-microscopic study of cell fragments derived by homogenization and centrifugation, J. Anat. (Lond.) 96:79 (1962).
5. R. Fried, "Methods of Neurochemistry," Marcel Dekker, Inc., New York (1972).

6. H. Schoemaker, E. Itoga, R. G. Boles, W. R. Roeske, F. J. Ehlert, S. Kito, and H.I. Yamamura, Temperature dependence and allosteric modulation by verapamil and diltiazem of [$^3$H]nitrendipine binding in the rat brain, Proceedings of the Workshop on Nitrendipine Meeting in New York, (1982).
7. O. H. Lowry, J. N. Rosebrough, A. L. Farr, and R. J. Randall, Protein measurement with the Folin phenol reagent, J. Biol. Chem. 193:265 (1951).
8. A. B. Cachelin, J. E. de Peyer, S. Kokubun, and H. Reuter, $Ca^{2+}$ channel modulation by 8-bromocyclic AMP in cultured heart cells, Nature, 304:462 (1983).
9. P. J. Marangos, J. Patel, and J. Stivers, Ontogeny of adenosine binding sites in rat forebrain and cerebellum, J. Neurochem. 39:267 (1982).
10. W. A. Himwich, Biochemical and neurophysiological development of the brain in the neonatal period, Int. Rev. Neurobiol. 4:117 (1962).
11. Y. Nomura, K. Oki, and T. Segawa, Ontogenetic development of the striatal [$^3$H]spiperone binding: Regulation by sodium and guanine nucleotide in rats, J. Neurochem. 38:902 (1982).
12. J. Fukuda and M. Kameyama, Enhancement of Ca spikes in nerve cells of adult mammals during neurite growth in tissue culture, Nature (Lond.) 279:546 (1979).
13. H. Meiri, E. M. Spira, and I. Parnas, Membrane conductance and action potential of a regenerating axonal tip, Science 211:709 (1981).

# BIOCHEMICAL PROPERTIES OF THE GABA/BARBITURATE/BENZODIAZEPINE RECEPTOR-CHLORIDE ION CHANNEL COMPLEX

R.W. Olsen, E.H.F. Wong, G.B. Stauber, D. Murakami, R.G. King, and J.B. Fischer

Department of Biochemistry and Division of Biomedical Sciences, University of California, Riverside, and Department of Pharmacology, UCLA School of Medicine Los Angeles, California

## INTRODUCTION

The inhibitory neurotransmitter γ-aminobutyric acid (GABA) acts by increasing chloride permeability in the postsynaptic cell membrane. Chloride ion channels are regulated via GABA binding to its receptor. The cellular response to GABA has been shown to be enhanced by benzodiazepine and barbiturate drugs. This potentiation of GABA-mediated inhibition may explain much of the nervous system depressant action of these clinically important agents.

We have developed an assay for GABA function with slices of mammalian brain, in which radioactive $^{36}Cl^-$ is used to follow increased chloride permeability induced by GABA. Barbiturates are likewise able to increase $^{36}Cl^-$ permeability in brain slices, and also to potentiate the response to GABA.

The GABA receptor site, defined by suitable *in vitro* radioactive ligand binding assays, has been shown to contain modulatory sites for barbiturates and benzodiazepines. The three classes of sites show mutual chloride-dependent allosteric interactions in membrane fragments from mammalian brain, suggesting a model for the GABA receptor which includes three classes of drug receptor and the chloride ion channel in a single macromolecular complex. This three-receptor complex has been demonstrated to exist in detergent-solubilized extracts and has a molecular weight of about 350,000. The allosteric properties are preserved in solution, including stereospecific, chloride-dependent, picrotoxin-sensitive enhancement by barbiturates of both benzodiazepine and GABA receptor binding. Barbiturate/picrotoxin sites can be demonstrated directly by the binding of [$^{35}$S]t-butyl bicyclophosphorothionate (TBPS). The complex has been purified about 2000 fold on a benzodiazepine affinity column. The purified protein contains all three binding activities and allosteric interactions, and shows 4 bands on SDS gels, the smallest of which at a molecular weight of 56,000 is photoaffinity labeled with [$^{3}$H]flunitrazepam.

## THE BARBITURATE/PICROTOXIN RECEPTOR SITE ON THE GABA/BENZODIAZEPINE BINDING PROTEIN COMPLEX

Barbiturates and related CNS depressants such as the pyrazolopyridines (e.g., SQ 20009 = etazolate) and etomidate enhance the binding of benzodiazepines and GABA receptor agonists, and allosterically inhibit the binding of antagonist ligands for these receptors. The relative activities in these *in vitro* assays of a series of selected compounds are shown in Table 1.

### Barbiturates Enhance Benzodiazepine and GABA Receptor Agonist Binding

The anxiolytic pyrazolopyridines, such as etazolate, were surprisingly found to enhance rather than inhibit benzodiazepine receptor binding using [$^{3}$H]diazepam (1-3). This effect was shown to be chloride-dependent and inhibited by picrotoxin (4). Previous studies in our laboratory had demonstrated specific binding sites in brain membranes for the convulsant drug [$^{3}$H]α-dihydropicrotoxinin (DHP). These binding sites were inhibited by biologically active convulsant picrotoxin analogs in a manner related to their activity as antagonists of GABA postsynaptic responses (5,6). [$^{3}$H]DHP binding was also inhibited by other convulsant drugs known to block GABA function, including "cage" convulsants, the convulsant benzodiazepine Ro5-3663, and pyrethroid insecticides (7,8). [$^{3}$H]DHP binding was furthermore inhibited by convulsant and depressant barbiturates known to enhance GABA function (8,9), as well as by certain purines and pyrimidines (7).

We found that the pyrazolopyridines inhibited [$^{3}$H]DHP binding at concentrations similar to those which enhanced benzodiazepine receptor binding (10). Further, picrotoxin and related convulsants inhibited the enhancement of benzodiazepine receptor binding by pyrazolopyridines, at concentrations similar to those active in inhibiting the binding of [$^{3}$H]DHP. Finally, the barbiturates and related depressants which inhibit [$^{3}$H]DHP binding acted like etazolate to enhance benzodiazepine receptor binding (11).

The barbiturate enhancement of benzodiazepine binding was chemically specific and stereospecific, and correlated reasonably well with sedative-hypnotic activity of the barbiturates, e.g.,

Table 1. Relative Potencies of Barbiturates and Related Depressants to Allosterically Modulate Benzodiazepine and GABA Receptors

| | $EC_{50}$ (μM) | | | $EC_{20}$ (μM) |
|---|---|---|---|---|
| Compound | [$^3$H]GABA(17) | [$^3$H]BMC(31) | [$^3$H]Diazepam(6,12) | [$^3$H]βCCM(31) |
| Etazolate | 2 | 3 | 1 | 8 |
| (+)Etomidate | 10 | 10 | 4 | 28 |
| (−)Etomidate | >100 | 100 | (100)[a] | >1000 |
| (±)DMBB | 150 | 50 | 80 | 40 |
| (±)Secobarbital | 300 | 200 | 100 | 110 |
| (±)Pentobarbital | 300 | 210 | 130 | 170 |
| (+)Hexobarbital | 300 | 260 | 150 | 250 |
| (−)MPPB | 400 | 280 | 100 | 800 |
| (−)Mephobarbital | 400 | 310 | 50 | 220 |
| Phenobarbital | >1000 | 550 | (200)[b] | 2000 |
| (+)Mephobarbital | >1000 | 950 | >1000 | 440 |
| (−)Hexobarbital | >1000 | >1000 | >1000 | >1000 |
| (+)MPPB | >1000 | >1000 | >1000 | 2000 |

[a] (−)Etomidate fails to enhance but inhibits [$^3$H]diazepam binding, $IC_{50}$ = 100 μM.
[b] Phenobarbital at 1 mM does not enhance equilibrium binding but alters kinetics of [$^3$H]diazepam binding ($EC_{50}$ = 200 μM) and reverses pentobarbital enhancement.

pentobarbital, secobarbital, and (+)hexobarbital were more potent than (-)hexobarbital and phenobarbital (Table 1) (11-13).

Interestingly, the barbiturate enhancement was dependent on the presence of certain anions, notably physiological concentrations of chloride. The only anions which supported the interaction (11) were exactly those demonstrated by Eccles and colleagues to be permeable at membrane chloride ion channels which mediate GABA-activated postsynaptic inhibitory potentials (14), synaptic responses which are enhanced _in vivo_ by barbiturates (15). The barbiturate enhancement, like that of etazolate, was competitively inhibited by picrotoxin and related convulsants (10,11). In addition, the barbiturate effect was allosterically inhibited by the GABA receptor antagonist bicuculline (10).

The anion specificity, chemical specificity, sensitivity to picrotoxin, and sensitivity to bicuculline leave little doubt that this _in vitro_ phenomenon defines a barbiturate/picrotoxin receptor site on the benzodiazepine receptor-linked GABA receptor-chloride ion channel complex (6).

Pyrazolopyridines were also reported to enhance GABA receptor binding (16), and once the anion requirement for barbiturate-benzodiazepine interactions became appreciated, we found that the same barbiturates could enhance GABA receptor binding in a chloride-dependent, picrotoxin-sensitive manner (17). Other laboratories reported similar observations for barbiturate enhancement of benzodiazepine (18,20) and GABA receptor binding (21-25), as well as interactions with picrotoxin and related convulsants (23,24).

The same barbiturates that enhance benzodiazepine receptor binding are those that enhance GABA binding (Table 1). The pentobarbital analogue, 1,3-dimethylbutyl barbiturate (DMBB), was the most potent of those tested. This compound, like (+)pentobarbital (26), is known as an excitatory barbiturate (25), but it also has depressant activity and is able to enhance GABA responses including $^{36}Cl^-$ flux, as described below. The excitatory action of DMBB would appear to involve a second mechanism of action in addition to the depressant effect involving GABA receptors.

## Barbiturates Inhibit Benzodiazepine and GABA Receptor Antagonist Binding

Table 1 shows that barbiturates and related compounds produce an allosteric inhibition of the binding of radioactive antagonists to GABA ([$^3$H]bicuculline methochloride, BMC) or benzodiazepine receptors ([$^3$H]β-carboline-3-carboxylate methyl ester (βCCM), an "inverse agonist" (27)). The same specificity of barbiturates was seen as that observed for allosteric enhancement of agonist binding just described. Etazolate and etomidate were more potent than barbiturates, with (+)etomidate > (-)etomidate, and (+)hexobarbital > (-)hexobarbital, (-)mephobarbital > (+)mephobarbital, and (-)$N^1$-methyl, 5-phenyl, 5-propyl barbiturate (MPPB) > (+)MPPB.

As described by Möhler and Okada (28) for the binding of [$^3$H]bicuculline methiodide (BMI), the binding of [$^3$H]BMC appears to involve an antagonist-preferring state of the GABA receptor, which has low affinity for agonists (29). These two (or more) states of the GABA receptor are differentially affected by various assay conditions, e.g., physiological saline (especially chloride) and temperature favor the low affinity agonist state which is the high affinity antagonist state (30). We have described how the low affinity (micromolar Kd) GABA agonist sites are enhanced in affinity by barbiturates but stabilized by picrotoxin. The same state of the receptor shows allosteric inhibition of high affinity antagonist ([$^3$H]BMC) binding by barbiturates via a decrease in affinity (31). Just as anions enhance the binding of [$^3$H]BMI and [$^3$H]BMC dramatically, and barbiturates enhance the binding of GABA agonists dramatically (over 10-fold change in Kd), likewise barbiturates lower the affinity of [$^3$H]BMC over 10-fold and are capable of inhibiting totally the high affinity binding of [$^3$H]BMC measured at 2 nM ligand in 0.1 M KSCN buffer (31).

Benzodiazepine receptor binding with [$^3$H]diazepam or [$^3$H]flunitrazepam (generally termed "agonists") is modestly enhanced (with a 2-fold increase in affinity) by GABA receptor agonists (32-35) and also by barbiturates and related depressants (11). The affinities of benzodiazepine receptor "antagonists", such as certain β-carbolines and Ro15-1788, are not enhanced by GABA (33), and the affinities of pro-convulsant excitatory benzodiazepine receptor ligands (the "inverse agonists" such as βCCM or methyl-6, 7-dimethoxy-4-ethyl-β-carboline-3-carboxylate, DMCM) are _lowered_ by GABA agonists (27). Barbiturates and related depressants likewise do not enhance the affinity of "antagonists" such as Ro15-1788 (A.M. Snowman, L.M.F. Leeb-Lundberg, and R.W. Olsen, _unpublished observations_). They do, however, _lower_ the affinity for the binding of benzodiazepine "inverse agonists" such as [$^3$H]βCCM (Table 1). Since only a modest (less than 2-fold) change in Kd is involved, this inhibition by barbiturates shows a plateau effect of about 40% inhibition at saturating concentrations of barbiturates measured at one low concentration (2 nM) of [$^3$H]βCCM. Therefore, the relative potencies are given in the table as $IC_{20}$ values (concentrations needed to inhibit by 20%).

It is evident that a very similar specificity is observed for the effects of barbiturates on all four receptor binding assays. This suggests that a similar barbiturate binding site is associated with both GABA and benzodiazepine binding sites, which in turn are associated with each other as indicated by allosteric interactions

(6,32-36). The barbiturate-sensitive site can be assayed directly by the binding of the convulsant [$^3$H]DHP (5). Due to low affinity of this ligand, however, a low ratio of specific binding to nondisplaceable background is unavoidable, making it difficult to obtain accurate data. This problem has been assuaged by the introduction of a new ligand for the picrotoxin/barbiturate sites, a radioactive cage convulsant [$^{35}$S]TBPS (37,38). Convulsant/ barbiturate binding activity can be shown to co-purify with the detergent-solubilized GABA/benzodiazepine receptor protein complex as described below.

## BARBITURATE-ENHANCED GABA RECEPTOR FUNCTION CAN BE ASSAYED WITH $^{36}Cl^-$ FLUX IN BRAIN SLICES

As mentioned, the physiological response to GABA at inhibitory synapses involves an increase in membrane permeability to chloride ions. In addition to electrophysiological methods at various levels of sophistication, the GABA response can be assayed with radioactive isotope tracer methods. We observed GABA stimulation of $^{36}Cl^-$ uptake in crayfish muscle fibers that showed properties very similar to the responses measured by microelectrodes (39). For mammalian brain, the method of Teichberg and colleagues (40) was employed to measure GABA enhancement of chloride permeability in tissue slices as monitored by the relative efflux rate of $^{36}Cl^-$ (41).

Fig. 1 shows that when GABA was added to hippocampal slices after the relative efflux rate had reached a low but stable level, the rate increased by up to 50%. This response was demonstrated to be receptor-mediated on the basis of pharmacological specificity and ion specificity. GABA did not increase the permeability of $^{22}Na^+$ but did increase that of $^{125}I^-$. Glycine and glutamate, as well as the GABA uptake inhibitors, 2,4-diaminobutyric acid and nipecotic acid, did not augment $^{36}Cl^-$ efflux. The response to GABA occurred in the presence of tetrodotoxin and high $Mg^{2+}$ (to block nerve conduction and synaptic transmission). $^{36}Cl^-$ efflux was also increased by the GABA agonists muscimol and 3-aminopropane sulfonic acid and blocked by the antagonists bicuculline and picrotoxin. Similar responses to GABA and muscimol were observed in tissue slices from several brain regions including cortex, striatum, and cerebellum. The response was dose-dependent and saturable, with half-maximal effect at about 0.1 mM. The weak apparent potency could be due to diffusion barriers (41).

Barbiturates also increased $^{36}Cl^-$ permeability in brain slices (without added GABA) and caused a more-than-additive enhancement of the GABA response (41). The maximal effect of saturating pentobarbital (100% increase in relative efflux rate) was larger than the maximal response to GABA. The related depressants etazolate and etomidate also increased $^{36}Cl^-$ permeability. Among the barbiturates tested, DMBB, secobarbital, and (+)hexobarbital, but not (-)hexobarbital and phenobarbital were similar to pentobarbital in increasing the rate of $^{36}Cl^-$ efflux and enhancing the response to GABA (41). This specificity agreed perfectly with that described for modulation of GABA/benzodiazepine receptor binding in membranes, demonstrating that a functional barbiturate/picrotoxin receptor site resides on the physiological GABA receptor/ionophore complex.

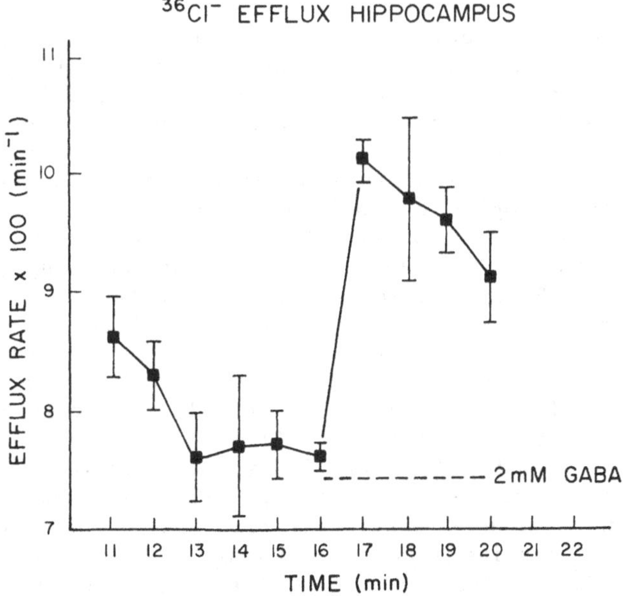


Fig. 1. Stimulation by GABA of $^{36}Cl^-$ efflux from slices of rat hippocampus. Slices of fresh hippocampus (230 μm) were incubated in 2 ml of oxygenated physiological saline at 37° in plastic baskets with sieve bottoms, and then loaded with $^{36}Cl^-$ (ref. 41). The wash-out of $^{36}Cl^-$ was measured by transferring the basket every minute at 22° through 2 ml of nonradioactive saline. The relative efflux rate plotted is the fraction of tissue $^{36}Cl^-$ lost for that one minute period. After the efflux had reached a low stable rate (16 min), supramaximal GABA was included in the saline, leading to increased efflux rate.

## PURIFICATION AND PROPERTIES OF THE GABA/BARBITURATE/BENZODIAZEPINE RECEPTOR-CHLORIDE ION CHANNEL COMPLEX

GABA receptor binding activity from bovine brain could be solubilized in high yield with the mild detergent deoxycholate, yielding a major peak of protein on size separation column chromatography with an apparent molecular weight of the protein-detergent complex of about 900,000 (42). Benzodiazepine receptor binding activity was likewise present in the deoxycholate extracts, and the two activities co-migrated on various protein separation procedures (43). The benzodiazepine binding of partially purified soluble receptor protein, like that in membranes, was still enhanced by GABA (43). Similar observations were likewise made by other groups (36,44,45). Gel filtration and sucrose gradient centrifugation in $H_2O$ and $D_2O$ (in Triton X-100) showed that the GABA/benzodiazepine receptor complex from bovine cortex and cerebellum had an apparent molecular weight of 350,000 (43).

Barbiturate enhancement of GABA and benzodiazepine receptor binding was unstable in extracts solubilized in deoxycholate and Triton X-100, but was reproducible in extracts solubilized with the detergent 3-[(3-cholamidopropyl)-dimethylammonio]propanesulfonate (CHAPS) (46). As in membranes, this barbiturate interaction was chloride-dependent and inhibited by picrotoxin. The CHAPS-soluble receptor protein complex had similar physicochemical properties as observed in the other detergents (46). CHAPS was also more effective than the other detergents in solubilizing the binding activity of [$^{35}$S]TBPS. Approximately 50% of the [$^{35}$S]TBPS binding in the membranes from rat cortex were solubilized in 1% CHAPS in 0.2 M KCl, 10 mM phosphate buffer, pH 7.5. The binding constant Kd in membranes and solution was 30 nM (38). Binding was inhibited by biologically active picrotoxin-like convulsants, cage convulsants, and the depressant drugs listed in Table 1. Stereospecificity was apparent, e.g., (+)etomidate and (+)hexobarbital were significantly more potent that the (-) isomers. [$^{35}$S]TBPS binding activity was observed to co-migrate on size separation column chromatography with GABA and benzodiazepine receptor binding activity (38), consistent with our proposal that all three sites are present on the same protein complex.

The benzodiazepine receptor protein could be bound to a benzodiazepine affinity column and subsequently eluted, with a high purification factor (36,47-51). We employed the benzodiazepine Ro7-1986/1 (52) as an affinity column ligand, with the purification summarized in Table 2. Sixty-five per cent of the GABA receptor activity and 97% of the benzodiazepine receptor activity applied to the column was retained and virtually all of the protein passed through. Negligible binding activity was eluted by extensive buffer

Table 2. Purification of the Rat Brain GABA/Benzodiazepine Receptor Complex

| Fraction | Protein | [$^3$H]Muscimol Binding | | | [$^3$H]Flunitrazepam Binding | | |
|---|---|---|---|---|---|---|---|
| | (total mg) | Total (pmol) | Yield (%) | Spec.Act. (pmol/mg) | Total (pmol) | Yield (%) | Spec.Act. (pmol/mg) |
| Membranes | 500 | 1000 | 100 | 2 | 400 | 100 | 0.8 |
| Deoxycholate sup. | 400 | 400 | 40 | 1 | 320 | 80 | 0.8 |
| Sepharose 6B (Triton X-100) | 50 | 250 | 25 | 5 | 200 | 50 | 4 |
| Affinity Column | 40 | 200 | 100 | 5 | 160 | 100 | 4 |
| Flow-through | 30 | 70 | 35 | -- | 5 | 3 | -- |
| Wash | 10 | -- | 2 | -- | -- | 1 | -- |
| Flurazepam | <0.05 | 80 | 40 | >1600 | -- | -- | -- |
| DEAE, G-25 | <0.05 | -- | -- | -- | 50 | 31 | >1000 |

Total rat brain (10 g) was homogenized and membranes prepared and solubilized in 0.5% deoxycholate in 50 mM Tris-HCl, 50 mM KCl, pH 8.0. This supernatant fraction was applied to a Sepharose 6B column (3 X 75 cm) in 0.5% Triton X-100; 2 ml fractions were assayed for receptor binding by the polyethylene glycol-bovine γ-globulin precipitation and centrifugation method (43). The peak of binding activity was pooled and 80% applied to a column (5 ml of Biorad Affi-gel 202 coupled to 0.5 micromoles of benzodiazepine Ro 7-1986/1 with 1-ethyl-3-(3-dimethylaminopropyl)carbodiimide). After washing with 100 ml of buffer, 10 ml of 5 mM flurazepam eluted the receptor. Free flurazepam was removed by passage through 5 ml columns of DEAE-cellulose, eluting with 0.4 M KCl, followed by Sephadex G-25.

washing, but over 40% of the bound [$^3$H]muscimol binding activity was eluted by millimolar free flurazepam. This sample has undetectably low protein content, indicating a specific activity of at least 1600 pmol/mg protein (compared to about 1 pmol/mg in crude membrane homogenates or initial detergent extracts; a totally purified receptor would be expected to have a specific activity of 3000-10000). Significant [$^3$H]flunitrazepam binding could be obtained following removal of the free flurazepam on a DEAE-cellulose column followed by G-25 Sephadex column chromotography. The purified receptor showed a 50% enhancement of [$^3$H]muscimol binding by pentobarbital (1 mM) and was able to bind [$^{35}$S]TBPS. SDS gel electrophoresis and Coomassie Blue staining revealed four peptide bands in the purified sample (Figure 2), at molecular weights of 56, 60, 61, and 66,000. Photoaffinity labeling of the sample with [$^3$H]flunitrazepam (as in ref. 53) and SDS gel electrophoresis

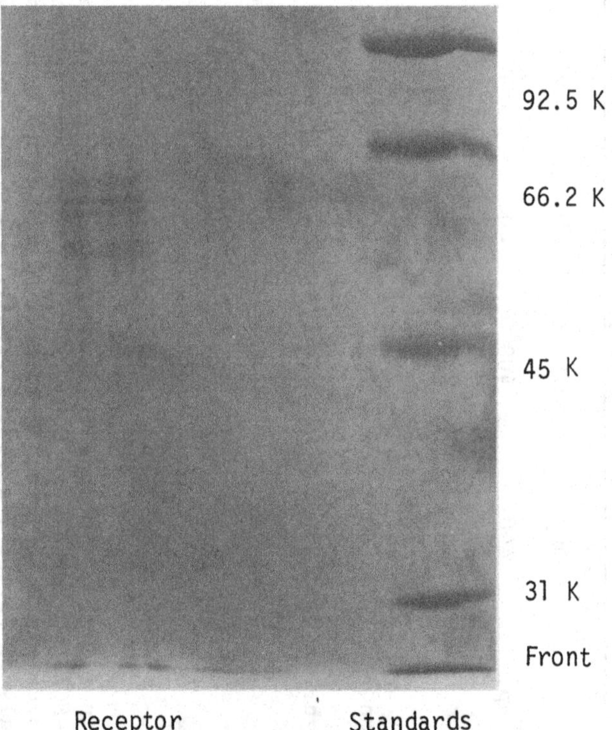


Fig. 2. SDS gel electrophoresis with Coomassie Blue stain of purified GABA/benzodiazepine/barbiturate receptor protein from rat brain. Ten to twenty micrograms of protein in the flurazepam eluant from the benzodiazepine affinity column (Table 2) were applied to the track at the left.

Table 3. Properties of Purified Rat Brain GABA/Benzodiazepine Receptor Complex.

---

-- Binds [$^3$H]Muscimol : Specific Activity > 1600 pmol/mg; Kd = 20 nM. Inhibited by muscimol, GABA, bicuculline, 3-amino-propane sulfonate, but not DABA (10 μM). Enhanced by pentobarbital (1 mM): 47-56%.

-- Binds [$^{35}$S]TBPS : Specific Activity > 300 pmol/mg; Kd = 30 nM. Inhibited by ethyl bicyclophosphate, picrotoxinin, pentobarbital, GABA.

-- Binds [$^3$H]Flunitrazepam : Specific Activity > 1000 pmol/mg; Kd = 4 nM. Inhibited 100% by flunitrazepam and clonazepam, but < 30% by Ro5-4864 (all at 300 nM). Photoaffinity labeled: 56 K band.

-- Subunit Composition: 56, 60, 61, 66 K bands

-- Native Molecular Weight : Stokes Radius $\simeq$ Catalase

---

resulted in a single radioactive band at 55,000 molecular weight, corresponding to the bottom stained band. The properties of the purified protein are summarized in Table 3.

In summary, following purification of the benzodiazepine receptor binding activity to near or total homogeneity, a single protein complex can be isolated that also contains GABA and convulsant/barbiturate receptor binding activity with allosteric interactions between the three. The chloride dependence of some of these interactions indicates that in all likelihood the receptor complex also includes the macromolecular components making up the associated chloride ion channel. Thus a major category of GABA binding protein contains chloride-sensitive modulatory sites for the two classes of drugs, the benzodiazepines, and the barbiturate/ picrotoxin compounds, and can be isolated intact. Since the purified protein contains the bulk of the measurable high affinity specific benzodiazepine receptor binding sites in brain, it is likely to include those GABA binding sites involved in regulating physiological chloride channels at synapses which can be modulated *in situ* by these drugs, and thus at least a portion of the physiological GABA receptor sites.

ACKNOWLEDGEMENTS

Supported by NIH Grant NS 20704 and Training Grant AM 07310, and NSF Grant NS 80-19722.

REFERENCES


1. P. Supavilai and M. Karobath, Action of pyrazolpyridines as modulators of [$^{3}$H]flunitrazepam binding to the GABA/ benzodiazepine receptor complex of the cerebellum, Eur. J. Pharmacol. 70:183 (1981).
2. B. Beer, C. A. Klepner, A. S. Lippa, and R. F. Squires, Enhancement of [$^{3}$H]diazepam binding by SQ 65396: a novel anti-anxiety agent, Pharmacol. Biochem. Behav. 9:849 (1978).
3. M. Williams and E. A. Risley, Enhancement of the binding of [$^{3}$H]diazepam to rat brain membranes in vitro by SQ 20009, a novel anxiolytic, gamma-aminobutyric acid (GABA) and muscimol, Life Sci. 24:833 (1979).
4. P. Supavilai and M. Karobath, Stimulation of benzodiazepine receptor binding by SQ 20009 is chloride-dependent and picrotoxin-sensitive, Eur. J. Pharmacol. 60:111 (1979).
5. M. K. Ticku, M. Ban, and R. W. Olsen, Binding of [$^{3}$H]α-dihydropicrotoxinin, a γ-aminobutyric acid synaptic antagonist, to rat brain membranes, Mol. Pharmacol. 14:391, (1978).
6. R. W. Olsen, Drug interactions at the GABA receptor ionophore complex, Ann. Rev. Pharmacol. Toxicol. 22:245 (1982).
7. R. W. Olsen, F. Leeb-Lundberg, and C. Napias, Picrotoxin and convulsant binding sites in mammalian brain, Brain Res. Bull. 5, Suppl. 2:217 (1980).
8. M. K. Ticku and R. W. Olsen, Cage convulsants inhibit picrotoxinin binding, Neuropharmacol. 18:315 (1979).
9. M. K. Ticku and R. W. Olsen, Interaction of barbiturates with dihydropicrotoxinin binding sites related to the GABA receptor-ionophore system, Life Sci. 22:1643 (1978).
10. F. Leeb-Lundberg, A. Snowman, and R. W. Olsen, Perturbation of benzodiazepine receptor binding by pyrazolpyridines involves picrotoxinin/barbiturate receptor sites, J. Neurosci. 1:471 (1981).
11. F. Leeb-Lundberg, A. Snowman, and R. W. Olsen, Barbiturate receptors are coupled to benzodiazepine receptors, Proc. Natl. Acad. Sci. USA 77:7468 (1980).
12. F. Leeb-Lundberg and R. W. Olsen, Interactions of barbiturates of various pharmacological categories with benzodiazepine receptors, Mol. Pharmacol. 21:320 (1982).
13. F. Leeb-Lundberg and R. W. Olsen, Heterogeneity of benzodiazepine receptor interactions with GABA and barbiturate receptor sites, Mol. Pharmacol. 23:315 (1983).
14. J. Eccles, R. A. Nicoll, T. Oshima, and F. J. Rubia, The anionic permeability of the inhibitory postsynaptic membrane of hippocampal pyramidal cells, Proc. R. Soc. Lond. B. 198:345 (1977).
15. R. A. Nicoll, J. C. Eccles, T. C. Oshima, and F. Rubia, Prolongation of hippocampal inhibitory post-synaptic potentials by barbiturates, Nature 258:625 (1975).

16. P. Placheta and M. Karobath, In vitro modulation by SQ 20009 and SQ 65396 of GABA receptor binding in rat CNS membranes, Eur. J. Pharmacol. 62:225 (1980).
17. R. W. Olsen and A. M. Snowman, Chloride-dependent enhancement by barbiturates of GABA receptor binding, J. Neurosci. 2:1812 (1982).
18. P. Skolnick, K. C. Rice, J. L. Barker, and S. M. Paul, Interaction of barbiturates with benzodiazepine receptors in the central nervous system, Brain Res. 233:143 (1982).
19. M. K. Ticku, Interaction of depressant, convulsant and anticonvulsant barbiturates with [$^3$H]diazepam binding site at the benzodiazepine-GABA-receptor-ionophore complex, Biochem. Pharmacol. 30:1573 (1981).
20. T. Asano and N. Ogasawara, Chloride-dependent stimulation of GABA and benzodiazepine receptor binding by pentobarbital. Brain Res. 225:212 (1981).
21. M. Willow and G. A. R. Johnston, Enhancement by anesthetic and convulsant barbiturates of GABA binding to rat brain synaptosomal membranes, J. Neurosci. 1:364 (1981).
22. S. R. Whittle and A. J. Turner, Differential effects of sedative and anticonvulsant barbiturates on specific [$^3$H]GABA binding to membrane preparations from rat brain cortex, Biochem. Pharmacol. 31:2891 (1982).
23. P. Supavilai, A. Mannonen, and M. Karobath, Modulation of GABA binding sites by CNS depressants and CNS convulsants, Neurochem. Int. 4:259 (1982).
24. P. Supavilai, A. Mannonen, J. F. Collins, and M. Karobath, Anion-dependent modulation of [$^3$H]muscimol binding and GABA-stimulated [$^3$H]flunitrazepam binding by picrotoxin and related CNS convulsants, Eur. J. Pharmacol. 81:687 (1982).
25. M. Willow and G. A. R. Johnston, Pharmacology of barbiturates: electrophysiological and neurochemical studies, Int. Rev. Neurobiol. 24:15 (1983).
26. J. L. Barker and D. A. Mathers, GABA receptors and the depressant action of pentobarbital, Trends in Neuroscience, p. 10, January (1981).
27. C. Braestrup and M. Nielsen, GABA reduces binding of $^3$H-methyl β-carboline-3-carboxylate to brain benzodiazepine receptors, Nature 292:472 (1981).
28. H. Möhler and T. Okada, Properties of γ-aminobutyric acid receptor binding with (+)[$^3$H]bicuculline methiodide in rat cerebellum, Mol. Pharmacol. 14:256 (1978).
29. R. W. Olsen and A. M. Snowman, [$^3$H]Bicuculline methochloride binding to low affinity GABA receptor sites, J. Neurochem. 41:1653 (1983).
30. R. W. Olsen, E. H. F. Wong, G. B. Stauber, and R. G. King, Biochemical Pharmacology of the GABA Receptor/Ionophore Protein, Fed. Proc., in press.

31. E. H. F. Wong, A. M. Snowman, L. M. F. Leeb-Lundberg, and R. W. Olsen, Barbiturates allosterically inhibit GABA-benzodiazepine receptor antagonist binding, manuscript submitted.
32. J. F. Tallman, S. M. Paul, P. Skolnick, and D. W. Gallager, Receptors for the age of anxiety: pharmacology of the benzodiazepines, Science 207:274 (1980).
33. F. J. Ehlert, W. R. Roeske, K. W. Gee, and H. I. Yamamura, An allosteric model for benzodiazepine receptor function, Biochem. Pharmacol. 32:2375 (1983).
34. M. Karobath, P. Placheta, M. Lippitsch, and P. Krogsgaard-Larsen, Is stimulation of benzodiazepine receptor binding mediated by a novel GABA receptor? Nature 278:748 (1979).
35. C. Braestrup, M. Nielsen, P. Krogsgaard-Larsen, and E. Falch, Partial agonists for brain GABA/benzodiazepine receptor complex, Nature 280:331 (1979).
36. M. Gavish and S. H. Snyder, γ-Aminobutyric acid and benzodiazepine receptors: copurification and characterization, Proc. Natl. Acad. Sci. USA 78:1939 (1981).
37. R. F. Squires, J. E. Casida, M. Richardson, and E. Saederup, [$^{35}$S]t-Butylbicyclophosphorothionate binds with high affinity to brain-specific sites coupled to γ-aminobutyric acid-A and ion recognition sites, Mol. Pharmacol. 23:326 (1983).
38. R. G. King and R. W. Olsen, Solubilization of cage convulsant/picrotoxin/barbiturate receptors, Trans. Am. Soc. Neurochem. (1984).
39. M. K. Ticku and R. W. Olsen, γ-Aminobutyric acid stimulated chloride permeability in crayfish muscle, Biochim. Biophys. Acta 464:519 (1977).
40. A. Luini, O. Goldberg, and V. I. Teichberg, Distinct pharmacological properties of excitatory amino acid receptors in the rat striatum: study by $Na^+$ efflux assay, Proc. Natl. Acad. Sci. USA 78:3250 (1981).
41. E. H. F. Wong, L. M. F. Leeb-Lundberg, V. I. Teichberg, and R. W. Olsen, γ-Aminobutyric acid activation of $^{36}Cl^-$ flux in rat hippocampal slices and its potentiation by barbiturates, Brain Res., in press (1984).
42. D. V. Greenlee and R. W. Olsen, Solubilization of gamma-aminobutyric acid receptor protein from mammalian brain, Biochem. Biophys. Res. Comm. 88:380 (1979).
43. F. A. Stephenson, A. E. Watkins, and R. W. Olsen, Physicochemical characterization of detergent-solubilized γ-aminobutyric acid and benzodiazepine receptor proteins from bovine brain, Eur. J. Biochem. 123:291 (1982).
44. T. Asano, Y. Yamada, and N. Ogasawara, Characterization of the solubilized GABA and benzodiazepine receptors from various regions of bovine brain, J. Neurochem. 40:209 (1983).
45. L. -R. Chang and E. A. Barnard, The benzodiazepine/GABA receptor complex: molecular size in brain synaptic membranes and in solution, J. Neurochem. 39:1507 (1982).

46. F. A. Stephenson and R. W. Olsen, Solubilization by CHAPS detergent of barbiturate-enhanced benzodiazepine-GABA receptor complex, J. Neurochem. 39:1579 (1982).
47. C. Martini, A. Lucacchini, G. Ronca, S. Hrelia, and C. A. Rossi, Isolation of putative benzodiazepine receptors from rat brain membranes by affinity chromatography, J. Neurochem. 38:15 (1982).
48. E. Sigel, C. Mamalaki, and E. A. Barnard, Isolation of a GABA receptor from bovine brain using a benzodiazepine affinity column, FEBS Lett. 147:45 (1982).
49. E. A. Barnard, F. A. Stephenson, E. Sigel, C. Mamalaki, G. Bilbe, A. Constanti, T.G. Smart, and D. A. Brown, Structure and properties of the brain GABA/benzodiazepine receptor complex, in: "Neurotransmitter Receptors: Mechanisms of Action and Regulation," S. Kito, T. Segawa, K. Kuriyama, H. I. Yamamura, and R. W. Olsen, eds., Plenum, New York (1984).
50. K. Kuriyama and J. Taguchi, Purification of γ-aminobutyric acid (GABA) and benzodiazepine receptors from rat brain using benzodiazepine-affinity column chromatography, in: "Neurotransmitter Receptors: Mechanisms of Action and Regulation," S. Kito, T. Segawa, K. Kuriyama, H. I. Yamamura, and R. W. Olsen, eds., Plenum, New York (1984).
51. R. W. Olsen, Biochemical properties of GABA receptors, in: "The GABA Receptors," S. J. Enna, ed., Humana Press, Clifton, New Jersey, p. 63 (1983).
52. R. W. Olsen, E. H. F. Wong, G. Stauber, D. Murakami, and J. Sussman, Progress in purification of the benzodiazepine/GABA receptor protein, Trans. Am. Soc. Neurochem. 14:194 (1983).
53. H. Möhler, M. K. Battersby, and J. G. Richards, Benzodiazepine receptor protein identified and visualized in brain tissue by a photoaffinity label, Proc. Natl. Acad. Sci. USA 77:1666 (1980).

# PURIFICATION OF γ-AMINOBUTYRIC ACID (GABA) AND BENZODIAZEPINE RECEPTORS FROM RAT BRAIN USING BENZODIAZEPINE-AFFINITY COLUMN CHROMATOGRAPHY

Kinya Kuriyama and Jun-ichi Taguchi

Department of Pharmacology, Kyoto Prefectural University of Medicine, Kawaramachi-Hirokoji, Kamikyo-Ku
Kyoto 602, Japan

γ-Aminobutyric acid (GABA) has been established as a major inhibitory neurotransmitter in the mammalian central nervous system (1-4) as well as invertebrate nervous systems (5,6). GABA acts through a physiologically relevant receptor protein, the GABA receptor, which is labeled specifically by [$^{3}$H]GABA (7) and various GABA agonists such as [$^{3}$H]muscimol (8,9). On the other hand, it has been demonstrated that specific and pharmacologically relevant benzodiazepine receptors exist in the mammalian central nervous system (10,11).

Recent pharmacological studies have shown that GABA agonists facilitate benzodiazepine receptor binding (12-15), while benzodiazepines stimulate GABA receptor binding (16,17) in the brain. In addition, it has been reported that both GABA and benzodiazepine receptors are co-solubilized from cerebral synaptic membranes by various detergents, and both binding sites appear in the same fraction when the solubilized fractions are subjected to various column chromatographic procedures (17-26). These results strongly suggest that the GABA receptor may be coupled with the benzodiazepine receptor in cerebral synaptic membranes, and the association of both receptors may be important for maintaining the function and/or the integrity of these receptors at central synapses. The above results also suggest that the affinity gels synthesized with either GABA (or GABA drugs) or benzodiazepines as immobilized ligands may be useful for the separation of GABA and/or benzodiazepine receptors.

In this chapter, we demonstrate the results of the purification of solubilized GABA/benzodiazepine receptors from rat brain by affinity column chromatography using a new benzodiazepine, 1012-S, as an immobilized ligand.

## MATERIALS AND METHODS

### Preparation of Synaptic Membranes and Solubilized Receptor Fractions

Crude synaptic membrane fractions were obtained from the brain of male Wistar rats weighing 180-200 g according to the method of Zukin et al. (7).

After repeating the washing of the synaptic membrane fraction with 50 mM Tris-citrate buffer (pH 7.1) three times, the solubilization of the synaptic membrane fraction by Nonidet P-40 was performed as described by Ito and Kuriyama (17) with the following modifications: synaptic membrane fractions were treated with 10 volumes of 50 mM Tris-citrate buffer (pH 7.1) containing 1% (v/v) Nonidet P-40 and 40 μg/ml of bacitracin for 90 min at 2°C. The Nonidet P-40 treated synaptic membrane fraction was centrifuged at 105,000 X g for 90 min at 2° C, and the supernatant thus obtained was used for affinity column chromatography as "solubilized receptor preparation".

### Synthesis of Benzodiazepine Affinity Gel

According to the methods of Cuatrecasas (27), 25 ml of CNBr-activated Sepharose 4B (Pharmacia, Uppsala) was incubated with 0.1 M sodium bicarbonate buffer (pH 9.0) containing 1.8 g of adipic acid dihydrazide for 16 hr at 4°C. After filtration of the mixture on a glass filter, the gel was suspended in 25 ml of 26.8 mM sodium iodoacetate and the pH was adjusted to 5.0 with HCl. 1-Ethyl-3-(3-dimethylaminopropyl)carbodiimide hydrochloride (1.2 g) was then added to catalyze the reaction, and the pH was readjusted to 5.0. After shaking for 3 hr at 25° C, the gel was filtered and washed with distilled water. The washed gel was resuspended in 12.5 ml of 50 mM sodium bicarbonate buffer (pH 9.0), and an equal volume of dimethylformamide containing 0.1 mmol of each benzodiazepine as a ligand was added. The mixture was then incubated for 24 hr at 25° C. The gel coupled with each benzodiazepine was treated with 1 M 2-aminoethanol (pH 8.0) for 16 hr at 25° C and was then filtered. After successive washings with 0.1 M sodium bicarbonate buffer (pH 8.5) and 0.1 M sodium acetate buffer (pH 4.5) containing 0.5 M NaCl, the resulting gel was finally washed with distilled water.

### Procedures for Affinity Column Chromatography

After pre-equilibrating the column (1.2 X 18 cm) packed with benzodiazepine affinity gel with 50 mM Tris-citrate buffer (pH 7.1) containing 0.1% Nonidet P-40, the receptor preparation was applied and washed at the flow rate of 10 ml/hr with the same medium used for pre-equilibrating the column until protein was not coming out. Following the washing with 100 ml of the same medium containing NaCl, the receptor protein retained in the column was bio-specifically eluted by chlorazepate or flurazepam dissolved in the same medium or by 1012-S dissolved in the same medium containing 10% dimethylsulfoxide, respectively. Subsequently, the receptor protein remaining in the column was nonspecifically eluted with sodium thiocyanate (1-2 M) dissolved in the same medium. Bio-specifically and nonspecifically eluted fractions were immediately dialysed overnight against the same medium at 4°C. All procedures for the affinity column chromatography were carried out at 4°C.

## Ion Exchange Chromatography for the Removal of Benzodiazepines

Ion exchange chromatography for removing benzodiazepines used for the bio-specific elution of the receptor protein was performed as described by Sigel et al. (26) with the following modifications: the fractions eluted from the affinity column (30 ml) were pooled, and the pH of the fraction was adjusted to 5.5 with 100 mM phosphoric acid. The pooled fraction was then applied to a DEAE-Sepharose CL-6B (Pharmacia, Uppsala) column (1.2 X 18 cm) pre-equilibrated with 50 mM potassium phosphate buffer (pH 5.5) containing 2 mM magnesium acetate, 10% (w/v) sucrose and 0.1% Nonidet P-40. After washing with 70 ml of the same medium at the flow rate of 20 ml/hr, the receptor protein was eluted with the same medium supplemented with 0.8 M potassium chloride and the pH readjusted to 7.1 with 100 mM potassium hydroxide.

## Assay of GABA and Benzodiazepine Receptor Bindings

The bindings of [$^3$H]muscimol ([methyl-$^3$H]-3-hydroxy-5-amino-ethyl-isoxazole, Spec. Act.: 29.4 Ci/mmol, New England Nuclear, Boston) and [$^3$H]flunitrazepam ([N-methyl-$^3$H]flunitrazepam, Spec. Act.: 76.9 Ci/mmol, New England Nuclear, Boston) to the receptor preparations were determined as described by Ito and Kuriyama (17). In brief, 0.5 ml of the solubilized or purified receptor preparation were incubated for 30 min at 2°C with 10 nM [$^3$H]muscimol or for 60 min at 2°C with 1 nM [$^3$H]flunitrazepam in 50 mM Tris-citrate buffer. Following the incubation, 0.3 ml of 50% (w/v) polyethyleneglycol (PEG, M.W.: 6000, Kanto Kagaku, Tokyo) were added to a final concentration of 15%. After allowing to stand for 5 min, the mixture was filtered rapidly under vacuum through a GF/B glass fiber filter. The filter was then washed three times with 3 ml of 50 mM Tris-citrate buffer containing 8% (w/v) PEG at 2°C. The specific binding of [$^3$H]muscimol and [$^3$H]flunitrazepam was defined as total binding of each [$^3$H]ligand minus the binding obtained in the presence of 1 mM GABA and 10 μM clonazepam, respectively.

The radioactivity trapped on each filter was measured by liquid scintillation spectrometer using Triton-toluene scintillator [Triton X-100 : toluene (containing 5 g PPO and 0.3 g POPOP per liter)=2:1] at a counting efficiency of 40-50%.

### Procedures for Photoaffinity Labeling and SDS-Polyacrylamide Gel Electrophoresis

Each protein sample was precipitated with 10% (w/v) trichloroacetic acid and the resultant pellet was washed twice with diethyl ether. The precipitated protein was dissolved in the sample buffer (containing 2.5 (w/v) SDS, 10 mM Tris-HCl (pH 8.0), 1 mM disodium EDTA, 5% 2-mercaptoethanol, 4% glycerol and 0.2% (w/v) Bromophenol blue) and heated for 5 min at 100° C. SDS-polyacrylamide gel electrophoresis was performed according to the method of Laemmli (28) using 10% (w/v) polyacrylamide slab gels. Photoaffinity labeling of the purified receptor fraction was employed as described by Sieghart and Karobath (31). Following incubation of the purified receptor fraction with 5 nM [$^3$H]flunitrazepam in 50 mM Tris-citrate buffer (pH 7.1), the sample was exposed to UV light for 5 min. After the irradiation, the sample was subjected to SDS-polyacrylamide gel electrophoresis as described above.

### Determination of Protein

To avoid the formation of protein precipitates with detergent and phenol reagent in the method of Lowry et al. (29), protein content was determined by the method of Bradford (30) using Coomassie Brilliant Blue G-250.

## RESULTS AND DISCUSSION

### Purification of the GABA Receptor

In the previous studies, we attempted to purify cerebral GABA and benzodiazepine receptors using Sephadex G-200, hydroxylapatite and DEAE-Sepharose CL-6B column chromatography, but these procedures were found to be unsatisfactory for the purificaiton of both types of receptors from the rat brain (32).

Recent pharmacological evidence has indicated that the cerebral GABA receptor is coupled with benzodiazepine receptor in synaptic membranes as well as in various solubilized fractions (17, 20-26). These facts strongly suggest that the isolation of the cerebral GABA receptor may be achieved by the use of an immobilized benzodiazepine affinity gel, while that of benzodiazepine receptor may be achieved by an immobilized GABA analogue affinity gel, respectively. In fact, the success of such a trial using the immobilized Ro 7-1986/1 affinity gel for the purification of GABA receptor from bovine brain has been reported recently (26). We have attempted, therefore, the purification of solubilized cerebral GABA receptor by benzodiazepine affinity column chromatography.

We have used in the present study acetamide adipic hydrazide Sepharose 4B synthesized from CNBr-activated Sepharose 4B, instead of that from AH-Sepharose 4B, since the latter gel had a higher nonspecific absorption of protein in the solubilized fraction with Nonidet P-40 than that found in the former gel.

In the case of nitrazepam-acetamide adipic hydrazide Sepharose 4B (Fig. 1), 94% of the protein applied to the column, 93% of total [$^3$H]muscimol binding and 102% of total [$^3$H]flunitrazepam binding were recovered in the run-through fraction. These results indicate that this gel does not retain the GABA and benzodiazepine receptors. When lorazepam was used as an immobilized ligand (Fig. 1), 97.2% of the protein applied to the column was recovered in the run-through fraction, in which the yields of [$^3$H]muscimol and [$^3$H]flunitrazepam binding were found to be 42.1% and 9.5%, respectively (Table 1).

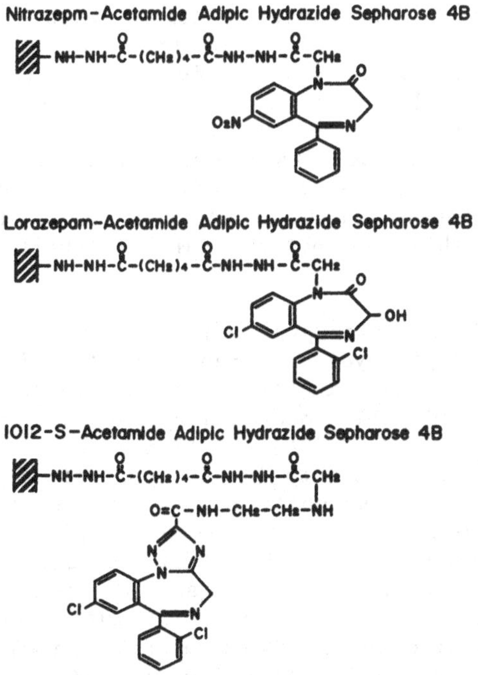

Nitrazepm-Acetamide Adipic Hydrazide Sepharose 4B

Lorazepam-Acetamide Adipic Hydrazide Sepharose 4B

1012-S-Acetamide Adipic Hydrazide Sepharose 4B

Fig. 1. Structures of immobilized benzodiazepine affinity gels used.

Table 1. Summary on Affinity Chromatographic Separation of Solubilized Cerebral GABA Receptor by Lorazepam-Acetamide Adipic Hydrazide Sepharose 4B.

| | Volume (ml) | Protein Content (mg) | $^{3}$H-Muscimol Binding (at 10 nM) Specific Binding (pmol/ mg prot.) | Purification Fold | Total Binding (pmol) | $^{3}$H-Flunitrazepam Binding (at 1 nM) Specific Binding (pmol/ mg prot.) | Purification Fold | Total Binding (pmol) |
|---|---|---|---|---|---|---|---|---|
| Solubilized fraction | 180 | 153.76 | 0.29 | 1.0 | 44.42 | 0.43 | 1.0 | 65.66 |
| Run-through | 220 | 149.51 (97.2%) | 0.13 | 0.4 | 18.73 (42.1%) | 0.04 | 0.1 | 6.29 ( 9.5%) |
| NaCl (1 M) | 60 | 0.57 ( 0.4%) | 6.43 | 22.3 | 3.67 ( 8.3%) | 1.24 | 2.9 | 0.71 ( 1.1%) |
| Chlorazepate (0.05-5 mM) | 30 | 2.40 ( 1.6%) | 0.66 | 2.3 | 1.62 ( 3.7%) | 0.01 | 0.1 | 0.02 ( 0.1%) |
| Flurazepam (0.05-5 mM) | 30 | 0.15 ( 0.1%) | 1.37 | 4.7 | 0.21 ( 0.5%) | 0.11 | 0.2 | 0.02 ( 0.1%) |
| NaSCN (1 M) | 30 | 1.41 ( 0.9%) | 2.46 | 8.5 | 3.47 ( 7.8%) | 0 | 0 | 0 |
| Recovery | | 154.04 (100.1%) | | | 27.69 (62.3%) | | | 7.04 (10.7%) |

Furthermore, the bio-specific elution with chlorazepate and flurazepam (0.05-5 mM) resulted in the overall purification for [$^{3}$H]muscimol binding of only 2.3 fold and 4.7 fold, respectively. In addition, the application of NaSCN (1 M) resulted in a similar purification fold as the bio-specific elution with these benzodiazepines. These results clearly indicate that the affinity column with lorazepam-acetamide adipic hydrazide Sepharose 4B retains GABA and benzodiazepine receptors, at least to some extent, but is not suitable to obtain further purification of these receptors.

As the third benzodiazepine affinity gel, we have synthesized 1012-S-acetamide adipic hydrazide Sepharose 4B (Fig. 1). The 1012-S has an aliphatic primary amino group and a higher affinity (IC$_{50}$=6.0 X $10^{-11}$ M) to solubilized benzodiazepine receptor than those of chlorazepate (IC$_{50}$=7.5 X $10^{-7}$ M) and flurazepam (IC$_{50}$=2.0 X $10^{-8}$ M). Similar to the results obtained in the lorazepam affinity gel (Table 1), the application of solubilized cerebral fraction to this gel exhibited the recovery of 96% of protein applied to the column in the run-through fraction, in which the yields of $^{3}$H-muscimol and $^{3}$H-flunitrazepam binding were 16.0% and 0%, respectively (Table 2, Fig. 2). In contrast, the bio-specific elution with 1012-S (1 mM) resulted in the elution of 25.6% of total muscimol binding, and the highest purification fold obtained was 4576 (specific activity: 0.99 nmole/mg protein). This purification fold for the GABA receptor was obviously superior to that reported previously in the bovine cerebral cortex (26). Furthermore, the successive Table 2.

Table 2. Summary on Affinity Chromatographic Separation of Solubilized Cerebral Receptor by 1012-S-Acetamide Adipic Hydrazide Sepharose 4B.

| | Volume (ml) | Protein Content (mg) | $^3$H-Muscimol Binding (at 10 nM) Specific Binding (pmol/mg prot.) | Purification Fold | Total Binding (pmol) | $^3$H-Flunitrazepam Binding (at 1 nM) Specific Binding (pmol/mg prot.) | Purification Fold | Total Binding (pmol) |
|---|---|---|---|---|---|---|---|---|
| Solubilized fraction | 180 | 162.57 | 0.22 | 1.0 | 35.32 | 0.38 | 1.0 | 61.97 |
| Run-through | 240 | 156.23 (96.0%) | 0.04 | 0.1 | 5.74 (16.0%) | 0 | 0 | 0 |
| NaCl (50 mM) | 100 | 0 | 0 | 0 | 0 | 0 | 0 | 0 |
| 1012-S (1 mM) | 40 | 0.03 ( 0.1%) | 348.23 | 1604.7 | 9.05 (25.6%) | 0.65 | 1.7 | 0.01 ( 0.1%) |
| [ I No. 8-11 | 8 | 0.02 | 344.93 | 1589.5 | 5.17 | 0.06 | 0.2 | 0.01 |
| II No.12-15 | 8 | 0.01 | 993.00 | 4576.0 | 2.98 | 1.66 | 4.4 | 0.01 |
| III No.16-17 | 4 | 0.02 | 383.33 | 1766.5 | 0.69 | 3.89 | 10.2 | 0.01 ] |
| NaSCN (1-2 M) | 40 | 0.19 ( 0.1%) | 73.71 | 339.6 | 14.08 (39.9%) | 0.03 | 0.1 | 0.01 ( 0.1%) |
| [ No. 7- 9 | 6 | 0.01 | 410.00 | 1889.4 | 3.69 | 0 | 0 | 0 ] |
| Recovery | | 156.45 (96.2%) | | | 29.18 (82.6%) | | | 0.18 ( 0.3%) |

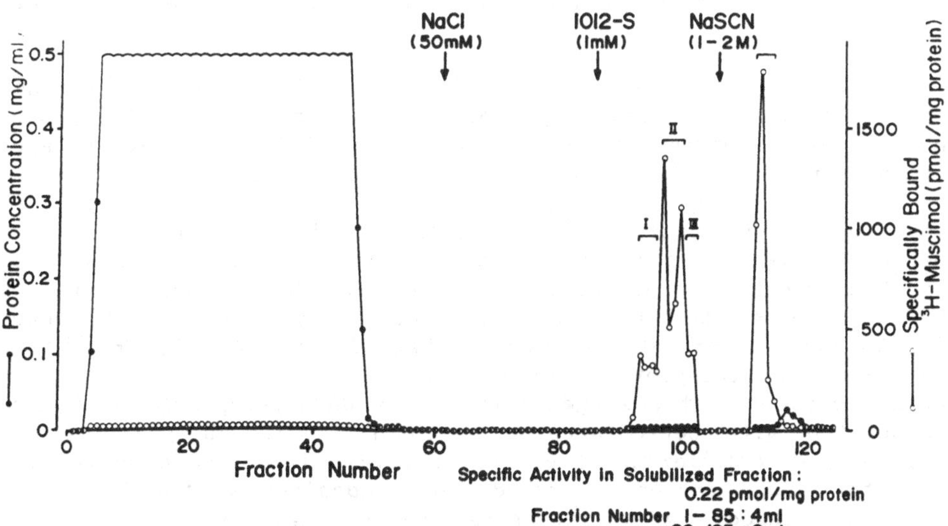


Fig. 2. Affinity chromatography using 1012-S-acetamide adipic hydrazide sepharose 4B of Nonidet P-40 (1%)-solubilized fraction.

application of NaSCN (1-2 M) resulted in the elution of 39.9% of total [$^3$H]muscimol binding, and the highest purification fold obtained under these experimental conditions was 1889 (specific activity: 0.41 nmole/mg protein). These results clearly indicate that 1012-S-acetamide adipic hydrazide Sepharose 4B is a suitable affinity gel for purifying cerebral GABA receptor. It should be mentioned, however, that [$^3$H]flunitrazepam binding was scarcely detected in these fractions highly enriched with GABA receptors, possibly due to the contamination of a benzodiazepine, 1012-S, used for the bio-specific elution.

## Coupling of Cerebral GABA and Benzodiazepine Receptors

Recent study using a benzodiazepine affinity gel indicated that the interference by the benzodiazepine used for affinity elution on the [$^3$H]flunitrazepam binding was prevented by the use of an ion-exchange chromatography (26), although the recovery of [$^3$H]flunitrazepam binding was significantly lower than that of [$^3$H]muscimol binding. In this study, we have also employed the same ion-exchange chromatographic procedures to remove benzodiazepines used for bio-specific elution and have found a small amount of [$^3$H]flunitrazepam binding indeed occurs in the same fraction in which a highly purified GABA receptor protein is detected. To confirm these results, we have also examined the binding of both [$^3$H]muscimol and [$^3$H]flunitrazepam to various fractions obtained from the affinity column of 1012-S-acetamide adipic hydrazide Sepharose 4B following the nonspecific elution with sodium thiocyanate. In this case, bio-specific elution with 1012-S was not used.

As shown in Table 3, 10.9% of total [$^3$H]muscimol binding and 0.7% of total [$^3$H]flunitrazepam binding were nonspecifically eluted under these experimental conditions. The purification fold for each receptor protein obtained was 165.7 and 10.0, respectively.

The above results clearly indicate that cerebral GABA and benzodiazepine receptors reside on the same macromolecular complex which is readily solubilized by various detergents and purified by benzodiazepine affinity column chromatography. It has also been found that the per cent recovery of benzodiazepine receptors in the purified GABA receptor fraction is considerably less than that of GABA receptors. This suggests that some GABA receptors may not be associated with benzodiazepine receptors, although the possible presence of a differential effect of sodium thiocyanate on the two receptor sites cannot be ruled out. Considering the fact that cerebral benzodiazepine receptors can be classified into type 1 (GABA-independent type) and type 2 (GABA-dependent type) (33), nonspecific elution with sodium thiocyanate may remove both of them from the affinity gel.

Table 3. Binding of [$^3$H]Muscimol and [$^3$H]Flunitrazepam in Sodium Thiocyanate-Eluted Fractions from 1012-S Affinity Column.

| | Percent of Total Protein | $^3$H-Muscimol Binding (at 10 nM) | | $^3$H-Flunitrazepam Binding (at 1 nM) | |
|---|---|---|---|---|---|
| | | Purification Fold | Percent of Total Binding | Purification Fold | Percent of Total Binding |
| Solubilized fraction | 100.0 | 1.0 | 100.0 | 1.0 | 100.0 |
| Run-through | 102.1 | 0.2 | 17.9 | 0.1 | 1.3 |
| NaCl (50 mM) | 0 | 0 | 0 | 0 | 0 |
| NaSCN (1-2 M) | 0.1 | 165.7 | 10.9 | 10.0 | 0.7 |

## Molecular Characteristics of GABA and Benzodiazepine Receptors

SDS-polyacrylamide gel electrophoretic profiles of the purified GABA/benzodiazepine receptor complex eluted bio-specifically with 1012-S showed the existence of two major bands (Fig. 3).

One band had a molecular weight of approximately 55,000, while that of the other band was approximately 49,000. It was also found that the benzodiazepine binding sites identified by photoaffinity labeling with [$^3$H]flunitrazepam were exclusively localized in the latter band. In addition, it was found that the purified GABA/benzodiazepine receptor complex eluted nonspecifically with NaSCN contained four additional minor bands, possibly due to a contamination (Fig. 3).

The above results, coupled with the previous findings (17,22) that the GABA/benzodiazepine receptor complex solubilized from synaptic membrane fraction of rat brain with Nonidet P-40 had a molecular weight of 250,000-270,000, suggest that the GABA/benzodiazepine receptor complex from the rat brain may consist of two subunits having the molecular weights of 55,000 and 49,000, respectively. Biochemical and pharmacological characteristics of these two major bands, however, remain to be analyzed in future studies.

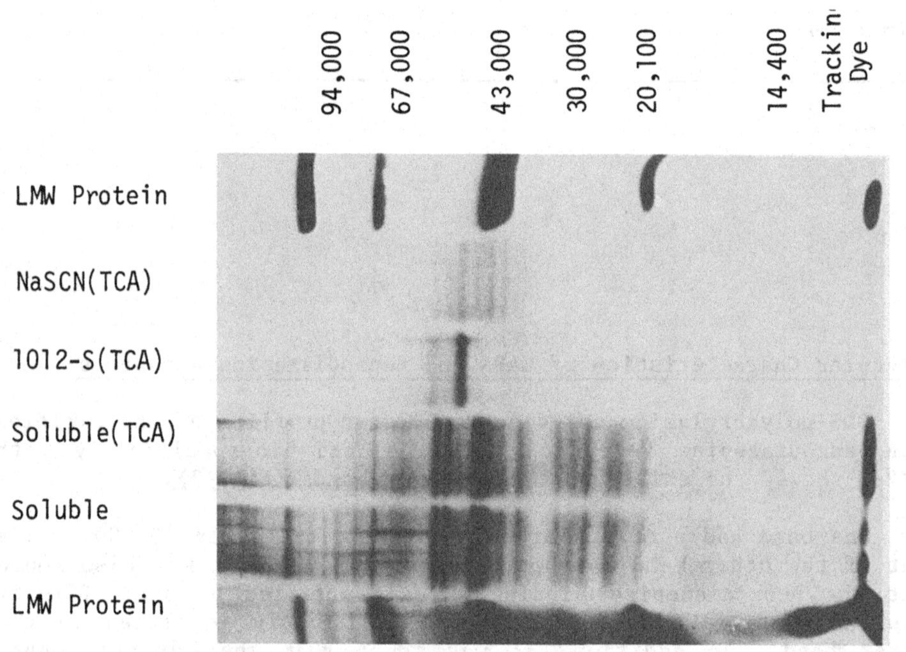


Fig. 3. SDS-polyacrylamide gel electrophoresis of solubilized and purified receptor fractions.

The experimental results obtained in this study have indicated that the GABA receptor of rat brain is readily purified by the use of a highly specific benzodiazepine affinity gel, 1012-S-acetamide adipic hydrazide Sepharose 4B. The purified GABA receptor is coupled, at least in part, with the benzodiazepine receptor. This GABA/benzodiazepine receptor complex seems to consist of two subunits (molecular weights: 55,000 and 49,000) as judged from the data obtained by SDS-polyacrylamide electrophoresis.

Further studies are obviously needed to clarify the biochemical and pharmacological characteristics of the purified GABA/ benzodiazepine receptor complex including those of subunit components of GABA receptor. Possible interactions of the purified GABA/benzodiazepine receptor complex with gated chloride ion channels should be also clarified before examining the function of the purified receptor at the membrane level.

## ACKNOWLEDGEMENTS

We would like to express our thanks to Drs. T. Okabayashi and K. Yamamoto, Shionogi Research Laboratories, Osaka, for kindly providing 1012-S used in this study. Their kind technical advice is also greatly appreciated.

The work was supported in part by Grant-in-Aid for Scientific Research (No 56480104, 1983) and Grant-in-Aid for Developmental Scientific Research (No 57870019, 1983) from the Ministry of Education, Science and Culture, Japan.

## REFERENCES

1. K. Kuriyama, B. Haber, B. Sisken, and E. Roberts, The γ-aminobutyric acid system in rabbit cerebellum, Proc. Natl. Acad. Sci. U.S.A. 55:846 (1966).
2. L. L. Iversen and F. E. Bloom, Studies on the uptake of [$^3$H]GABA and [$^3$H]glycine in slices and homogenates of rat brain and spinal cord by electron microscopic autoradiography, Brain Res. 41:131 (1972).
3. Y. Miyata and M. Otsuka, Distribution of γ-aminobutyric acid in cat spinal cord and the alteration produced by local ischemia, J. Neurochem. 19:1833 (1972).
4. Y. Yoneda and K. Kuriyama, A comparison of microdistribution of taurine and cysteine sulphinate decarboxylase activity with those of GABA and L-glutamate decarboxylase, J. Neurochem. 30:821 (1978).

5. E. A. Kravitz, P. B. Molinoff, and Z. W. Hall, A comparison of the enzymes and substrates of gamma-aminobutyric acid metabolism in lobster excitatory and inhibitory axons, Proc. Natl. Acad. Sci. U.S.A. 54:778 (1965).
6. M. Otsuka, L. L. Iversen, Z. W. Hall, and E. A. Kravitz, Release of gamma-aminobutyric acid from inhibitory nerves of lobster, Proc. Natl. Acad. Sci. U.S.A. 56:1110 (1966).
7. S. R. Zukin, A.B. Young, and S. H. Snyder, Gamma-aminobutyric acid binding to receptor sites in the rat central nervous system, Proc. Natl. Acad. Sci. U.S.A. 71:4802 (1974).
8. K. Beaumont, W. Chilton, H. I. Yamamura, and S. J. Enna, Muscimol binding in rat striatum: association with synaptic GABA receptors, Brain Res. 148:153 (1978).
9. S. R. Snodgrass, Use of [$^3$H]muscimol for GABA receptor studies, Nature (Lond.) 273:392 (1978).
10. H. Möhler and T. Okada, Benzodiazepine receptors: Demonstration in the central nervous system, Science 198:849 (1977).
11. C. Braestrup and R. F. Squires, Specific benzodiazepine receptors in rat brain characterized by high affinity [$^3$H]diazepam binding, Proc. Natl. Acad. Sci. U.S.A. 74:3805 (1977).
12. M. S. Briley and S. Z. Langer, Influence of GABA receptor agonists and antagonists on the binding of [$^3$H]diazepam to the benzodiazepine receptor, Eur. J. Pharmacol. 52:129 (1978).
13. G. J. Wastek, R. C. Speth, T. D. Reisine, and H. I. Yamamura, The effect of gamma-aminobutyric acid on $^3$H-flunitrazepam binding in the rat brain, Eur. J. Pharmacol. 50:445 (1978).
14. J. F. Tallman, J. W. Thomas, and D. W. Gallager, GABAergic modulation of benzodiazepine binding site sensitivity, Nature (Lond.) 274:383 (1978).
15. M. Karobath and G. Sperk, Stimulation of benzodiazepine receptor binding by γ-aminobutyric acid, Proc. Natl. Acad. Sci. U.S.A. 76:1004 (1979).
16. A. Guidotti, G. Toffano, and E. Costa, An endogenous protein modulates the affinity of GABA and benzodiazepine receptors in rat brain, Nature (Lond.) 275:553 (1978).
17. Y. Ito and K. Kuriyama, Some properties of solubilized GABA receptor, Brain Res. 236:351 (1982).
18. M. Gavish, R. S. L. Chang, and S. H. Snyder, Solubilization of histamine H-1, GABA and benzodiazepine receptors, Life Sci. 25:783 (1979).
19. T. Asano and N. Ogasawara, Soluble gamma-aminobutyric acid and benzodiazepine receptors from rat cerebral cortex, Life Sci. 29:193 (1981).
20. M. Gavish and S. H. Snyder, γ-Aminobutyric acid and benzodiazepine receptors: Copurification and characterization, Proc. Natl. Acad. Sci. U.S.A. 78:1939 (1981).

21. T. Asano, Y. Yamada, and N. Ogasawara, Soluble GABA/benzodiazepine receptors from bovine cerebral cortex, in: "Problems in GABA Research," Y. Okada and E. Roberts, eds., Excerpta Medica, Amsterdam-Oxford-Princeton (1982).
22. Y. Ito, K. Kuriyama, E. Ueno, C. Nishimura, and Y. Yoneda, Solubilization and partial purification of cerebral GABA receptors, in: "Problems in GABA Research," Y. Okada and E. Roberts, eds., Excerpta Medica, Amsterdam-Oxford-Princeton (1982).
23. F. A. Stephenson and R. W. Olsen, Solubilization by CHAPS detergent of barbiturate-enhanced benzodiazepine-GABA receptor complex, J. Neurochem. 39:1579 (1982).
24. F. A. Stephenson, A. E. Watkins, and R. W. Olsen, Physicochemical characterization of detergent-solubilized γ-aminobutyric acid and benzodiazepine receptor proteins from bovine brain, Eur. J. Biochem. 123:291 (1982).
25. T. Asano, Y. Yamada, and N. Ogasawara, Characterization of the solubilized GABA and benzodiazepine receptors from various regions of bovine brain, J. Neurochem. 40:209 (1983).
26. E. Sigel, F. A. Stephenson, C. Mamalaki, and E. A Barnard, A γ-aminobutyric acid/benzodiazepine receptor complex of bovine cerebral cortex: Purification and partial characterization, J. Biol. Chem. 258:6965 (1983).
27. P. Cuatrecasas, Protein purification by affinity chromatography: Derivations of agarose and polyacrylamide beads, J. Biol. Chem. 245:3059 (1970).
28. U.K. Laemmli, Cleavage of structural proteins during the assembly of the head of bacteriophage T4, Nature (Lond.) 227:680 (1970).
29. O. H. Lowry, N. J. Rosebrough, A. L. Farr, and R. J. Randall, Protein measurement with the Folin phenol reagent, J. Biol. Chem. 183:265 (1951).
30. M. M. Bradford, A rapid and sensitive method for the quantitation of microgram quantities of protein utilizing the principle of protein-dye binding, Anal. Biochem. 72:248 (1976).
31. W. Sieghart and M. Karobath, Molecular heterogeneity of benzodiazepine receptors, Nature (Lond.) 286:285 (1980).
32. K. Kuriyama and Y. Ito, Some characteristics of solubilized and partially purified cerebral GABA and benzodiazepine receptors, in: "CNS Receptors - From Molecular Pharmacology to Behavior," P. Mandel and F. V. DeFeudis, eds., Raven Press, New York, p. 59 (1983).
33. C. A Klepner, A. S. Lippa, D. I. Benson, M C. Sano, and B. Beer, Resolution of two biochemically and pharmacologically distinct benzodiazepine receptors, Pharmacol. Biochem. Behav. 11:457 (1979).

# STRUCTURE AND PROPERTIES OF THE BRAIN GABA/BENZODIAZEPINE RECEPTOR COMPLEX

Eric A. Barnard, F. Anne Stephenson, Erwin Sigel, Cleanthi Mamalaki and Graeme Bilbe

Department of Biochemistry, Imperial College of Science and Technology, London SW7 2AZ, England

Andrew Constanti, Trevor G. Smart and David A. Brown

Department of Pharmacology, The School of Pharmacy University of London, London, WC1N 1AX, England

Two classes of receptor for γ-aminobutyrate (GABA) in the brain have been clearly defined, although there are hints of the existence of others. The first is the GABA-A receptor, at which benzodiazepines potentiate the electrophysiological activity of GABA, bicuculline is a strong antagonist and muscimol, isoguvacine and certain other ligands are strong agonists. The second is the GABA-B receptor (1) which has none of the properties just mentioned but which, in contrast to the GABA-A receptor, requires $Ca^{2+}$ for the binding of GABA and has baclofen as a characteristic agonist. The GABA-A receptor, exclusively, will be dealt with here.

At this receptor GABA controls the opening of a channel for chloride ions. The response to stimulation shows no appreciable lag and typical rapidly-gated channels are opened, with no evidence for involvement of any second messenger system (2-4); hence all the available evidence indicates that the binding of a GABA agonist to this receptor opens a chloride channel in the receptor complex itself. From a variety of pharmacological and ligand binding studies it has been concluded (as reviewed by Olsen, 5) that this receptor complex in situ contains binding sites for GABA agonists, benzodiazepines (BZ), β-carbolines, barbiturates and picrotoxin, and also a chloride channel.

The β-carboline site is closely related to the BZ site and they may partly overlap: each of these types of drug can displace the binding of the other. However, this interaction, e.g. between the flunitrazepam and ethyl-β-carboline 3-carboxylate binding activity, is not a simple competitive one (6). This and other evidence (7) suggests that the BZ and β-carboline binding sites are not identical.

Another site on this receptor complex is the barbiturate/picrotoxin site. Anesthetic barbiturates potentiate GABA post-synaptic responses and, in fact, at the ion channel level it is now known that they prolong the lifetime of the channel opened by GABA (3). They also enhance BZ binding and low-affinity GABA binding (8). These enhancement effects are blocked competitively by picrotoxinin, are stereospecific and are dependent on the presence of an anion which can permeate the receptor channel. All of the available evidence (reviewed by Olsen, 5; see also Olsen et al., this volume) indicates that these barbiturates act as agonists and picrotoxin as antagonist, at a site (distinct from the GABA or BZ sites) which regulates the ion channel. Bicyclophosphate cage convulsants act like picrotoxin but more potently and with higher specificity, and a $^{35}$S-labelled analogue, t-butyl bicyclophosphorothionate (TBPS), is a superior ligand for this site (9). In rat brain membranes its binding (competitive with picrotoxin and with barbiturates and requiring the specific anions) gives a linear Scatchard plot, with one set of sites (at least up to 20 nM ligand concentration) being found (9), with $K_d$ = 20 nM at 21° C, as we have confirmed on our membrane preparations.

Since barbiturates bound at this site lengthen the open channel lifetime and since TBPS competitively blocks their effect, we have employed the specific binding of [$^{35}$S]TBPS and the effect of barbiturates as probes for this regulatory site related to the anion channel. We report here on the receptor complex as an isolated protein, in which the GABA-binding site can be probed by the binding of the agonist [$^{3}$H]muscimol, and the two types of regulatory site by (i) the binding of [$^{3}$H]flunitrazepam or (ii) by the binding of [$^{35}$S]TBPS and the effect of a barbiturate.

## PURIFICATION OF THE RECEPTOR AND RETENTION OF BINDING SITES

Initially we used bovine brain cortex extracted in sodium deoxycholate. The protease inhibitors phenylmethane sulphonyl fluoride (0.3 mM), benzamidine (1 mM), soybean trypsin inhibitor (10 mg/L), eggwhite ovomucoid/ovoinhibitor (10 mg/L) and EDTA (1 mM) were present, together with 300 mM sucrose, in the extraction stages. Column affinity chromatography using an immobilized BZ formed the main purification stage. The BZ Ro 7-1986/1 (generously provided by the Hofmann LaRoche Co.) was bound to agarose via a spacer of 14 atoms length (10), with the features imposed that a positive charge is maintained therein and that the bound ligand concentration is low, features which reduce nonspecific hydrophobic binding to the column. Extensive washing in a medium including 0.2% Triton X-100, 2 mM $Mg^{2+}$ and 10% sucrose removed much protein but no receptor. A water-soluble BZ, chlorazepate (10 mM), applied in the same medium, eluted the receptor (11). Ion-exchange chromatography on DEAE-Sephacel removed all of the chlorazepate. The product showed receptor activity in binding both [$^{3}$H]muscimol and [$^{3}$H]flunitrazepam (Fig. 1; Table 1). The fact that affinity purification on a BZ column yielded a purified muscimol-binding protein is particularly significant, showing that the GABA and BZ sites are in one molecular complex.

This purified receptor also could undergo the photo-affinity reaction, which in the membrane-bound state incorporates [$^{3}$H]flunitrazepam covalently into the receptor protein when UV-illuminated (12). It also showed (Fig. 2) displacement of muscimol binding by a characteristic GABA antagonist, bicuculline methobromide, and the enhancement of this effect by the thiocyanate anion which is known (13) to occur with GABA antagonists in the membrane-bound receptor. Further, a β-carboline binds to the pure receptor with a single high affinity, and at about 1:1 stoichiometry with [$^{3}$H]flunitrazepam (Fig. 3). The antagonist-type BZ ligand, Ro 15-1788 (14), also binds similarly (Table 1). The pharmacological specificity, tested with a series of relevant ligands (Table 1), is similar to that seen in the corresponding membrane-bound state (15) or in the crude solubilized (16,17) state.

However, three properties which monitor regulatory interactions in the complex, as discussed above, are lost in this form of the pure receptor. These are the binding of the cage convulsant ligand [$^{35}$S]TBPS, the stimulation of BZ binding by GABA agonists and the stimulation of GABA and BZ binding by pentobarbital. The receptor in this preparation is present finally in a medium containing 0.2% Triton X-100 as the detergent: if 0.5% deoxycholate is maintained throughout, instead, the pure receptor is very unstable and rapidly loses all of its binding properties. In contrast, the binding activities shown in Table 1 are stable if the detergent used from the affinity column elution onwards is 0.2% Triton X-100. However, the GABA enhancement of BZ binding is detectable only at a low level in the crude solubilized state if Triton is used in the extraction, and it is lost in the pure state in Triton medium whatever the initial extraction procedure. Likewise the other regulatory effects mentioned are absent in the receptor purified in these deoxycholate and Triton media.

Table 1. The pharmacological specificity of the binding sites of the GABA/benzodiazepine receptor purified from bovine cortex. (For experimental details, see Sigel et al. (11)).

| Ligand | $K_d$ or $K_i$ (nM) | Bmax nmol/mg protein) |
|---|---|---|
| [$^3$H]Muscimol | 12 | 4.3 |
| GABA | 50 | |
| 3 Aminopropanesulphonic acid | 200 | |
| L-2,4, Diaminobutyric acid | >$10^4$ | |
| Baclofen (with 2 mM $Ca^{2+}$) | >$10^4$ | |
| [$^3$H]Flunitrazepam | 10 | 1.2 |
| [$^3$H]Ro15 1788 | 9 | 1.2 |
| [$^3$H]Propyl -β-carboline 3-carboxylate | 20 | 1.0 |
| Clonazepam | 4 | |
| Ro7 1986/1 | 8 | |
| Diazepam | 40 | |
| Ro5 4864 | > $10^3$ | |
| Ethyl-β-carboline 3-carboxylate | 25 | |

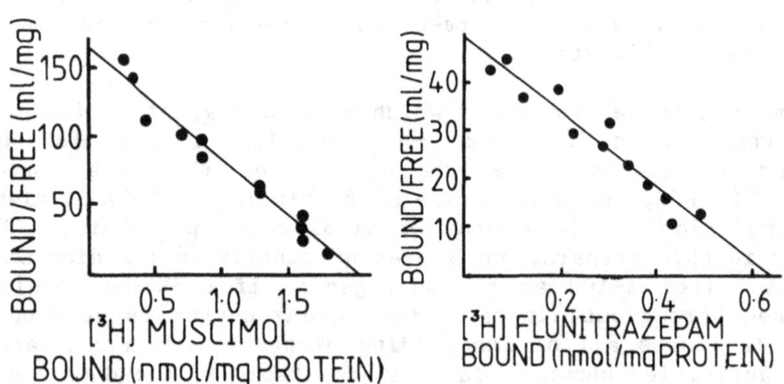


Fig. 1. Scatchard plots of the binding of [$^3$H]flunitrazepam and [$^3$H]muscimol to the purified GABA/benzodiazepine receptor complex. Each binding was measured by the polyethylene glycol (PEG) precipitation/filtration method. Non-specific binding was measured in the presence of 30 μM unlabelled flunitrazepam or muscimol, respectively, and has been subtracted.

Table 2. Stimulation of [$^3$H] flunitrazepam binding by GABA and by pentobarbital in the CHAPS-purified receptor.

| Addition | Concentration | Change in binding |
|---|---|---|
| GABA | $10^{-6}$M | +9% |
| | $10^{-5}$M | +23% |
| pentobarbital | 2 x $10^{-4}$M | +9% |
| | $10^{-3}$M | +20% |

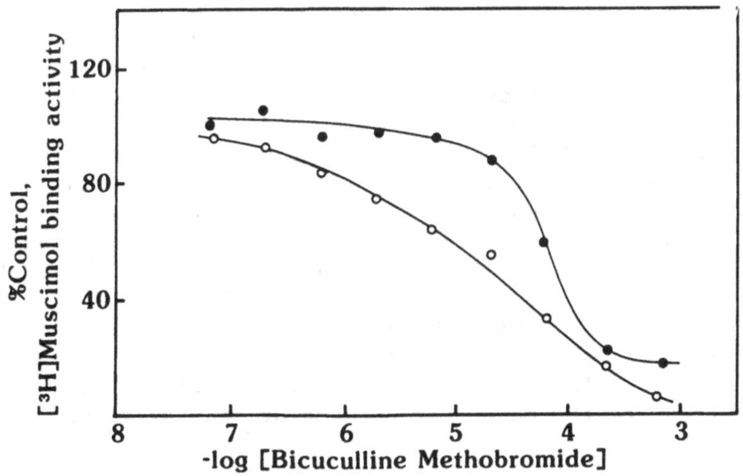


Fig. 2. Inhibition of the muscimol binding of the purified receptor by bicuculline methobromide. [$^3$H]Muscimol binding activity was measured (by a PEG filtration assay) in the presence of increasing concentrations of bicuculline methobromide, in the absence (•) or the presence (o) of 50 mM sodium thiocyanate. The $K_i$ value (determined from the $IC_{50}$ value) is 24 μM in the absence, and 4.4 μM in the presence, of 50 mM sodium thiocyanate (mean of two determinations).

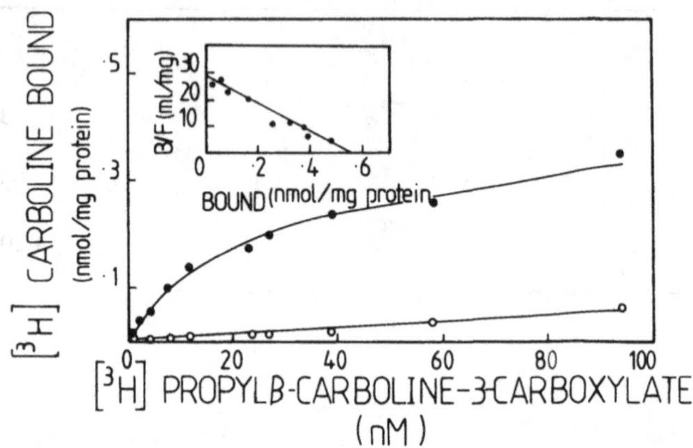


Fig. 3. Binding (by PEG filtration assay) of [$^{3}$H]propyl-β-carboline 3-carboxylate to the isolated GABA/benzodiazepine receptor complex. Non-specific binding (o) was measured in the presence of 10 μM ethyl-β-carboline 3-carboxylate. •, specific binding. The inset shows a Scatchard plot of the saturation curve. A typical best-fit plot showed a single high affinity site, $K_d$= 20 nM and Bmax = 0.56 nmol sites/mg protein. This corresponds to 78% of the total [$^{3}$H]-labelled flunitrazepam binding sites of the purified receptor which were determined simultaneously as in Fig. 1. Similarly, the antagonist Ro 15-1788 also showed saturable binding to the purified complex, and a best-fit Scatchard plot showed a single high affinity site, $K_d$ = 9 nM and Bmax = 0.71 nmol specific binding sites/mg protein. Thus the number of agonist binding sites (i.e. for [$^{3}$H]flunitrazepam) is equal to the number of antagonist binding sites.

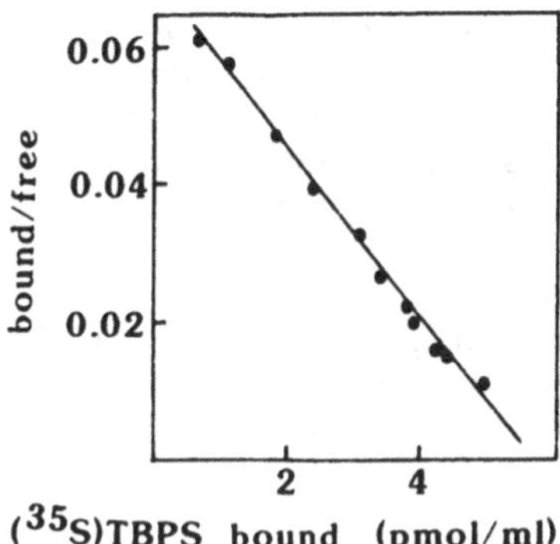


Fig. 4. Scatchard plot of the [$^{35}$S]TBPS binding to the CHAPS purified receptor. The purified receptor was incubated (90 min, 30° C) with [$^{35}$S]TBPS in a medium containing 20 mM K phosphate (pH 7.4), 200 mM KCl, 0.1 mM EDTA and 0.6% CHAPS. Ligand binding was measured by PEG precipitation/filtration. Nonspecific binding was determined in the presence of 500 μM picrotoxin and was, at its maximum, 24% of the total binding.

## MODIFIED PURIFICATION PROCEDURE

An alternative procedure was, therefore, adopted. The zwitterionic detergent CHAPS (18), which has been found satisfactory for retaining these receptor sites in a crude extract (19), was employed. 1.5% CHAPS/0.15% Asolectin (i.e., soybean lipids) was used instead of deoxycholate for the extraction stage, and 0.6% CHAPS/0.06% Asolectin replaced 0.2% Triton in the subsequent stages. Otherwise the purification procedures were as described above. Extraction of the receptor in CHAPS was only about half as efficient as in deoxycholate, but the purification was as high as before.

The CHAPS-purified receptor binds flunitrazepam as before, the Scatchard plot (for binding in 0.3% CHAPS medium) being essentially as in Fig. 1. It also binds muscimol, the Scatchard plot (for binding in 0.1% Triton medium) being again linear, but with Bmax about one-half of that found in the previous purification. However, the stimulations of BZ binding by GABA and by pentobarbital were consistently found in the receptor when purified and assayed in CHAPS medium (Table 2). Their magnitudes are lower than is seen in the original membranes, or in a crude extract in CHAPS medium (19). We do not know if this represents a partial inactivation of the regulatory linkage during the purification procedure.

The CHAPS-purified receptor also binds the cage convulsant ligand TBPS, at a single high-affinity site (Fig. 4). Hence all of the sites detectable on the GABA receptor complex in the membrane are in this case still present in the purified protein.

## CHEMICAL AND PHYSICAL PROPERTIES

The purified receptor was shown to be a single protein species by isoelectric focussing (Fig. 5). In SDS gel electrophoresis it showed 2 subunits at 53,000 (α) and 57,000 (β) daltons (Fig. 6). However, two other faint bands of about 62,000 and 65,000 daltons could sometimes be discerned and it is possible that these represent other subunits which have been proteolytically nicked during the preparation, remaining present up to the stage of SDS denaturation when they are largely lost as fragments. The same pattern was seen in the deoxycholate/Triton and the CHAPS preparations. The subunit composition of the purified GABA/BZ receptor is not yet firmly established.

In photo-affinity labelling by [$^3$H]flunitrazepam, the α-subunit reacts very strongly (as it does in the membranes). The β-subunit also labels, but to a much lesser extent; whether this labelling is equally significant has not yet been determined.

In ultracentrifugation on a sucrose density gradient, the purified receptor (obtained by either method and analyzed in the appropriate detergent) was heterogeneous; the main peak was broad, centered at about 11 S (11). Similar physical heterogeneity, with the main peak in the same range, has been reported in 0.5% Triton X-100/0.1 mM mercaptoethanol medium for the unpurified receptor (21,22). In gel filtration an equivalent heterogeneity was found in the purified receptor (11). However, this descrepancy between the hydrodynamic heterogeneity and the isoelectric homogeneity of the pure receptor disappeared when the ionic strength of the medium was raised to 0.5 M or above; the receptor then migrated as a single peak (Fig. 7). This effect was ionic strength dependent and is

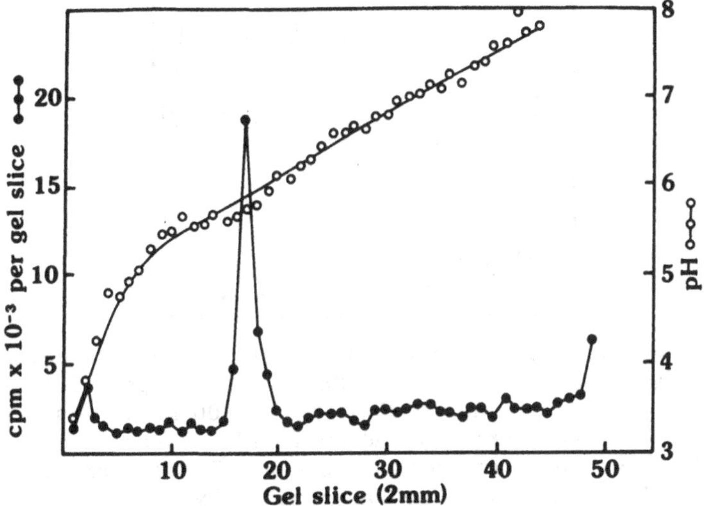


Fig. 5. Purity of the purified receptor as demonstrated by isoelectric focussing after [$^{125}$I]labelling. The receptor, fully purified in deoxycholate/Triton media (11), was iodinated to a specific radioactivity of 1750 Ci/mmol muscimol binding sites (mean value) by the mild chloramine-T method of Froehner et al. (20). Disc gels (10 X 0.5 cm) were prepared from 4% acrylamide and contained 1% (v/v) Triton X-100 with 0.05 (v/v) Ampholines (pH 4-6.5 and pH 6.5-9, from Pharmacia). Gels were pre-run at 1 mA/gel for 20 min at 4° C with 10 mM $H_3PO_4$ (pH 2.0)/0.2% Triton X-100 in the anode chamber and 20 mM NaOH (pH 11.0)/0.2% Triton X-100 in the cathode chamber. Samples (20 µl [$^{125}$I] receptor or 1 mg myoglobin) were focussed for 5 h at 4° C at a constant voltage of 300 V. Gels were sliced at 2 mm intervals and the discs were either suspended in degassed 10 mM KCl and the pH measured or were counted in a gamma-counter. It is seen that all of the protein present focussed as a single sharp symmetrical band, with pI 5.6 (mean of 2 determinations). The radioactivity that migrated to the cathode was shown to be only iodinated detergent.

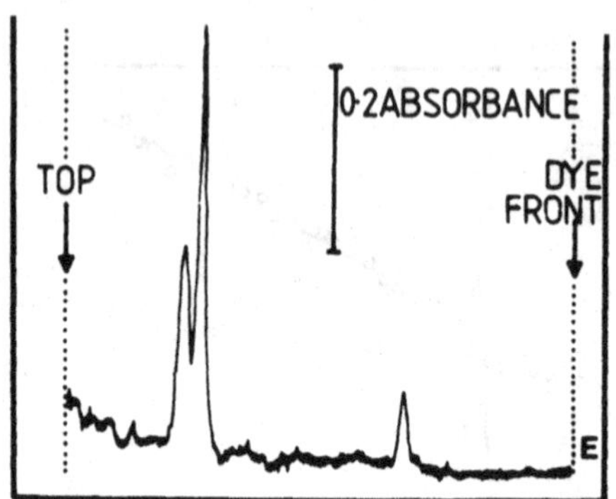


Fig. 6. Polyacrylamide gel electrophoresis in the presence of sodium dodecyl sulphate (SDS). The receptor fully purified in deoxycholate and 0.2% Triton X-100 media was denatured in SDS/dithiothreitol and analyzed. A scan at 595 nm after Coomassie blue staining is shown. Details are given in Sigel et al. (11). By calibration by a set of standards run in parallel, the major band (α) is at 53,000 $M_r$ and the overlapping band (β) is at 57,000 $M_r$. (the very minor band at 21,500 is added soybean trypsin inhibitor).

interpreted as due to a reversible aggregation of the receptor molecules in detergent media at low ionic strength, the inter-molecular attraction being weakened by salt effects. Some residual aggregation is still detectable on the heavy side of the peak (Fig. 7). Likewise in gel filtration (on an Ultrogel (LKB) AcA-22 column) the purified receptor was heterogeneous in low-salt media of either detergent but much more homogeneous in 1 M NaCl in either detergent, with a Stokes' radius then of 7.3 nm.

In all conditions of sedimentation or gel filtration, the muscimol binding and the BZ binding co-migrated, as is also illustrated in Fig. 7. This is further evidence that in the purified state a single protein complex bears both binding sites.

The apparent sedimentation constant of the non-aggregated pure receptor in 0.5% Triton X-100/1 M NaCl (Fig. 7) is 7.4 S. This value has been considerably depressed by the buoyancy effect of the binding of detergent to the protein, which is enhanced in the high-salt medium. Such a buoyancy distortion can be corrected by increasing the density of the medium using $D_2O$ (method reviewed by Barnard, 23). By means of parallel sedimentations in the 0.5% Triton/1 M NaCl medium in either $H_2O$ or $D_2O$, the true sedimentation constant of the receptor-detergent complex in that medium was found to be 8.46 S. Taking the partial specific volume of the protein itself as 0.730, and using also the Stokes' radius of 7.3 nm obtained in the same conditions, a value of the protein molecular weight has been obtained of 230,000.

This corrected molecular weight of the receptor can be compared with the size estimated for it in the membrane by the method of radiation inactivation. This method has been applied in an improved form (25) in which internal markers of known molecular weight are used. Brain membranes were irradiated and the losses of binding sites for muscimol, BZ and β-carboline were monitored throughout. It was found that all three activities decayed together (24). The rate of decay corresponded to a target size of 220,000 daltons. These findings, and the agreement of that size with the molecular weight found for the pure receptor protein, are interpreted as meaning that a single complex in the membrane carries all of the sites of this receptor, and remains as a single entity upon extraction and purification.

## THE MESSENGER RNA FOR THE RECEPTOR COMPLEX

As the first stage in the application of molecular genetics to this receptor, the mRNAs coding for it were identified. As with the nicotinic ACh receptor (26), mRNA for such a membrane receptor protein cannot be fully processed and translated in a conventional cell-free system. Translation was achieved by micro-injection of mRNA (extracted from chick or rat brain and isolated as poly $(A)^+$ mRNA) into _Xenopus_ oocytes (Fig. 8). As found first there with mRNA coding for the nicotinic ACh receptor (27), so here a production from the foreign mRNA of the receptor/ion channel complex was observed. GABA receptors are absent in the control oocyte, but they develop soon after the injection of the appropriate mRNA. Application of GABA to the exterior of the injected oocyte produces an increase in chloride conductance. This increase is dose-dependent (Fig. 8). The dose-conductance curve (Fig. 8) is the same as that for GABAergic brain neurons _in situ_. The response is found with mRNA extracted from the brain of newly-hatched chicks (Fig. 8) and likewise from the brain of rats about 14 days after birth (Fig. 9),

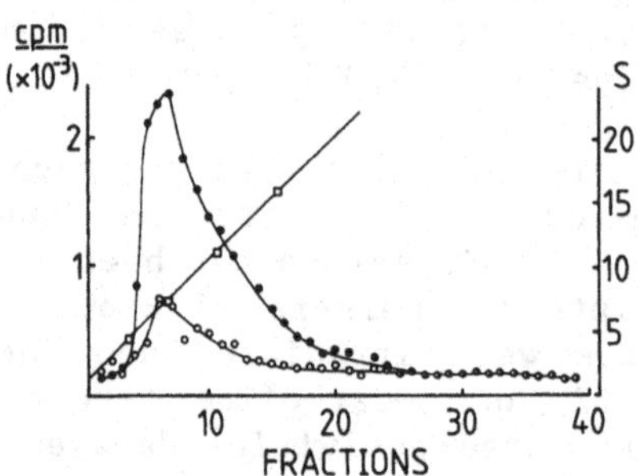


Fig. 7. Sucrose density gradient centrifugation of the purified receptor complex. The receptor was purified (Sigel et al., 11) in the deoxycholate/Triton media (using bovine brain membranes which had been pre-extracted with 0.05% Triton X-100 as described by Chang and Barnard (24)). The sample and the gradient were in 0.5% Triton X-100 1 M NaCl/0.02 M K phosphate (pH 7.4)/0.02% $NaN_3$ /0.001 M 2-mercaptoethanol. In the case illustrated $D_2O$ was used (replacing $H_2O$ in all the solutions). Other methods were as given by Sigel et al. (11). Each fraction was assayed for muscimol (o) or flunitrazepam (•) specific binding activity: note the coincidence throughout of these two activities. A plot is superimposed ( □ ), right-hand scale) of the calibration by marker proteins, these being (from left to right) haemoglobin, yeast alcohol dehydrogenase, catalase and β-galactosidase.

corresponding to the ontogeny of the GABA-A/BZ system.

The receptor produced in the cell membrane of the oocyte shows the characteristic pharmacological specificity of the GABA-A receptor in the brain neuron. Muscimol is more potent than GABA (Fig. 9c). Other GABA agonists show the predicted series of potency (28). Desensitization is observed in the continued presence of 40 μM GABA (Fig. 8b). Bicuculline shows antagonism to GABA (Fig. 10d). The regulatory effects of BZ and barbiturate ligands are also exerted (Fig. 10a,b). Picrotoxin blocks the new channels, but only when opened by a GABA agonist (Fig. 10c). All of these effects are exhibited at the same drug concentrations as are active on neuronal GABA-A receptors.

These findings demonstrate that mRNA species from brain can be translated to form polypeptides which are assembled to form the GABA receptor in the cell membrane, with all of the functional sites and the chloride channel intact and correctly oriented. The fact that all of these properties are developed together in this system further supports the view that they are present on a single macromolecular complex, which is independent of the type of membrane in which it is present.

## MOLECULAR GENETIC APPROACH TO ANALYSIS OF THE GABA RECEPTOR

One application of this receptor mRNA translation in the oocyte will be the fuller study of the ion channel of this receptor in an accessible form and isolated from other influences in neuronal preparations. Another will be the study of the biosynthesis of the receptor and the requirements for formation of the ion channel.

However, a wider perspective is opened up by these findings. The identification of mRNA species for this receptor will allow the formation of cDNAs coding for it. These can be recognized either in an expression system using anti-receptor antibodies or by hybridization with complementary oligonucleotide probes. The latter can be constructed when the subunits, separated as in Fig. 6, are subjected to N-terminal micro-sequencing. The former is feasible since specific antibodies have been prepared against the receptor and its subunits (Fig. 11). The application of these techniques should in time lead to the cloning of cDNAs for the α and β subunits, the determination of their entire amino acid sequences, and of the full subunit complement of the receptor complex. The isolation of the receptor protein has been a necessary first stage in the application to this system of the powerful methods of modern molecular biology.

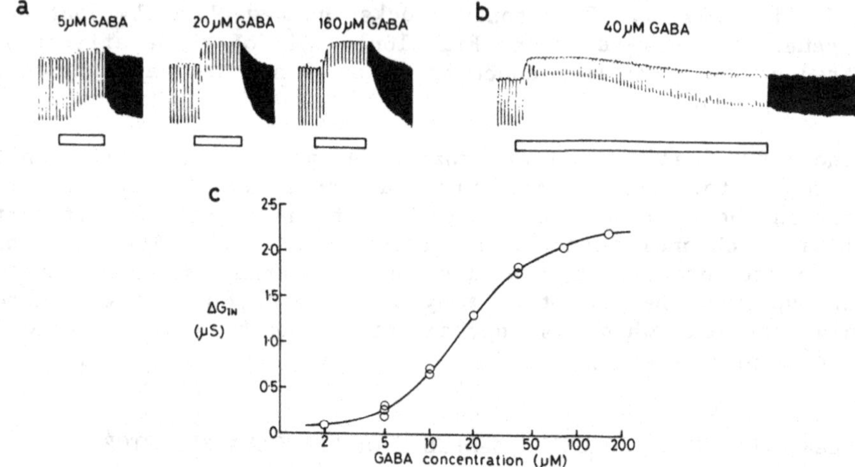


Fig. 8. GABA receptors induced in Xenopus oocyte by the microinjection of brain mRNA. a: membrane conductance changes evoked by bath-applied GABA in a single Xenopus oocyte, recorded 1 day after injecting chick brain mRNA. Downward pen deflections are electrotonic potentials evoked by intracellularly applied hyperpolarizing current pulses (40 nA, 1 s, 0.2 Hz). Upward baseline shifts are membrane depolarizations produced by GABA (applied during periods indicated by bars). Note chart speed was slower during the offset of the responses. b; fading response produced by a 5 min application of 40 μM GABA in the same cell. c: corresponding GABA log-dose/conductance relationship. Ordinate, change in input conductance (μS); abscissa, GABA concentration (μM). From Smart et al. (28). Voltage scale is as shown in Fig. 10.

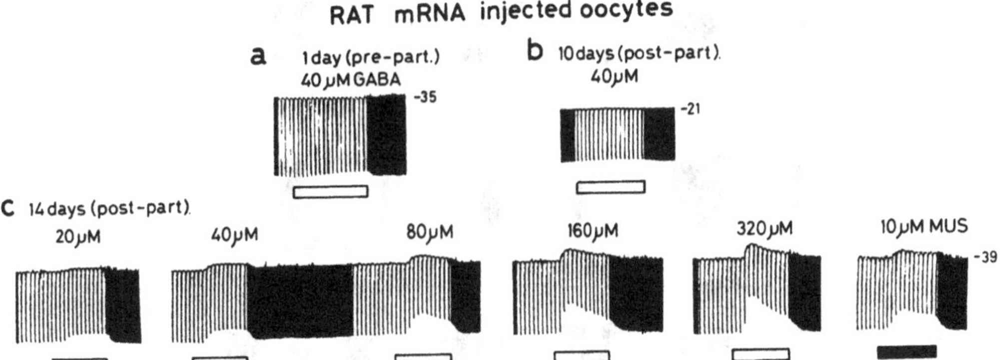


Fig. 9. The GABA-dependent conductance change in oocytes injected with mRNA extracted from the whole rat brain. a: from rats 1 day preparturition. b: 10 days postparturition, with dose-dependent responses to the GABA concentrations shown and to muscimol (MUS). Note that only the oocytes injected with the 14-day brain mRNA exhibit a significant response to GABA. Note also that muscimol (applied at 10 μM, filled bar) is more potent than GABA, as in the rat brain, and that the conductance change desensitizes at 80 μM GABA concentration and at 10 μM muscimol. (Scales and other details as in Fig. 10).

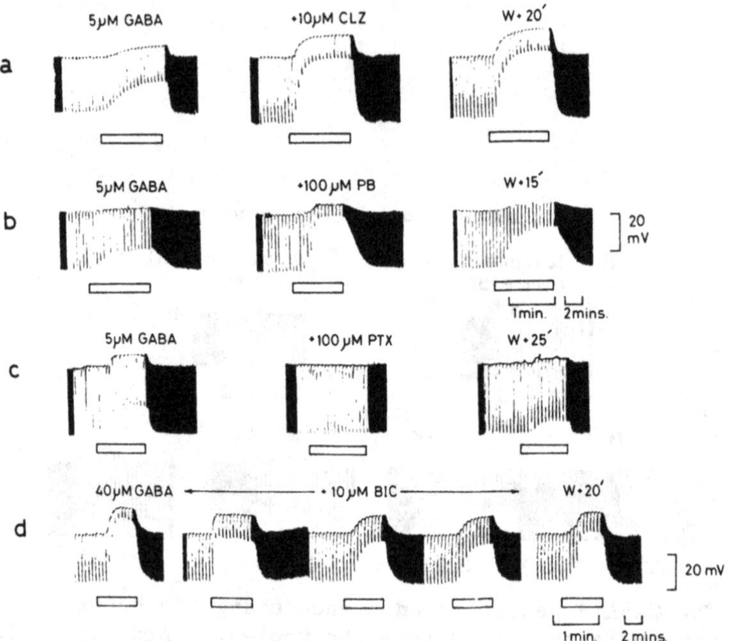


Fig. 10. Enhancement and blockade of GABA-evoked conductance change in chick mRNA-injected oocytes. a: responses to GABA recorded in control solution, after 12 min in 10 μM chlorazepate (CLZ), and then after 20 min wash (W) in normal Ringer solution. b: GABA response in control solution, after 10 min in pentobarbital (PB), then after a 15 min wash. Note partial recovery of GABA response after PB but not after 10 μM CLZ exposure. c: blockade by 100 μM picrotoxin (PTX) shows reversal occurs on washing (W). d: partial blockade by a sub-optimal dose of bicuculline methobromide (BIC). Full recovery occurs after washing. In all cases, the open bars indicate the periods during which GABA (at the concentration shown) was applied. Other details as in Fig. 8.

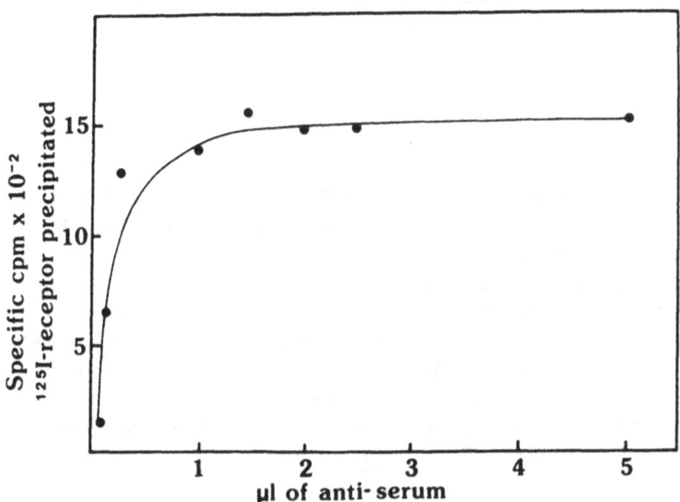


Fig. 11. Immuno-precipitation of the purified [$^{125}$I]labelled receptor. Polyclonal antibodies were raised by the immunisation of rabbits with the receptor (approximately 30 μg protein) purified from bovine brain in the deoxycholate/Triton media. The primary immunisations were in complete Freund's adjuvant at six sites intradermally. All subsequent injections were in Freund's incomplete adjuvant at six sites subcutaneously. For immunoassay, iodinated pure receptor (prepared as noted for Fig. 5: 43 pM) was incubated for 5h at 4° C with increasing amounts of anti-serum, diluted in a pre-immune rabbit serum to give a fixed amount of serum in the assay. Anti-(rabbit IgG Fc), 10 μl, was added and the [$^{125}$I]labelled precipitate collected and counted. Maximum precipitation of the antigen is seen at 1.5 μl of added anti-serum.

ACKNOWLEDGEMENTS

This work was conducted by the Medical Research Council (MRC) Molecular Neurobiology Research Group at Imperial College. The work at the School of Pharmacy was also supported by the MRC. We also thank Dr. B.M. Richards (G.D. Searle Company) for help and encouragement. F.A.S. holds a Royal Society Research Fellowship, E.S. held an EMBO Fellowship and G.B holds an SERC postgraduate studentship.

REFERENCES

1. D. R. Hill and N. G. Bowery, [$^3$H]Baclofen and [$^3$H]GABA bind to bicuculline-insensitive GABA sites in rat brain, Nature 290:149 (1981).
2. A. Nistri, A. Constanti, and K. Krnjević, Electrophysiological studies of the mode of action of GABA on vertebrate central neurons, Adv. Biochem. Psychopharmacol. 21:81 (1980).
3. R. E. Study and J. L. Barker, Diazepam and (-)pentobarbital: Fluctuation analysis reveals different mechanisms for potentiation of γ-aminobutyric acid responses in cultured central neurons, Proc. Natl. Acad. Sci. 78:7180 (1981).
4. W. Haefely and P. Polc, Electrophysiological studies on the interaction of anxiolytic drugs with GABAergic mechanisms, in: "Anxiolytics: Neurochemical, Behavioral and Clinical Perspectives," J. B. Malick, S. J. Enna and H.I. Yamamura, eds., p. 113, Raven Press, New York (1983).
5. R. W. Olsen, Drug interactions at the GABA receptor-ionophore complex, Ann. Rev. Pharmacol. Toxicol. 22:245 (1982).
6. M. Nielsen and C. Braestrup, Ethyl-β-carboline 3-carboxylate shows differential benzodiazepine receptor interaction, Nature 286:606 (1980).
7. C. L. Brown and I. L. Martin, Photoaffinity labelling of the benzodiazepine receptor cannot be used to predict ligand efficacy, Neurosci. Lett. 35:37 (1983).
8. R. W. Olsen and A. M. Snowman, Chloride-dependent enhancement by barbiturates of GABA receptor binding, J. Neurosci. 2:1812 (1982).
9. R. F. Squires J. E. Casida, M. Richardson, and E. Saederup, [$^{35}$S]-Butyl bicyclophosphorothionate binds with high affinity to brain-specific sites coupled to γ-aminobutyric acid-A and ion recognition sites, Mol. Pharmacol. 23:326 (1983).
10. E. Sigel, C. Mamalaki, and E. A. Barnard, Isolation of a GABA receptor from bovine brain using a benzodiazepine affinity column, FEBS Lett. 147:45 (1982).

11. E. Sigel, F. A. Stephenson, C. Mamalaki, and E. A. Barnard, A γ-aminobutyric acid/benzodiazepine receptor complex of bovine cerebral cortex. Purification and partial characterization, J. Biol. Chem. 258:6965 (1983).
12. H. Möhler, M. K. Battersby, and J. G. Richards, Benzodiazepine receptor protein identified and visualized in brain tissue by a photoaffinity label, Proc. Natl. Acad. Sci. USA 77:1666 (1980).
13. S. J. Enna and S. H. Snyder, Influence of ions, enzymes and detergents on γ-aminobutyric acid-receptor binding in synaptic membranes of rat brain, Mol. Pharmacol. 13:442 (1977).
14. H. Möhler and J. G. Richards, Agonist and antagonist benzodiazepine receptor interaction in vitro, Nature 294:763 (1981).
15. C. Braestrup and R. F. Squires, Specific benzodiazepine receptors in rat brain characterized by high affinity [$^3$H]diazepam binding, Proc. Natl. Acad. Sci. USA 74:3805 (1977).
16. D. V. Greenlee and R. W. Olsen, Solubilization of γ-aminobutyric acid receptor protein from mammalian brain, Biochem. Biophys. Res. Commun. 88:380 (1979).
17. R. Sherman-Gold and Y. Dudai, Solubilization and properties of a benzodiazepine receptor from calf cortex, Brain Res. 198:485 (1980).
18. W. F. Simonds, G. Kuski, R. A. Streaty, L. M. Hjelmeland, and W. A. Klee, Solubilization of active opiate receptors, Proc. Natl. Acad. Sci. USA 77:4623 (1980).
19. F. A. Stephenson and R. W. Olsen, Solubilization by CHAPS detergent of barbiturate-enhanced benzodiazepine-GABA receptor complex, J. Neurochem. 39:1579 (1982).
20. S. C. Froehner, C. T. Reiness, and Z. W. Hall, Subunit structure of the acetylcholine receptor from denervated rat skeletal muscle, J. Biol. Chem. 252:8589 (1977).
21. F. A. Stephenson, A. E. Watkins, and R. W. Olsen, Physicochemical characterization of detergent-solubilized γ-aminobutyric acid and benzodiazepine receptor proteins from bovine brain, Eur. J. Biochem. 123:291 (1982).
22. T. Asano and N. Ogasawara, Soluble γ-aminobutyric acid and benzodiazepine receptors from rat cerebral cortex, Life Sci. 29:193 (1981).
23. E. A. Barnard, Molecular weight of receptors in solution and in the membrane, in: "Cell Membrane Receptors," E. H. Reid, ed., Pitman, Bath, England, in press.
24. L. -R. and E. A. Barnard, The benzodiazepine/GABA receptor complex: Molecular weight in brain synaptic membranes and in solution, J. Neurochem. 39:1507 (1982).

25. M. M. S. Lo, E. A. Barnard, and J. O. Dolly, Size of acetylcholine receptors in the membrane. An improved version of the radiation inactivation method, Biochemistry 21:2210 (1982).
26. K. Sumikawa, M. Houghton, J. S. Emtage, B. M. Richards, and E. A. Barnard, Active multi-subunit ACh receptor assembled by translation of heterologous mRNA in Xenopus oocytes, Nature 292:862 (1981).
27. E. A. Barnard, R. Miledi, and K. Sumikawa, Translation of exogenous messenger RNA coding for nicotinic acetylcholine receptors produces functional receptors in Xenopus oocytes, Proc. R. Soc. Lond. B. 215:241 (1982).
28. T. G. Smart, A. Constanti, G. Bilbe, D. A. Brown, and E. A. Barnard, Synthesis of functional chick brain GABA-benzodiazepine-barbiturate/receptor complexes in mRNA-injected Xenopus oocytes, Neurosci. Lett. 40:55 (1983).

# USE OF AUTORADIOGRAPHIC TECHNIQUES FOR THE LOCALIZATION OF NEUROTRANSMITTER RECEPTORS IN BRAIN AND PERIPHERY: RECENT APPLICATIONS

D.R. Gehlert, H.I. Yamamura and J.K. Wamsley

Departments of Psychiatry, Anatomy and Pharmacology
University of Utah School of Medicine, Salt Lake City
Utah 84132 and Departments of Pharmacology,
Biochemistry, Psychiatry and the Arizona Research
Laboratories, University of Arizona Health Science
Center, Tucson, Arizona 85724

## INTRODUCTION

The role of the receptor in mediating drug and neurotransmitter action has become increasingly more well-defined. Studies of a compound's interaction with its receptor has led to the concept of rational drug design and subsequent *in vitro* testing procedures. Anatomical localization of these drug and neurotransmitter receptors can lead to an increased understanding of their potential role in physiological and behavioral responses, as well as helping to define mechanisms of drug action.

The study of neurotransmitter receptors has understandably lagged behind the study of their respective chemical transmitters. Since the development of labeled compounds to identify receptors in brain homogenate preparations (1), the field has rapidly expanded in participation and technology. These binding techniques have seen a variety of applications from identifying specific receptor populations to monitoring changes which occur during chronic drug treatment or disease processes.

Though the studies performed in membrane preparations have contributed a great deal of information about disease states and drug action, they have provided limited resolution of the anatomical location of receptors. Previous studies attempting to use receptor binding techniques to localize receptor populations in individual brain regions, have required gross dissection of the area of interest or isolation of individual nuclear regions by micropunching the area from a tissue slice. These techniques are laborious and time-consuming, and oftentimes have required pooling of the desired area from several animals, thereby reducing statistical capabilities. Some regions, such as the cortical laminae, do not lend themselves to dissection and require the higher resolution afforded by the light microscope. Consequently, in vitro autoradiographic techniques for diffusible ligands (2,3) were developed to bridge this gap. Using this technology, resolution of receptor populations at the light microscopic level is in the micrometer range. Receptor populations can also be detected with greater sensitivity, and discrete brain regions from individual animals can be examined in their anatomical positions. The use of in vitro labeling techniques for autoradiography, as opposed to labeling in vivo, allows the researcher to carefully control the binding conditions and prevent the metabolism of the ligand (4). Compounds which do not readily cross the blood-brain barrier, such as tritiated peptides, can be used to characterize central receptor populations. Human post-mortem tissue samples can also be used with in vitro techniques. By using serial sections, different receptor populations or subtypes of a single receptor can be localized. Receptor autoradiographic techniques are also ideally suited to measure small or opposing receptor changes in experimental models which could not be detected in whole brain homogenates. Though this autoradiographic methodology has been primarily applied to the brain, other organs are also well suited for autoradiographic receptor localization as well. In this communication, we will briefly discuss the methodology and present examples of recent applications of in vitro receptor autoradiography.

## METHODOLOGY

### Initial Experiments

In vitro labeling of tissue sections for autoradiography is performed in a manner similar to receptor binding in tissue homogenates. Initial experiments are performed to determine rinse time, incubation time and binding saturation. Using this data, appropriate competition curves can be generated to determine the pharmacological specificity of the ligand. The tissue sections can be wiped from the slides using microfiber glass filter discs, and counted using conventional liquid scintillation counting methods. This provides a means of replicating work previously performed in homogenate studies in order to verify that the appropriate receptor is being labeled in the tissue slice. This methodology also provides a means of determining the optimal binding conditions which afford the highest specific to nonspecific (signal to noise) ratios for the labeling of a particular receptor type. A $K_D$ and $B_{max}$ can be determined from a Scatchard plot of the saturation data and cooperativity can be assessed with a Hill Plot, just as these parameters are studied in homogenate preparations (5). Once these parameters are verified and the appropriate conditions defined, receptor autoradiographic localization of the actual labeled site can be attempted.

Generation of Autoradiograms

Tissue slices (usually 6 to 10 microns in thickness) are obtained, by cryostat sectioning, and thaw-mounted onto cold, "subbed" (chrome-alum/gelatin coated) microscope slides. These slide-mounted tissue sections are then incubated (using the conditions defined in the previous experiments), rinsed (to remove the unbound radioactive ligand) and given a dip in distilled water (to remove the buffer salts). The labeled sections are next placed on metal pans over ice and individually dried under a stream of dry, cool air in order to minimize ligand diffusion. The slides should be refrigerated with desiccant overnight to remove any traces of moisture.

Autoradiograms can be generated in two ways. Perhaps the more versatile is the use of tritium-sensitive film. For this technique, the slides are affixed to photographic mounting board and apposed to tritium-sensitive film (LKB Ultrofilm, LKB Products; Rockville, MD) in X-ray cassettes. After an appropriate exposure period, the film is then removed and developed. The autoradiograms can then be visualized microscopically under brightfield illumination as a pattern of dark autoradiographic grains.

The second method involves the use of emulsion-coated coverslips. With this method, long, thin coverslips (25x77mm, Corning #0, Corning Glass Works; Corning, NY) are dipped in liquid photographic emulsion (Kodak NTB-3 diluted 1:1 with water) and allowed to dry. The coverslip is then affixed to the frosted end of the slide with cyanoacrylate glue so the emulsion-coated part of the coverslip is positioned over the tissue section. The coverslip is then clamped onto the slide using a piece of Teflon and a binder clip. After an exposure period, in light tight desiccant-filled boxes, the coverslip is separated from the slide with a piece of toothpick and the emulsion subsequently developed. The tissue section is fixed, stained and dried for several hours. The coverslip is then mounted on the slide using permanent mounting fluid. The autoradiogram can then be visualized under low power using darkfield illumination. Alternatively, photographic grains can be visualized directly under high power with brightfield illumination. The depth of focus is such that the desired tissue area should be identified and the focus adjusted to visualize the grains. This allows for localization of autoradiographic grains directly over the tissue area. In most instances, it is difficult to visualize the tissue and the autoradiographic grains at the same time. A highly detailed explanation of these techniques is available elsewhere (6).

Both the film and coverslip techniques yield autoradiograms that are highly reproducible. Yet each method possesses distinct advantages. With the use of tritium brain paste standards, the autoradiographic images on tritium-sensitive film can be quantitated using microdensitometric techniques (7). The image can also be scanned with the use of a microdensitometer and color-coded computer reconstruction can be obtained (3,6). This method allows for the rapid localization of changes in certain receptor populations in small, discrete brain nuclei or changes after experimental manipulation.

The use of the coverslip technique provides higher resolution and localization of photographic grains over the desired tissue area. For instance, this higher resolution and the ability to focus on the tissue is of indispensable value in accurately localizing kidney receptor populations (8). Unfortunately, at the present time, quantitation using coverslips is difficult and is impractical to use with standards. Therefore, the tritium sensitive film has the distinct advantage of being rapidly quantifiable, as well as being relatively easy and simple to use.

## APPLICATIONS

_In vitro_ receptor autoradiography can play an important role in localizing potential sites of drug action. Coupling this information with known regions underlying the physiological actions of a compound can contribute to drug design. Thus, beyond the benefits of neurotransmitter receptor localization in the brain, creative application of these techniques can contribute to the fundamental questions of drug selectivity and response.

One of the most useful recent applications of receptor autoradiography is the co-localization of neurotransmitter receptors. Serial sections can be labeled with different ligands to compare receptor distributions in a single animal. Specific grain densities, representing the specifically bound ligand, can then be quantitated to compare receptor populations labeled by the respective compounds. Receptor subtypes can be localized _indirectly_ by labeling serial sections with a ligand which binds to all the subtypes with a single high affinity. The binding to adjacent sections is then displaced by including an appropriate concentration of an unlabeled compound, in the incubation medium, which displays selectivity for the desired subtype. Alternatively, the receptor subtype can be labeled directly, if appropriate labeled ligands are available which specifically bind to the receptor subtype of interest. Subtypes of muscarinic cholinergic receptors have been identified using both of these methodologies. Discrete areas of high and low affinity muscarinic agonist binding have been identified within the brain by the ability of the muscarinic agonist carbachol to displace the binding of the antagonist [$^3$H]-N-methylscopolamine (9, Wamsley et al., in preparation).

More recently, direct labeling of the high affinity agonist receptor subtype has been accomplished using a radiolabeled form of the muscarinic agonist cis methyldioxolane (Yamamura et al., in preparation). In most regions, albeit not all, these studies provided complementary localizations of the high affinity agonist sites. The high affinity sites predominate in such regions as lamina IV of the cerebral cortex, nucleus tractus diagonalis, several thalamic nuclei, superior colliculus, facial nerve nucleus, hypoglossal nucleus and the ventral horn of the spinal cord. Low affinity agonist sites predominate in the nucleus accumbens, caudate-putamen, most laminae of the cerebral cortex (except lamina IV), olfactory tubercle, pontine nuclei and in the dorsal horn of the spinal cord.

The functional significance of the high vs. low affinity agonist sites has yet to be determined, but there has been some indication that the high and low affinity sites are different conformations of the same receptor (10-12) and the low affinity conformation is associated with the physiological response (13-15). In the peripheral nervous system, the high affinity agonist sites appear to be the conformation undergoing orthograde axonal transport (these can be converted to a conformation with lower affinity by incubating in the presence of guanine nucleotides), while the low affinity sites appear to be undergoing transport in a retrograde direction (10,16).

In addition, muscarinic antagonist receptor subtypes have been proposed on the basis of the binding of the selective anti-ulcer drug, pirenzepine. While classical muscarinic antagonists bind to the receptor with equal high affinity, pirenzepine appears to recognize a subpopulation of muscarinic antagonist sites (17-19). Sites binding pirenzepine with high affinity have been termed the $M_1$ receptor while those recognizing pirenzepine with low affinity have been termed the $M_2$ receptor (17,20). In order to localize these subtypes within the brain, serial tissue sections were directly labeled with either the classical antagonist [$^3$H]-quinuclidinyl benzilate (21) or a tritiated form of pirenzepine (22,23). Brain areas which label with both quinuclidinyl benzilate ([$^3$H]-QNB) and pirenzepine ([$^3$H]-PZ) would be considered to contain the $M_1$ receptor subtype while those areas binding [$^3$H]-QNB but not [$^3$H]-PZ would be considered to contain the $M_2$ muscarinic receptor subtype. Autoradiographic grain densities, corresponding to specifically bound [$^3$H]-QNB, have been localized to several brain nuclei which did not bind [$^3$H]-PZ. Interestingly, most of these regions correspond to those showing predominantly the high affinity agonist receptor subtype described above (i.e., nucleus tractus diagonalis, lamina IV of the parietal cortex, superior colliculus, facial nerve nucleus, etc.). The reasons underlying these observations are unknown at this time, but suggest some relationship between the high affinity agonist site and the $M_2$ receptor.

These same binding principles can be used to co-localize binding sites for ligands which recognize different receptor types. This is of particular benefit to localize receptors for drugs which are believed to act by similar mechanisms. Benzodiazepines and barbiturates are believed to act by an enhancement of the putative inhibitory neurotransmitter, gamma-amino butyric acid (GABA) in the brain (24-26). The binding of benzodiazepines in membrane preparations is enhanced by the presence of GABA in a chloride-dependent process. This GABA effect is not directly mediated by the benzodiazepine recognition site and therefore a benzodiazepine-GABA receptor ionophore complex has been postulated (27,28). The barbiturates can also enhance the binding of benzodiazepines and GABA leading to the suggestion that recognition sites for these drugs are also coupled to this multireceptor complex (27,28). In order to determine if the barbiturate/convulsant receptor complex exists in some regions of the brain independent of the presence of the benzodiazepine sites, serially sectioned brain regions were prepared for autoradiographic localization of the convulsant bicyclophosphate [$^{35}$S] t-butyl bicyclophosphorothionate (29), a ligand which is highly specific for the barbiturate/picrotoxin binding site (30), or for the localization of benzodiazepine receptors using [$^3$H]-flunitrazepam (31,32). The distribution of binding sites for these two ligands alligned very closely in many areas of the brain (Figure 1). However, in the cerebellum a different localization of the two sites was found. The barbiturate/convulsant sites were concentrated in the granule cell layer (Figure 1), while the benzodiazepine receptors predominated in the molecular layer (33). Interestingly, it has been demonstrated that barbiturates and related compounds do not enhance the binding of benzodiazepines in the cerebellum as effectively as they do in other regions (34,35). These results, taken together, supply evidence that the receptors for the barbiturate/convulsant sites in the cerebellum may exist in a conformation independent of the benzodiazepine component of the

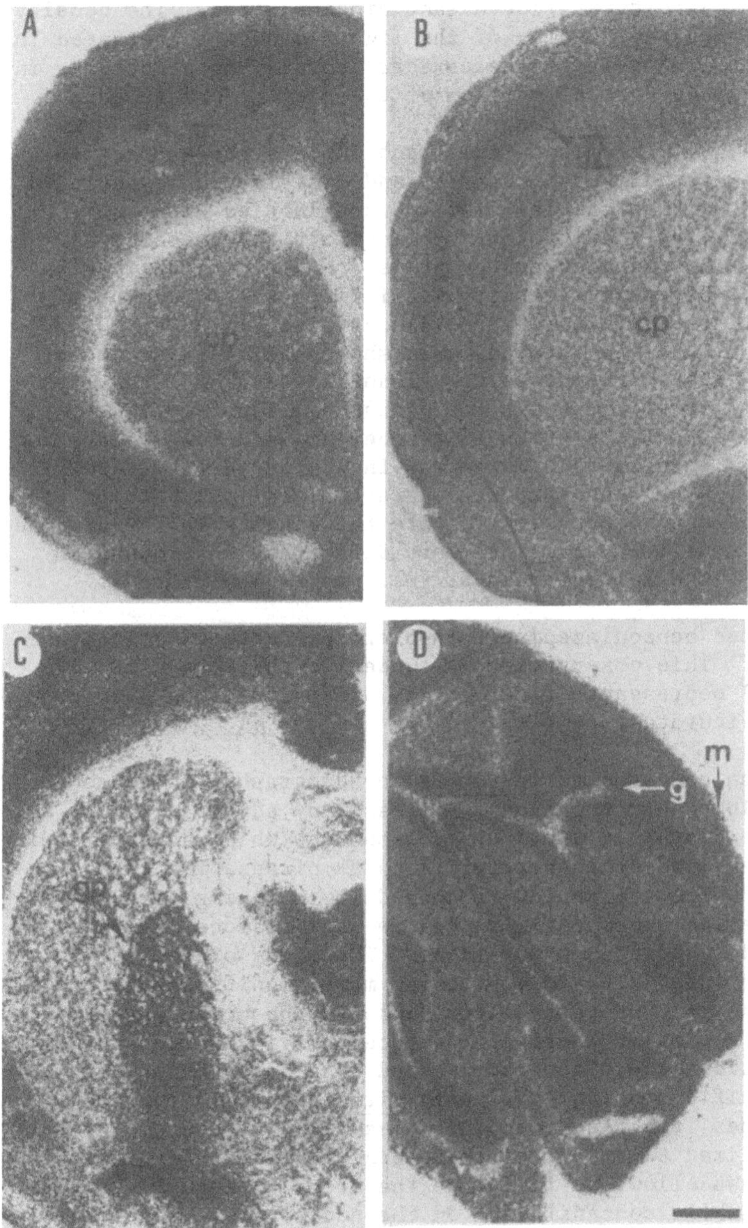


Fig. 1. A. The autoradiogram shown in this photomicrograph was generated by a tissue section specifically labeled with [$^{35}$S]t-butyl bicyclophophorothionate. Thus, the distribution of the autoradiographic grains (black dots against a

(continued)

light background) indicate the presence of the barbiturate/convulsant binding sites. The density of the binding of the ligand can be appreciated in lamina IV of the cerebral cortex, in structures in the ventral portion of the section and in the caudate-putamen (cp).
B. This photomicrograph depicts the autoradiographic grain density and distribution associated with a section of forebrain labeled with [$^3$H]-flunitrazepam to indicate the presence of benzodiazepine receptors. Note the distribution of these receptors is similar to that shown for the barbiturate/convulsant sites in A.
C. This autoradiogram shows the distribution of the barbiturate/convulsant sites associated with a section through the globus pallidus (gp).
D. The section of cerebellum used to generate this autoradiogram was incubated in [$^{35}$S]t-butyl bicyclophosphorothionate to label the barbiturate/convulsant sites. Note the binding in the granule cell layer (g) is more dense than that found in the molecular layer (m). Bar = 500 microns.

postulated benzodiazepine-GABA-picrotoxin-chloride ionophore complex. This observation may underlie the more selective CNS neuronal depressant effect of the benzodiazepines over that of the barbiturates.

When compared to in vivo autoradiography, in vitro receptor autoradiographic techniques possess many advantages. In addition to those discussed previously, compounds which do not readily pass the blood-brain barrier, such as pirenzepine (36), and ligands which are rapidly broken down in living tissues, such as the putative neuropeptides, can be used to localize their respective receptor populations. The study of receptors for the neuropeptides is rapidly becoming more significant as the potential roles for these compounds in the brain are expanded. Identification of neuropeptide receptors in homogenate membrane preparations are often complicated by a high degree of non-specific binding. Using in vitro autoradiographic techniques, the receptors for these putative neuromodulators can be localized to small, discrete brain nuclei. Autoradiographic techniques allow one to ignore the areas of high nonspecific binding while concentrating on the nuclei where specific binding exists. For example, the binding sites for [$^3$H]-arginine vasopressin ([$^3$H]-AVP) have been localized to very small brain nuclei (37-39). Figure 2 illustrates the specific binding of [$^3$H]-AVP to the magnocellular division of the paraventricular nucleus of the hypothalamus. Note the high degree of nonspecific binding throughout the remainder of the autoradiogram.

The use of receptor autoradiographic techniques has also provided for the localization of receptor sites for the neurohormone angiotensin II (40). Many of these sites occur in regions of the brainstem known to be involved in central mechanisms of blood pressure regulation and where a putative role for this peptide as a neurotransmitter has been hypothesized (Figure 2).

Receptors for yet another putative peptide neurotransmitter, substance P, have recently been localized autoradiographically in the central nervous system (41). Using tritiated substance P itself, Yamamura et al. (in preparation) have localized these receptors in the lateral horn of the thoracic spinal cord gray matter (Figure 2). The latter localization would have been impossible by any other means currently available, and defines a region where substance P containing neurons projecting from the ventral medulla to the preganglionic sympathetic neurons in the intermediolateral cell column (42) may be playing an important role in influencing vasomotor activity.

The utilization of in vitro receptor autoradiography to detect receptor populations in discrete areas of the brain can also be used to detect alterations induced by lesions or to demonstrate changes which occur in experimental models of human disease. As alluded to earlier, autoradiographic techniques are also ideal for the evaluation of axonal transport of receptors in the nervous system. The accumulation of muscarinic cholinergic (43) or serotonergic (44) receptors has been visualized after interruption of white matter pathways by electrolytic or surgical lesioning (Figure 2). Alterations in muscarinic receptors have also been characterized autoradiographically in severely thiamine deficient animals, a model of the Wernicke-Korsakoff's syndrome (Gehlert et al., in preparation). In the latter experiments, increases in muscarinic receptor densities, presumably due to a reduction in acetylcholine levels (45,46), were seen in several areas believed to be involved in memory. This includes several laminae of the hippocampus and cerebral cortex. In the same autoradiographic experiments, a marked reduction in muscarinic binding was seen in the ventro-medial hypothalamus, an area believed to be involved in satiety (47,48). Interestingly, a sign of severe thiamine deficiency is anorexia and these results may indicate a role for the muscarinic receptor in appetite control. The increase in muscarinic receptor binding seen in one region with a concomitant decrease in receptor binding in an adjacent region would be difficult to measure in membrane homogenate preparations and demonstrates another distinct advantage of autoradiographic localization of receptor binding.

These techniques are not limited to the brain; other organs are well suited for the use of receptor autoradiography. For

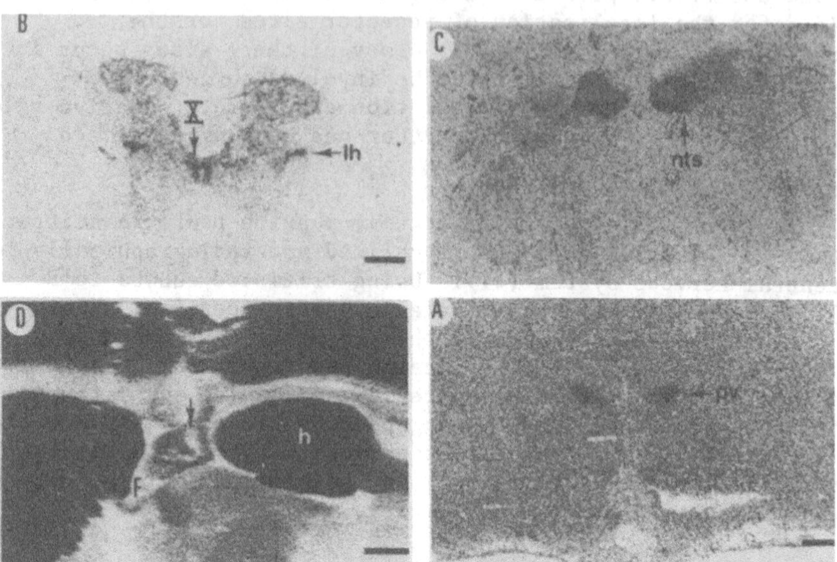


Fig. 2. A. This photomicrograph shows the grain density on a region of film overlying a section through the hypothalamus which was labeled with [$^3$H]-arginine vasopressin. Dense labeling of sites in the magnocellular division of the paraventricular nucleus (pv) can be seen (Bar = 250 microns).

B. A section of thoracic spinal cord labeled with [$^3$H]-substance P was used to produce this autoradiogram. High grain densities, associated with sites specifically binding [$^3$H]-substance P, can be seen in the lateral horn (lh) and in lamina X (the gray matter surrounding the central canal). Lower densities of substance P receptors occur in most of the rest of the gray matter regions (Bar = 400 microns).

C. This autoradiogram shows the angiotensin II receptor binding associated with the nucleus tractus solitarius (nts) in the brainstem.

D. The autoradiogram shown in this photomicrograph depicts the phenomenon of axonal transport of muscarinic receptors in the brain. A microelectrode was introduced stereotaxically into the brain and a small radiofrequency lesion (arrow) was created in the fimbria (F) of the hippocampal formation (h). The animal was allowed to survive for 24 hours post-lesion before sagittal sections of brain were taken through the lesioned area and prepared for the autoradiographic localization of muscarinic receptors using [$^3$H]-pirenzepine. An accumulation of these sites can be seen surrounding the lesion indicating that $M_1$ receptors are undergoing axonal transport in the fimbria. Bar = 500 microns.

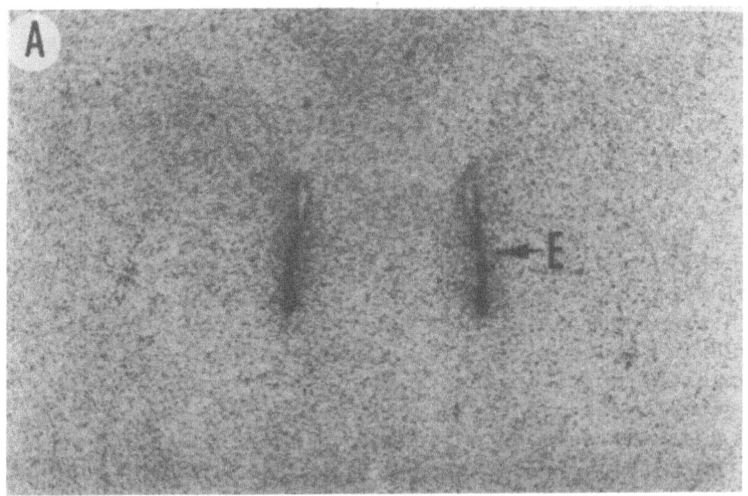


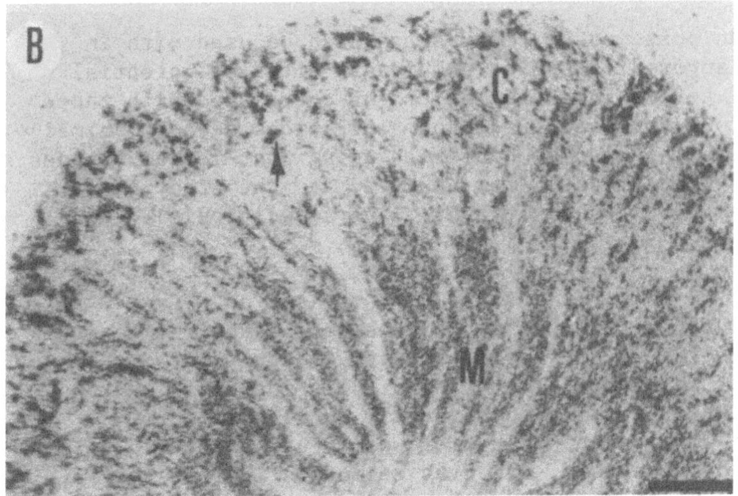


Fig. 3. A. The autoradiographic grain density associated with a section of brain labeled with [$^3$H]-R05-4864 is shown in this photomicrograph. Under the conditions used in this experiment (i.e., a long (continued)

rinse time) the specific binding is limited to the ependyma cells (E) lining the ventricles.
B. The autoradiogram shown in this photomicrograph was made from a section of kidney labeled with [$^3$H]-R05-4864. Thus, the distribution and density of the "peripheral" type benzodiazepine receptors in the kidney can be appreciated. Specific labeling is found in both the renal cortex (C) and renal medulla (M). The arrow points out the specific binding of [$^3$H]-R05-4864 associated with a distal convoluted tubule. Bar = 500 microns.

instance, peripheral-type benzodiazepine binding sites have been reported in several tissues including the brain and kidney (49). The labeled compound [$^3$H]-R05-4864 has been demonstrated to specifically bind to these sites. The binding sites for [$^3$H]-R05-4864 have recently been autoradiographically localized in the kidney and brain (8). Binding in the kidney was associated with the ascending loop of Henle and distal tubule while peripheral-type binding sites in the brain, using conditions which provided the highest specific to nonspecific ratio, were found in the chorioid plexis and ependyma cells (Figure 3). While the physiological role of the peripheral-type benzodiazepine receptor is presently unknown, the localization of these sites can add considerable insight into their potential effects and define particular regions for future studies to focus on.

Since human post-mortem tissues can also be used with *in vitro* receptor autoradiographic techniques, sites of potential pharmacological action in the human can be correlated with those seen in laboratory animals. Receptor alterations seen in animal models of disease can also be confirmed using post-mortem tissues from victims of that disease (for example, see the study of Whitehouse et al. [50] on amyotrophic lateral sclerosis).

## CONCLUSION

Receptor autoradiographic techniques can be applied for the microscopic localization of a wide variety of drug and neurotransmitter receptors. Autoradiographic localization provides the highest resolution currently attainable for determining the emplacement of neurotransmitter receptor subtypes, peptide receptors, and drug receptors in both normal and experimentally- or diseased-altered tissues. *In vitro* receptor autoradiography thus provides a sensitive and quantitative tool with which to study potential sites of receptor function in drug and neurotransmitter action.

ACKNOWLEDGEMENTS

The authors wish to thank Jane Stout for her excellent secretarial assistance. Portions of this work were supported by grants from the NIMH (MH-36563) and DOD (DAMD17-83-C-3023) to JKW; and by grants from the NIMH (MH-2725, MH-30626 and RSDA MH-00095) to HIY.

REFERENCES

1. S.H. Snyder, Overview of neurotransmitter binding, in: "Neurotransmitter Receptor Binding," S.J. Enna and M.J. Kuhar, eds., Raven Press, New York, p. 1 (1978).
2. W.S. Young and M.J. Kuhar, A new method for receptor autoradiography: $^3$H-opioid receptor labeling in mounted tissue sections, Brain Res. 179:255 (1979).
3. J.M. Palacios, D.L. Niehoff and M.J. Kuhar, Receptor autoradiography with tritium sensitive film: potential for computerized densitometry, Neurosci. Lett. 25:101 (1981).
4. J.K. Wamsley and J.M. Palacios, Receptor mapping by histochemistry, in: "Handbook of Neurochemistry," Vol. 2, A. Lajtha, ed., Plenum Press, New York, p. 27 (1982).
5. J.P. Bennett, Jr., Methods in binding studies, in: "Neurotransmitter Receptor Binding," H.I. Yamamura, S.J. Enna and M.J. Kuhar, eds., Raven Press, New York, p. 57 (1978).
6. J.K. Wamsley and J.M. Palacios, Apposition techniques of autoradiography for microscopic receptor localization, in: "Current Methods in Cellular Neurobiology," J. Barker and J. McKelvy, eds., John Wiley and Sons, New York, p. 241 (1983).
7. J.R. Unnerstall, D.L. Niehoff, M.J. Kuhar and J.M. Palacios, Quantitative receptor autoradiography using [$^3$H] ultrofilm: Application to multiple benzodiazepine receptors, J. Neurosci. Methods 6:59 (1982).
8. D.R. Gehlert, H.I. Yamamura and J.K Wamsley, Autoradiographic localization of peripheral benzodiazepine binding sites in the rat brain and kidney using [$^3$H]-R05-4864, Eur. J. Pharmacol., in press.
9. J.K. Wamsley, M.A. Zarbin, N.J.M. Birdsall and M.J. Kuhar, Muscarinic cholinergic receptors: Autoradiographic localization of high and low affinity agonist binding sites, Brain Res. 200:1 (1980).
10. M.A. Zarbin, J.K. Wamsley and M.J. Kuhar, Axonal transport of muscarinic cholinergic receptors in rat vagus nerve: High and low affinity agonist receptors move in opposite directions and differ in nucleotide sensitivity, J. Neurosci. 2:934 (1982).
11. E. Burgisser, A. DeLeon and R.J. Lefkowitz, Reciprocal modulation of agonist and antagonist binding to muscarinic cholinergic receptor by guanine nucleotide, Proc. Natl.

Acad. Sci. 79:1732 (1982).
12. F.J. Ehlert, W.R. Roeske and H.I. Yamamura, Muscarinic receptor: Regulation by guanine nucleotides, ions and N-ethylmaleimide, Fed. Proc. 40:153 (1981).
13. N.J.M. Birdsall, A.S.V. Burgen and E.C. Hulme, Correlation between the binding properties and pharmacological responses of muscarinic receptors, in: "Cholinergic Mechanisms and Psychopharmacology," D.J. Jenden, ed., New York, Plenum Press, p. 25 (1977).
14. N.J.M. Birdsall and E.C. Hulme, Biochemical studies on muscarinic acetylcholine receptors, J. Neurochem. 27:7 (1976).
15. P.G. Strange, N.J.M. Birdsall and A.S.V. Burgen, Occupancy of muscarinic acetylcholine receptors stimulates a guanylate cyclase in neuroblastoma cells, Biochem. Soc. Trans. 5:189 (1977).
16. J.K. Wamsley, M.A. Zarbin and M.J. Kuhar, Muscarinic cholinergic receptors flow in the sciatic nerve, Brain Res. 217:155 (1981).
17. R. Hammer, C.P. Berrie, N.J.M. Birdsall, A.S.V. Burgen and E.C. Hulme, Pirenzepine distinguishes between different subclasses of muscarinic receptors, Nature 283:90 (1980).
18. N.J.M. Birdsall, A.S.V. Burgen, R. Hammer, E.C. Hulme and J. Stockton, Pirenzepine - a ligand with original binding properties to muscarinic receptors, Scand. J. Gastroenterol. 15, Suppl. 66:1 (1980).
19. M. Watson, W.R. Roeske and H.I. Yamamura, [$^3$H]-Pirenzepine selectively identifies a high affinity population of muscarinic cholinergic receptors in the rat cerebral cortex, Life Sci. 31:2019 (1982).
20. M. Watson, H.I. Yamamura and W.R. Roeske, A unique regulatory profile and regional distribution of [$^3$H]-pirenzepine in the rat provide evidence for distinct $M_1$ and $M_2$ muscarinic receptor subtypes, Life Sci. 32:3001 (1983).
21. J.K. Wamsley, M. Lewis, W.S. Young, III and M.J. Kuhar, Autoradiographic localization of muscarinic cholinergic receptors in the rat brainstem, J. Neurosci. 1:176 (1981).
22. H.I. Yamamura, J.K. Wamsley, P. Deshmukh and W.R. Roeske, Differential light microscopic autoradiographic localization of muscarinic cholinergic receptors in the brainstem and spinal cord of the rat using [$^3$H]-pirenzepine, Eur. J. Pharmacol. 91:147 (1983).
23. J.K. Wamsley, D.R. Gehlert, W.R. Roeske and H.I. Yamamura, Muscarinic antagonist binding site heterogeneity as evidenced by autoradiography after direct labeling with [$^3$H]-QNB and [$^3$H]-pirenzepine, Life Sci., in press.
24. E. Costa and A. Guidotti, Molecular mechanisms in the receptor action of benzodiazepines, Ann. Rev. Pharmacol. Toxicol. 19:531 (1979).

25. W. Haefely, P. Polc, R. Schaffner, H.H. Keller, L. Pieri and H. Möhler, Facilitation of GABA-ergic transmission of drugs, in: "GABA-Neurotransmitters," P. Krogsgaard-Larsen, J. Scheel-Kruger and H. Kofod, eds., Munksgaard, Copenhagen, p. 357 (1979).
26. W. Haefely, L. Pieri, P. Polc and R. Schaffner, in: "Handbook of Experimental Pharmacology," F. Hoffmeister and G. Stille, eds., Springer-Verlag, Berlin, p. 213 (1981).
27. R.W. Olsen, GABA-benzodiazepine-barbiturate receptor interactions, J. Neurochem. 37:1 (1981).
28. R.W. Olsen, Drug interactions at the GABA receptor ionophore complex, Ann. Rev. Pharmacol. Toxicol. 22:245 (1982).
29. K.W. Gee, J.K. Wamsley and H.I. Yamamura, Light microscopic autoradiographic identification of picrotoxin/barbiturate binding sites in rat brain with [$^{35}$S]-t-butyl-bicyclophosphorothionate, Eur. J. Pharmacol. 89:323 (1983).
30. R.F. Squires, J.E. Casida, M. Richardson and E. Saederup, [$^{35}$S]-t-butylbicyclophosphorothionate binds with high affinity to brain specific sites coupled to GABA-A and ion recognition sites, Mol. Pharmacol. 23:326 (1983).
31. W.S. Young and M.J. Kuhar, Autoradiographic localization of benzodiazepine receptors in the brains of humans and animals, Nature 280:393 (1979).
32. W.S. Young and M.J. Kuhar, Radiohistochemical localization of benzodiazepine receptors in rat brain, J. Pharmacol. Exp. Therap. 212:337 (1980).
33. J.K. Wamsley, K.W. Gee and H.I. Yamamura, Comparison of the distribution of convulsant/barbiturate and benzodiazepine receptors using light microscopic autoradiography, Life Sci. 33:2321 (1983).
34. R.W. Olsen and F. Leeb-Lundberg, Convulsant and anticonvulsant drug binding sites related to the GABA receptor/ionophore system, in: "Neurotransmitters, Seizures and Epilepsy," P.O. Morselli, K.G. Lloyd, W. Löscher, B.S. Meldrum and E.H. Reynolds, eds., Raven Press, New York, p. 151 (1981).
35. F. Leeb-Lundberg and R.W. Olsen, Heterogeneity of benzodiazepine receptor interactions with gamma-aminobutyric acid and barbiturate receptor sites, Mol. Pharmacol. 23:315 (1983).
36. R. Hammer and F.W. Koss, The pharmacokinetic profile of pirenzepine, Scand. J. Gastroenterol. 14, Suppl. 57:1 (1979).
37. H.I. Yamamura, K.W. Gee, R.E. Brinton, J.P. Davis, M. Hadley and J.K. Wamsley, Light microscopic autoradiographic visualization of [$^{3}$H]-arginine vasopressin binding sites in the rat brain, Life Sci. 32:1919 (1983).
38. D.G. Baskin, F. Petracca and D.M. Dorsa, Autoradiographic localization of specific binding sites for [$^{3}$H]-[Arg$^{8}$]

vasopressin in the septum of the rat brain with tritium sensitive film, Eur. J. Pharmacol. 90:155 (1983).
39. F.W. Van Leeuwen and P. Wolters, Light microscopic autoradiographic localization of [$^3$H]-arginine-vasopressin binding sites in the rat brain and kidney, Neurosci. Lett. 41:61 (1983).
40. D.R. Gehlert, R.C. Speth, D.P. Healy and J.K. Wamsley, Autoradiographic localization of angiotensin II receptors in the rat brainstem, Life Sci., in press.
41. R. Quirion, C.W. Shults, T.W. Moody, C.B. Pert, T.N. Chase and T.L. O'Donohue, Autoradiographic distribution of substance P receptors in rat central nervous system, Nature 203:714 (1983).
42. C.J. Helke, J.J. Neil, V.J. Massari and A.D. Loewy, Substance P neurons project from the ventral medulla to the intermediolateral cell column and ventral horn in the rat, Brain Res. 243:147 (1982).
43. J.K. Wamsley, Muscarinic cholinergic receptors undergo axonal transport in the brain, Eur. J. Pharmacol. 86:309 (1983).
44. E. Snowhill and J.K. Wamsley, Serotonin type-2 receptors undergo axonal transport in the medial forebrain bundle, Eur. J. Pharmacol., in press.
45. C.U. Vorhees, D.E. Schmidt and R.J. Barrett, Effects of pyrithiamin and oxythiamin on acetylcholine levels and utilization in rat brain, Brain Res. Bull. 3:493 (1978).
46. K.V. Speeg, D. Chen, D.W. McCandless and S. Scheuker, Cerebral acetylcholine in thiamine deficiency, Proc. Soc. Exp. Biol. Med. 135:1005 (1970).
47. G.A. Bray and D.A. York, Hypothalamic and genetic obesity in experimental animals: An autonomic and endocrine hypothesis, Physiol. Rev. 59(3):719 (1979).
48. S.C. Woods and D. Porte, The central nervous system, pancreatic hormones, feeding and obesity, Adv. Metab. Disord. 9:283 (1978).
49. H. Shoemaker, R.G. Boles, W.D. Horst and H.I. Yamamura, Specific high affinity binding sites for [$^3$H]-R05-4864 in rat brain and kidney, J. Pharmacol. Exp. Therap. 225:61 (1983).
50. P.J. Whitehouse, J.K. Wamsley, M.A. Zarbin, D.L. Price, W.W. Tourtellotte and M.J. Kuhar, Amyotrophic lateral sclerosis: Alteration in neurotransmitter receptors, Ann. Neurol. 14:8 (1983).

# THE STRUCTURE AND EVOLUTION OF NEUROTRANSMITTER RECEPTORS (α- AND β-ADRENERGIC, DOPAMINERGIC AND MUSCARINIC CHOLINERGIC)

J. Craig Venter and Claire M. Fraser

Department of Molecular Immunology
Roswell Park Memorial Institute
666 Elm Street
Buffalo, NY 14263

## INTRODUCTION

This laboratory has studies underway on the molecular characterization of the major receptors of the autonomic nervous system, as well as the slow inward calcium channel of cardiac and smooth muscle (1). The receptors under investigation include $\alpha_1$, $\alpha_2$, $\beta_1$ and $\beta_2$-adrenergic, muscarinic cholinergic and dopaminergic types. To date we have made substantial progress with the complete purification of $\alpha_1$, $\beta_1$ and $\beta_2$ adrenergic and muscarinic cholinergic receptors and have molecular size information on $\alpha_2$-adrenergic and $D_2$ dopaminergic receptors (Table I).

Technical advances such as the production of monoclonal antibodies to receptors (2-6), the use of covalent affinity labels such as NHNP-NBE for β-receptors (7,8), phenoxybenzamine for $\alpha_1$-adrenergic receptors (9,10) and propylbenzilylcholine mustard for muscarinic receptors (6,11), and the use of radiation inactivation/target size analysis (1,7,8,10-12) have singularly and in combination helped to provide data not obtainable prior to the late 1970's. The forthcoming structural data on neurotransmitter receptors has also permitted more informed speculation as to the origin and evolution of receptor structure and function (6,13).

TABLE I

MOLECULAR PROPERTIES OF NEUROTRANSMITTER RECEPTORS

| Receptor Class | $\beta_1$- | $\beta_2$- | $\alpha_1$- | $\alpha_2$- | $D_2$- | Muscarinic Cholinergic |
|---|---|---|---|---|---|---|
| Endogenous ligand | Epi-NE | Epi | Epi-NE | Epi-NE | DA | Ach |
| Isoelectric Point (pI) | 5.5 (turkey RBC) | 4.2 (dog lung) | 3.9 (rat liver) | --- | 5.0 (dog brain) | 5.9 (human, rat *Drosophila* brain) |
| Target Size (Radiation Inactivation) | 130 Kdal (dog heart) | 109 Kdal (dog lung) | 160 Kdal (rat liver) | 160 Kdal (human platelets) | 123 Kdal (human, dog brain) | 80 Kdal (same as above) |
| Hydrodynamic Mol wt. | 65 Kdal (dog heart, turkey RBC) | 90 Kdal (dog lung, liver) | 83 Kdal (rat liver) | --- | 116 Kdal (dog brain) | 80 Kdal (same as above) |
| SDS-PAGE Mol wt. | 65 Kdal (dog heart) | 58 Kdal (dog lung) | 85 Kdal (rat liver) | --- | --- | 80 Kdal (same as above) |
| Receptor Structure | dimer (heart) | dimer (lung) | dimer (rat liver) | | | monomer |

Data from 2,3,5,6,7,8,10,11,12,13,16,17,18,23,28.

Abbreviations: Epi: epinephrine; NE: norepinephrine; DA: dopamine; ACh: acetylcholine; RBC: erythrocyte.

## BETA-ADRENERGIC RECEPTOR STRUCTURE

### Beta$_1$-Adrenergic Receptors

Hydrodynamic properties of canine heart and turkey erythrocyte $\beta_1$-receptors, determined in 1978 (14), indicated a 65,000 dalton molecular weight for this receptor. Immunoaffinity chromatography studies with the first monoclonal antibodies developed against $\beta$-receptors also indicated a molecular weight of 65-70,000 daltons for the purified $\beta_1$ receptor (3). The model proposed (16) for the $\beta_1$ receptor is that of a monomeric protein of 65,000-70,000 daltons (Fig. 1), however, recent studies on radiation inactivation of canine heart $\beta_1$-receptors indicated an in situ membrane molecular weight of 130,000 daltons and suggested that the mammalian $\beta_1$-receptor may exist as a functional dimer (Fraser and Venter, in preparation).

Studies with sulfhydryl reagents have shown that there is a sulfhydryl group in the ligand binding site of the $\beta_1$-receptor that is apparently involved in agonist binding (19,20). $\beta_1$-Receptors also contain at least one disulfide bond which is important for maintaining the receptor in a conformation necessary for ligand binding (19,20). Furthermore, agonist binding produces a conformational change in the $\beta_1$-receptor via cleavage of a second disulfide bond (19).

For other aspects of $\beta_1$ receptor structure, many recent reviews are recommended (15-18).

### Beta$_2$-Adrenergic Receptors

Using a covalent affinity label and $\beta_2$-receptor specific autoantibodies (7,21-23), we isolated a 58,000 dalton subunit of the canine lung $\beta_2$-receptor. Using monoclonal antibody immunoaffinity chromatography, we also obtained a similar size subunit from bovine lung $\beta_2$-receptors (7). Evidence from immunoaffinity isolation and SDS-PAGE suggested a disulfide crosslinked $\beta_2$-receptor dimer with a molecular weight of 120,000 daltons. Consistent with this idea were data obtained from target size analysis of lung membrane $\beta_2$-receptors showing the presence of a molecular weight complex of approximately 110,000 daltons for the $\beta_2$ receptor in situ (7). A proposed model for the lung $\beta_2$-receptor is illustrated in Fig. 2.

Autoantibodies from allergic respiratory disease patients, specific for $\beta_2$-receptors, display no cross-reactivity with $\beta_1$-adrenergic receptors (21) supporting the concept originally advanced by us that $\beta_1$ and $\beta_2$ receptors differ structurally (14,16,17,21,24). However, monoclonal antibody data indicate some structural homology between $\beta_1$ and $\beta_2$ receptors, particularly in the region of the ligand binding site (2,3). Similar data on the homology of the

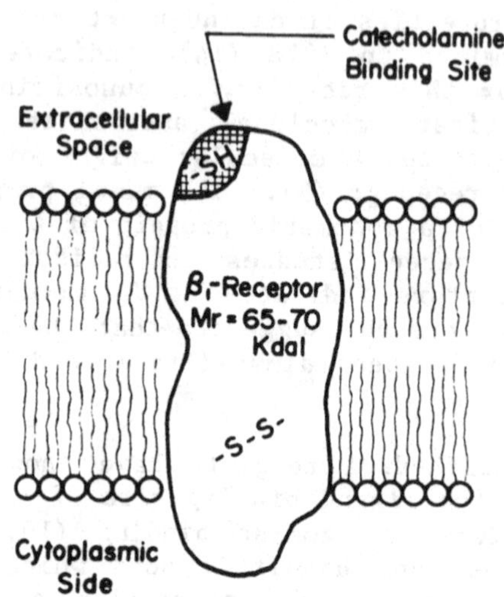


Fig. 1. Proposed membrane model for the $\beta_1$-adrenergic receptor (16).

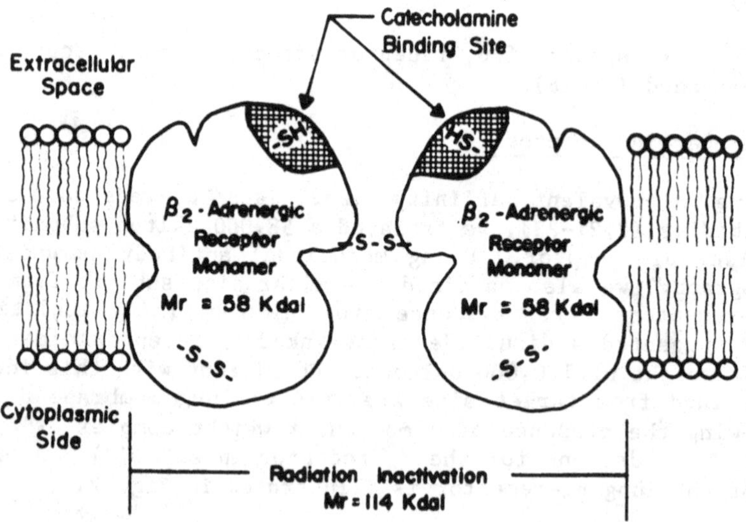


Fig. 2. Proposed membrane model for the $\beta_2$-adrenergic receptor (16).

ligand binding region of $\beta_1$- and $\beta_2$-receptors has also been obtained by Strosberg and coworkers (25).

## ALPHA ADRENERGIC RECEPTOR STRUCTURE

### Alpha$_1$ Adrenergic Receptor Structure

$\alpha_1$-Adrenergic receptors are found in most tissues including brain, blood vessel and airway smooth muscle, cardiac muscle and liver. The rat liver $\alpha_1$-adrenergic receptor has undergone extensive physiological and pharmacological characterization (26,27). In structural studies on the rat liver $\alpha_1$-adrenergic receptor we utilized a high specific activity [$^3$H]phenoxybenzamine (PCB) as an affinity reagent (9). The [$^3$H]POB was found to be highly specific for the rat liver $\alpha_1$-receptor and could be utilized in the 0.5 to 1.0 nM range to specifically and covalently label $\alpha_1$ receptors in the membrane (9).

The structure of the $\alpha_1$-receptor was investigated by comparing polypeptides identified by SDS polyacrylamide gel electrophoresis of [$^3$H]POB labeled receptors with the size of the intact receptor in the cell membrane as determined by target size analysis. The $\alpha_1$-receptor labeled in rat liver plasma membranes with [$^3$H]POB appeared as a single polypeptide with a molecular weight of 85,000 on SDS polyacrylamide gels (Fig. 3). In the absence of protease inhibitors, proteolytic fragments of the $\alpha_1$-receptor of 58,000, 45,000 and 40,000 daltons were also apparent (10). Similarly, limited proteolysis of the 85,000 dalton protein by trypsin, chymotrypsin or papain produced peptides of 62,000, 45,000 40,000, 27,000, 23, 000 and 18,000 daltons (Fig. 4). Limited proteolysis of the membrane bound receptor by chymotrypsin produced water soluble peptides of 45,000, 40,000, 27,000, 23,000 and 18,000 daltons, all of which contain the ligand binding site and protrude from the membrane into the extracellular space. This suggests that over 50 percent of the receptor protein is present in the aqueous phase on the extracellular surface of the membrane. Limited proteolysis of the membrane-associated receptor also produced a 58-62,000 dalton peptide; however, this protein required detergent for solubilization from the membrane. Taken together, these studies suggest the presence of a receptor tail of 22,000 daltons in the extracellular or cytoplasmic aqueous phase (10).

In order to determine whether the 85,000 dalton protein labeled by [$^3$H]phenoxybenzamine represented all or only a portion of the $\alpha_1$-receptor, radiation inactivation studies were undertaken. Radiation inactivation of the $\alpha_1$-receptor was measured by loss of specific [$^3$H]POB and [$^3$H]prazosin binding and by loss of affinity labeled $\alpha_1$-adrenergic receptors on SDS polyacrylamide gels (Fig. 5). Target size analysis of the rat liver $\alpha_1$-receptor indicated that the

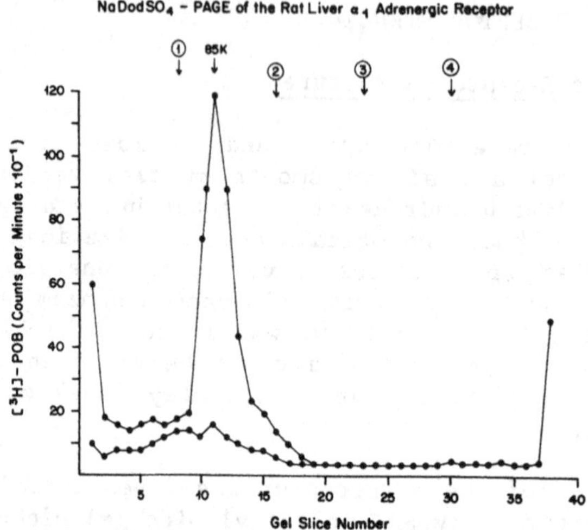


Fig. 3. NaDodSO$_4$-PAGE analysis of rat liver $\alpha_1$-adrenergic receptors (10).

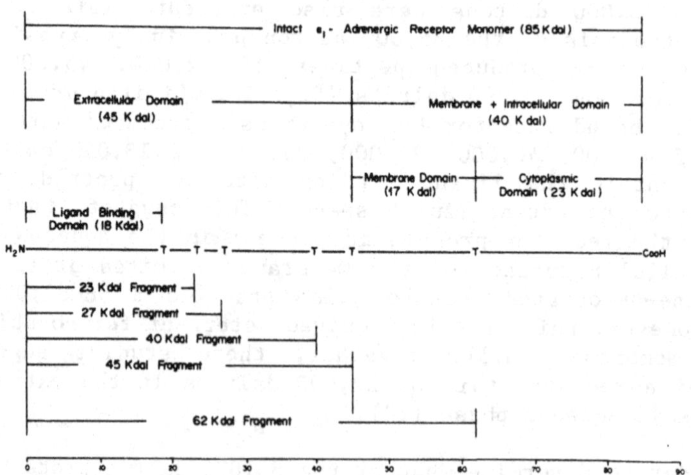


Fig. 4. Tryptic digest map of the $\alpha_1$-adrenergic receptor monomer (10).

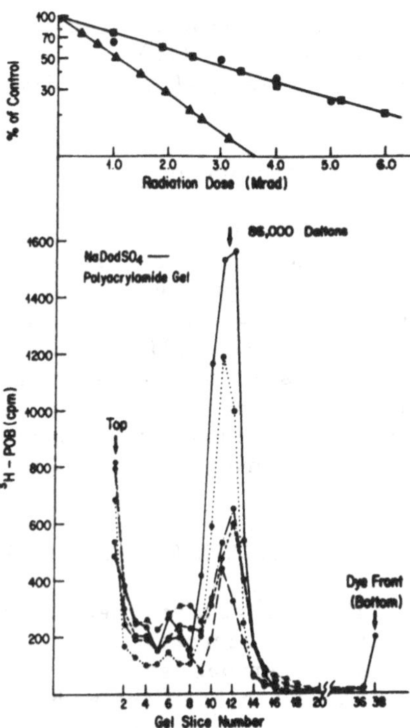


Fig. 5. Radiation inactivation of rat liver $\alpha_1$-adrenergic receptor (10).

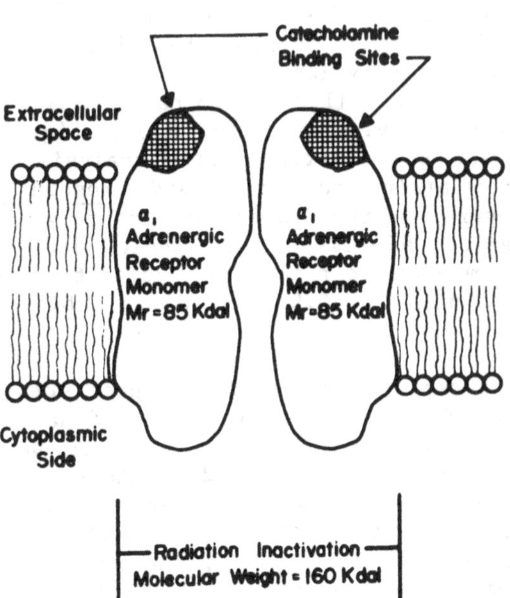


Fig. 6. Proposed membrane model for the $\alpha_1$-adrenergic receptor (16).

intact membrane-bound receptor has an average molecular mass of 160,000 daltons (10). This finding suggests that in situ the $\alpha_1$-receptor may exist as a dimer of two 85,000 dalton subunits (Fig. 6).

## $\alpha_2$-Adrenergic Receptor

To date structural studies on $\alpha_2$-adrenergic receptors have been more limited. We have recently characterized the molecular size of the $\alpha_2$-adrenergic receptor in the membrane of human platelets using target size analysis (28). When [$^3$H]yohimbine specific binding was used to assess radiation induced loss of $\alpha_2$-receptors (21), the molecular size of the $\alpha_2$-receptor in situ was found to be 160,000 daltons, a size identical to that of the $\alpha_1$-receptor of rat liver (Table 1). These data suggest possible structural similarities between $\alpha_1$- and $\alpha_2$-receptor subtypes.

## MUSCARINIC CHOLINERGIC RECEPTOR STRUCTURE

The muscarinic acetylcholine receptor is a principal neurotransmitter receptor and controls the cholinergic side of the autonomic nervous system. Muscarinic receptors are found in most tissues of the body and modulate, often by opposing the β-adrenergic receptor system, heart rate, airway and vascular smooth muscle contraction and glandular secretions.

We have isolated muscarinic receptors from a wide variety of sources including human brain, rat brain, dog brain, *Drosophila* heads, dog heart, rat heart, guinea pig ileum longitudinal smooth muscle, and monkey ciliary muscle. Using [$^3$H]propylbenzilylcholine mustard ([$^3$H]PrBCM) as a covalent affinity label, receptors isolated from the above tissues and species were all shown to have a molecular weight of 80,000 on SDS-polyacrylamide gels (11). Fig. 7 illustrates an SDS polyacrylamide gel electrophoretic analysis of muscarinic receptors from human brain, rat brain and *Drosophila* heads. Further evidence supporting the molecular identity of muscarinic receptors from a variety of tissues and species derives from the demonstration that all muscarinic receptors studied to date have an identical isoelectric point, pI = 5.9 (Fig. 8 and Table 2) (6).

As with the $\alpha_1$-adrenergic receptor, muscarinic receptor structure was further studied by limited proteolysis of the 80,000 dalton protein from different tissues. Trypsin and papain produced peptides of 64,000, 52,000, 42,000, 36,000, 23,000 and 18,000 daltons from all [$^3$H]PrBCM labeled receptors examined, further indicating the absence of multiple structural forms of the receptor (11). Limited proteolysis of the membrane bound receptor produced a major peptide of 42,000 daltons and minor peptides of 36,000, 23,000, and

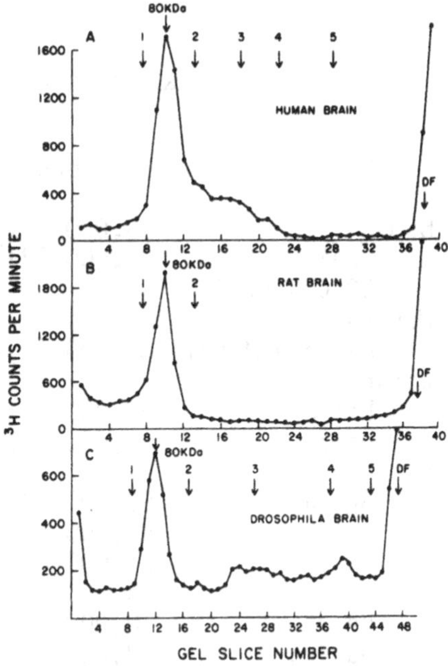


Fig. 7. NaDodSO$_4$-PAGE analysis of muscarinic acetylcholine receptors (6).

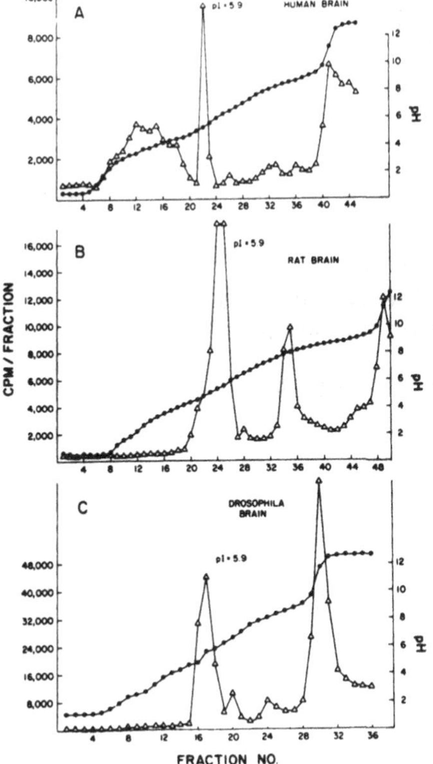


Fig. 8. Isoelectric focusing of muscarinic acetylcholine receptors (6).

TABLE 2

MUSCARINOC CHOLINERGIC RECEPTOR STRUCTURE

| Tissue | Molecular Weight | Isoelectric Point[a] |
|---|---|---|
| Human | 78,000 ± 1,200 (1)[b] | 5.9 |
| Brain | 82,000 (2)[b] | |
| 80,000 (1)[e] | | |
| Monkey | | |
| Ciliary Muscle | 80,000 (1)[e] | 5.9 |
| Canine | | |
| Brain | 82,000 ± 1,800 (1)[b] | |
| Heart | 81,000 ± 3,000 (1)[b] | 5.9 |
| Rat | | |
| Brain | 82,000 (2)[b] | 5.9 |
| | 86,000 (3)[c] | |
| | 80,000 ± 2,000 (1)[b] | |
| | 83,200 ± 2,500 (1)[d] | |
| Heart | 78,000 ± 1,800 (1)[b] | --- |
| Guinea Pig | 78,000 (2)[b] | --- |
| Ileum Smooth | 79,000 ± 4,200 (1)[b] | |
| Muscle | 77,600 ± 2,000 (1)[d] | |
| Brain | 83,200 ± 6,000 (1)[d] | --- |
| Frog | | |
| Brain | 80,000 (1)[d] | --- |
| Drosophila | | |
| Head | 80,000 (1)[e] | 5.9 |

Methods for molecular weight determination

(1) $NaDodSO_4$ polyacrylamide gel electrophoresis.
(2) Radiation inactivation/Target size analysis.
(3) Hydrodynamic.

[a] Isoelectric points were determined in Ref. 6.
[b] Data from Venter (11).
[c] Data from Haga (29).
[d] Data from Birdsall *et al.* (30).
[e] Data from Venter *et al.* (6).

18,000 daltons all of which contain the ligand binding site and protrude from the membrane into the extracellular space. Based on these data a structural map for the muscarinic receptor has been constructed (Fig. 9). Our proposed model for the muscarinic receptor is illustrated in Fig. 10. Target size data indicated that the receptor is a 80,000 dalton monomer in the membrane (11).

## Possible Relationship Between Muscarinic Cholinergic and $\alpha_1$-Adrenergic Receptors

The similarity in molecular weights and tryptic digest patterns between $\alpha_1$-adrenergic and muscarinic cholinergic receptors (Figs. 5, 9) suggested possible homologous structural relationships, an idea consistent with biological data indicating that these pharmacologically distinct proteins can activate common effectors such as the slow inward calcium channel, or guanine nucleotide-binding, negative regulatory proteins for adenylate cyclase.

We have produced monoclonal antibodies against the muscarinic cholinergic receptor and the $\alpha_1$-adrenergic receptor (6). In performing cross-reactivity studies, we discovered that one monoclonal antibody with specificity for the muscarinic receptor (antibody M1.30) was able to selectively immunoprecipitate partially purified $\alpha_1$-adrenergic receptors (6), whereas five other monoclonal antibodies specific for muscarinic receptors did not recognize antigenic determinants on $\alpha_1$-adrenergic receptors (Table 3). In contrast, two out of five monoclonal antibodies raised and screened against rat liver $\alpha_1$-adrenergic receptors (antibodies $\alpha$-IVC$_2$ and $\alpha_1$-IIIE$_3$) were able to immunoprecipitate muscarinic cholinergic receptors isolated from rat brain, dog heart and monkey ciliary muscle with equal efficacy (Table 3) (6). These data strongly suggest that muscarinic cholinergic and $\alpha_1$-adrenergic receptors contain identical antigenic determinants recognized by monoclonal antibodies.

Despite the fact that muscarinic cholinergic and $\alpha_1$-adrenergic receptors are associated with opposing sides of the autonomic nervous system and respond to unique neurotransmitters *in vivo*, these receptors do modulate a number of common effector proteins. Consequently, it would not be surprising to find similar or identical regions within each receptor which are involved in the specific interactions between receptor proteins and their effectors. We proposed that some degree of structural homology may be found among the diverse group of peptide hormone and neurotransmitter receptors that mediate responsiveness via the calcium channel and adenylate cyclase system (6). Monoclonal antibody cross-reactivity between these two diverse pharmacological classes of receptors has provided the first direct evidence for possible common structural determinants between these receptor classes.

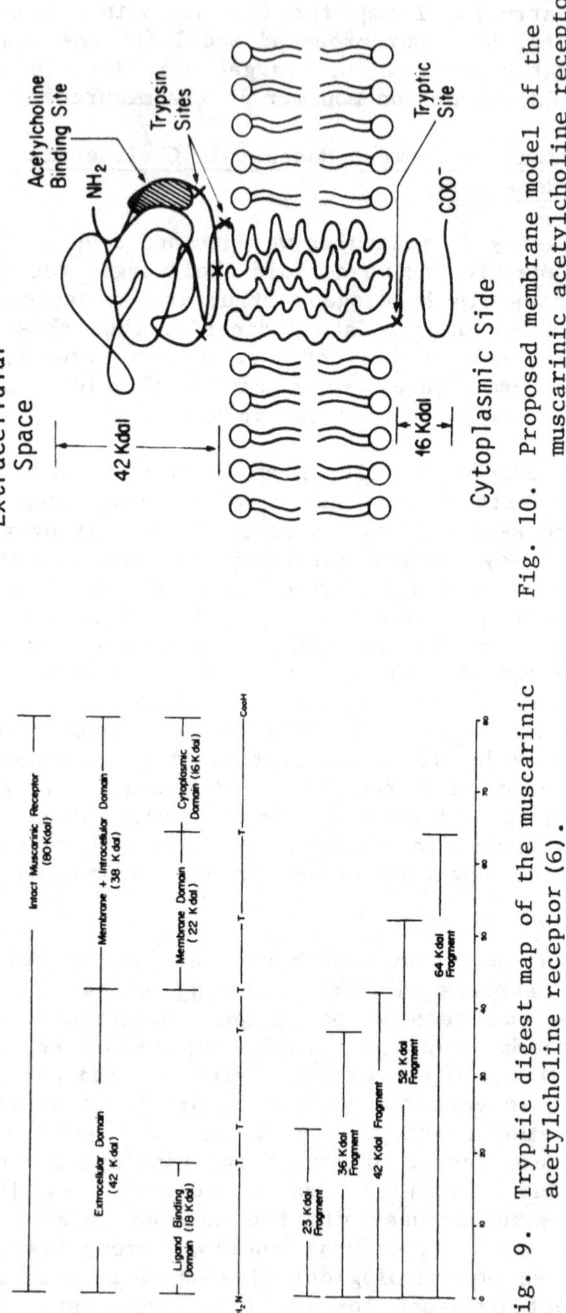


Fig. 9. Tryptic digest map of the muscarinic acetylcholine receptor (6).

Fig. 10. Proposed membrane model of the muscarinic acetylcholine receptor.

TABLE 3

IMMUNOPRECIPITATION OF $\alpha_1$-ADRENERGIC AND CHOLINERGIC RECEPTORS BY ANTI-MUSCARINIC CHOLINERGIC AND $\alpha_1$-ADRENERGIC RECEPTOR MONOCLONAL ANTIBODIES

| | | | Receptor | | | | | |
|---|---|---|---|---|---|---|---|---|
| Monoclonal Antibody | Isotype[c] | Antigen | Rat Brain Muscarinic | Human Brain Muscarinic | Dog Heart Muscarinic | Monkey Ciliary Muscle Muscarinic | Drosophila Head Muscarinic | Rat Liver $\alpha_1$-Adrenergic[b] |
| Musc. 1.1 | $IgG_{2a}$ | Purified Rat Brain Muscarinic Receptor | + | + | + | + | + | – |
| Musc. 1.6 | $IgG_{2b}$ | " | + | + | + | + | + | + |
| Musc. 1.30 | $IgG_{2a}$ | " | + | + | + | + | + | – |
| Musc. 1.42 | $IgG_1$ | " | + | + | + | + | + | – |
| Musc. 1.43 | $IgG_{2a}$ | " | + | + | + | + | + | – |
| Musc. 1.44 | $IgG_1$ | " | + | + | + | + | + | – |
| $\alpha_1$ IV $C_2$ | $IgG_1$ | Rat Liver Membranes | + | N.T. | + | + | N.T. | + |
| $\alpha_1$ IV $E_3$ | $IgG_{2b/2a}$ | " | + | N.T. | + | + | N.T. | + |
| $\alpha_1$ I $F_9$ | $IgG_{2a}$ | " | – | N.T. | – | – | N.T. | + |
| $\alpha_1$ III $F_3$ | $IgG_1$ | " | – | N.T. | – | – | N.T. | + |
| $\alpha_1$ I $D_8$ | $IgG_{2a}$ | " | – | N.T. | – | – | N.T. | + |

a Data indicate the ability of monoclonal antibodies to immunoprecipitate affinity labeled muscarinic receptors relative to the data obtained with rat brain receptors (which were used as a source of antigen). (+) indicates 100% cross reactivity with a fixed amount (1.0 μg IgG) of each antibody with 2.0 fmol of purified receptor.

b Data indicates the ability of monoclonal antibodies (1 μg IgG) to immunoprecipitate $^3$H-POB labeled $\alpha_1$-adrenergic receptors (2.0 fmol).

c Isotype determined using spent culture media with Ouchterlony's test.

NT - Indicates not tested.

Data from reference 6.

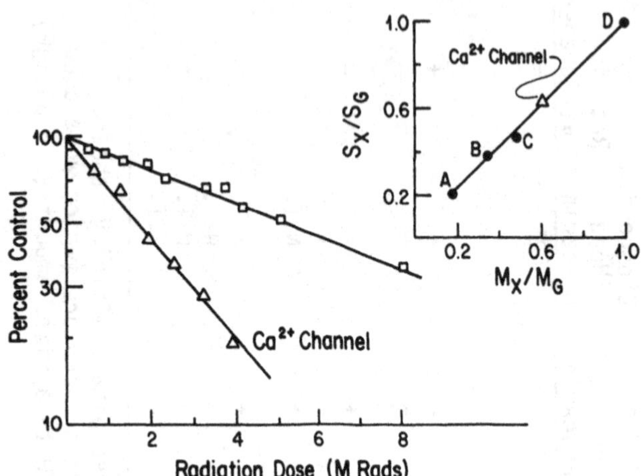


Fig. 11. Radiation inactivation of the slow inward calcium channel (1).

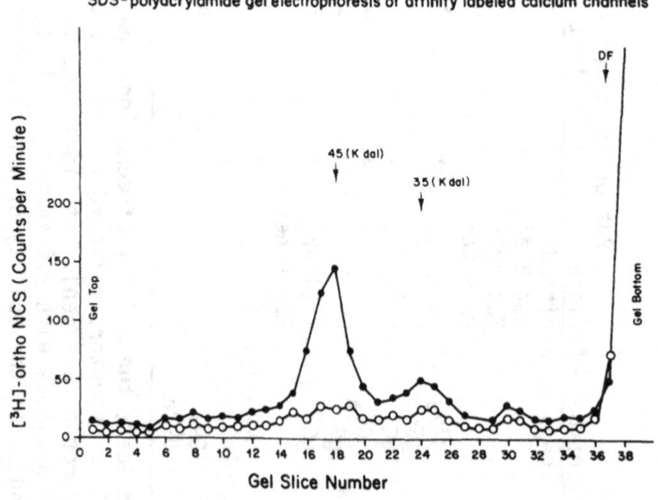


Fig. 12. SDS-PAGE of o-NCS affinity labeled calcium channels (1).

## SLOW INWARD CALCIUM CHANNEL STRUCTURE

The slow inward calcium channel is thought to be one important mediator of receptor action in tissues such as cardiac and smooth muscle. In order to better understand receptor structure and function, we have initiated studies to purify and reconstitute the calcium channel. Our studies to date using target size analysis (Fig. 11) and covalent affinity labeling (Fig. 12) indicate that the calcium channel is an oligomeric protein with a membrane molecular size of 278,000 daltons and a possible subunit of 45,000 daltons to which calcium channel antagonists bind (1). Studies are underway to investigate the interactions of receptor proteins with the calcium channel.

## RECEPTOR EVOLUTION

The muscarinic cholinergic receptor data indicate considerable structural homology between receptors from human brain to *Drosophila* (11). There is much evidence which indicates that muscarinic receptors also exist throughout much of the phylogenetic tree, in molluscs, annelids, and other athropods (6,11) as well as in insects, as our data illustrate. These findings suggest that the muscarinic cholinergic receptor has existed essentially unchanged for several hundred million years. Further studies into the evolution of other neurotransmitter receptors may also reveal a similar heritage for these proteins.

## ACKNOWLEDGEMENTS

This work was supported in part by AI-20863 and HL-31178 to JCV awarded by the National Institutes of Health, DHEW and AHA-82881 to CMF awarded by the American Heart Association.

## REFERENCES

1. J. C. Venter, C. M. Fraser, J. S. Schaber, C. Y. Jung, G. Bolger, and D. Triggle, Molecular properties of the slow inward $Ca^{++}$ channel, *J. Biol. Chem.* 258:9344 (1983).
2. C. M. Fraser and J. C. Venter, Monoclonal antibodies to β-adrenergic receptors: their use in purification and molecular characterization of β-receptors, *Proc. Natl. Acad. Sci. USA* 77:7034 (1980).
3. J. C. Venter and C. M. Fraser, The development of monoclonal antibodies to β-adrenergic receptors and their use in receptor purification and characterization, *in*: "Monoclonal Antibodies in Endocrine Research," R. E. Fellows and G. S. Eisenbarth, eds., Raven Press, New York (1981).

4. J. C. Venter, Monoclonal antibodies and autoantibodies in the isolation and characterization of neurotransmitter receptors: the future of receptor research, J. Molec. Cell Cardiol. 14:687 (1982).
5. C. M. Fraser, R. Greguski, B. Eddy, and J. C. Venter, Autoantibodies and monoclonal antibodies in the purification and molecular characterization of neurotransmitter receptors, J. Cell. Biochem., in press.
6. J. C. Venter, B. Eddy, L. Hall, and C. M. Fraser, Monoclonal antibodies detect the conservation of muscarinic cholinergic receptor structure from human brain to Drosophila and possible homology with $\alpha_1$-adrenergic receptors, Proc. Natl. Acad. Sci. USA, in press.
7. C. M. Fraser and J. C. Venter, The size of the mammalian lung $\beta_2$-adrenergic receptor as determined by target size analysis and immunoaffinity chromatography, Biochem. Biophys. Res. Commun. 109:21 (1982).
8. J. C. Venter, High efficiency coupling between $\beta$-adrenergic receptors and cardiac contractility: direct evidence for "spare" $\beta$-adrenergic receptors, Mol. Pharmacol. 16:429 (1979).
9. G. Kunos, W. H. Kan, R. Greguski, and J. C. Venter, Selective affinity labeling and molecular characterization of hepatic $\alpha_1$-adrenergic receptors with [$^3$H]phenoxybenzamine, J. Biol. Chem. 58:326 (1983).
10. J. C. Venter, P. Horne, B. Eddy, R. Greguski, and C. M. Fraser, $\alpha_1$-Adrenergic receptor structure, Mol. Pharmacol., submitted.
11. J. C. Venter, Muscarinic cholinergic receptor structure, J. Biol. Chem. 258:4842 (1983).
12. L. Lilly, C. M. Fraser, C. Y. Jung, P. Seeman, and J. C. Venter, Molecular size of the canine and human brain $D_2$ dopamine receptor as determined by radiation inactivation, Mol. Pharmacol. 24:10 (1983).
13. J. C. Venter, Neurotransmitter receptor structure and evolution, in: "Receptor Biochemistry and Methodology, Vol. 4," J. C. Venter, C. M. Fraser and J. Lindstrom, eds., Allan R. Liss, Inc., New York, in press.
14. W. L. Strauss, G. Ghai, C. M. Fraser, and J. C. Venter, Hydrodynamic properties and sulfhydryl reagent sensitivity of $\beta_1$ (cardiac) and $\beta_2$ (lung and liver) adrenergic receptors: Molecular evidence for isoreceptors, Fed. Proc. 38:843 (1979).
15. D. R. Jeffrey, R. R. Charlton, and J. C. Venter, Reconstitution of turkey erythrocyte $\beta$-adrenergic receptors into human erythrocyte acceptor membranes, J. Biol. Chem. 255:5015 (1980).
16. J. C. Venter and C. M. Fraser, The structure of alpha and beta-adrenergic receptors, TIPS 4:256 (1983).

17. J. C. Venter and C. M. Fraser, β-Adrenergic receptor isolation and characterization with immobilized drugs and monoclonal antibodies, Fed. Proc. 42:273 (1983).
18. C. M. Fraser, β-Adrenergic receptors: Monoclonal antibodies and receptor structure, in: "Receptor Biochemistry and Methodology, Vol. 4," J. C. Venter, C. M, Fraser, and J. Lindstrom, eds., Alan R. Liss, Inc., New York, in press.
19. W. L. Strauss, Molecular characterization of β-adrenergic receptors, Doctoral dissertation, State University of New York at Buffalo (1980).
20. W. L. Strauss, Sulfhydryl groups and disulfide bonds: Modification of amino acid residues in studies of receptor structure and function, in: "Receptor Biochemistry and Methodology, Vol. 1," J. C. Venter and L. C. Harrison, eds., Alan R. Liss, Inc., New York, in press.
21. J. C. Venter, C. M. Fraser, and L. C. Harrison, Autoantibodies to $\beta_2$-adrenergic receptors: A possible cause of adrenergic hyporesponsiveness in allergic rhinitis and asthma, Science 207:1361 (1980).
22. C. M. Fraser, J. C. Venter, and M. Kaliner, Autonomic abnormalities and autoantibodies to β-adrenergic receptors: their use in receptor purification and characterization, N. Engl. J. Med. 305:1165 (1981).
23. J. C. Venter, C. M. Fraser, A. I. Soiefer, D. R. Jeffrey, W. L. Strauss, R. R. Charlton, and R. Greguski, Autoantibodies and monoclonal antibodies to β-adrenergic receptors: their use in receptor purification and characterization, Adv. Cyclic Nucleotide Res. 14:135 (1981).
24. W. Strauss, G. Ghai, C. Fraser, and J. Venter, Detergent solubilization of mammalian cardiac and hepatic β-adrenergic receptors, Arch. Biochem. Biophys. 196:566 (1979).
25. P.O. Couraud, B. Z. Lu, A. Schmutz, O. Durieu-Trautmann, C. Klutchko-Delavier, J. Hoebeke, and A. D. Strosberg, Immunological studies of β-adrenergic receptors, J. Cell. Biochem., in press.
26. J. P. DeHaye, P. F. Blackmore, J. C. Venter, and J. H. Exton, Studies on the α-adrenergic activation of hepatic glucose output: α-adrenergic activation of phosphorylase by immobilized epinephrine, J. Biol. Chem. 255:3905 (1980).
27. P. F. Blackmore, B. P. Hughes, R. Charest, E. A. Shuman, and J. A. Exton, Time course of $\alpha_1$-adrenergic and vasopressin actions on phosphorylase activation, calcium efflux, pyridine nucleotide reduction, and respiration in hepatocytes, J. Biol. Chem. 258:10488 (1983).
28. J. C. Venter, J. S. Schaber, D. C. U'Prichard, and C. M. Fraser, Molecular size of the human platelet $\alpha_2$-adrenergic receptor as determined by radiation inactivation, Biochem. Biophys. Res. Comm., submitted.
29. T. Haga, Molecular size of muscarinic acetylcholine receptors of rat brain, FEBS Lett. 113:68 (1980).

30. N. J. M. Birdsall, A. S. V. Burgen, and E. Hulme, A study of the muscarinic receptor by gel electrophoresis, Br. J. Pharmacol. 66:337 (1979).

## CONTRIBUTORS

Klaus Aktories, Dr.med., Dr.rer.nat.
Pharmakologisches Institut
der Universitat Heidelberg
D-6900 Heidelberg
Federal Republic of Germany

Takeo Asakawa, M.D., Ph.D.
Department of Pharmacology
Saga Medical School, Nabeshimamachi
Saga 840-01, Japan

Eric A. Barnard, Ph.D., FRS
Department of Biochemistry
Imperial College
London SW7, England

Henry Bourne, M.D.
Department of Pharmacology
University of California
San Francisco, CA 94143

Dermot M.F. Cooper, Ph.D.
Department of Pharmacology
University of Colorado
School of Medicine
Denver, CO 80262

Motohatsu Fujiwara, M.D.
Department of Pharmacology
Faculty of Medicine
Kyoto University
Kyoto 606, Japan

Fusao Hirata, M.D., Ph.D.
Laboratory of Cell Biology
National Institute of Mental Health
Building 10, Room 2D-47
Bethesda, MD 20205

Sadahiko Ishibashi, Ph.D.
Department of Physiological Chemistry
Hiroshima University School of Medicine
1-2-3 Kasumi, Minamiku
Hiroshima 734, Japan

Shozo Kito, M.D., Ph.D.
Third Department of Internal Medicine
Hiroshima University School of Medicine
1-2-3 Kasumi, Minamiku
Hiroshima 734, Japan

Takayoshi Kuno, Ph.D.
Department of Pharmacology
Kobe University School of Medicine
7-5-1 Kusunoki cho, Chuoku
Kobe 650, Japan

Kinya Kuriyama, M.D.
Department of Pharmacology
Kyoto Prefectural University of Medicine
Kawaramachi, Hirokoji, Kamikyoku
Kyoto 602, Japan

Yoichiro Kuroda, Ph.D.
Department of Neurochemistry
Tokyo Metropolitan Institute for Neurosciences
2-6 Musashidai, Fuchu-shi
Tokyo 183, Japan

Norio Ogawa, M.D.
Department of Neurochemistry
Institute for Neurobiology
Okayama University Medical School
2-5-1 Shikatacho
Okayama 700, Japan

Richard W. Olsen, Ph.D.
Department of Pharmacology
University of California, Los Angeles
School of Medicine
Los Angeles, CA 90024

William R. Roeske, M.D.
Departments of Internal Medicine and Pharmacology
University of Arizona
Arizona Health Sciences Center
Tucson, AZ 85724

Elliot M. Ross, Ph.D.
Department of Pharmacology
University of Texas
School of Medicine
Dallas, TX 75235

Tomio Segawa, Ph.D.
Department of Pharmacology
Institute of Pharmaceutical Sciences
Hiroshima University School of Medicine
1-2-3 Kasumi, Minamiku
Hiroshima 734, Japan

Chikako Tanaka, M.D. Ph.D.
Department of Pharmacology
Kobe University School of Medicine
7-5-1 Kusunoki cho, Chuoku
Kobe 650, Japan

Takashi Taniguchi, Ph.D.
Department of Pharmacology
Faculty of Medicine
Kyoto University
Kyoto 606, Japan

Michio Ui, Ph.D.
Department of Physiological Chemistry
Faculty of Pharmaceutical Sciences
Hokkaido University
Nishi 6, Kita 12jo, Kitaku
Sapporo 060, Japan

J. Craig Venter, Ph.D.
Department of Molecular Immunology
Roswell Park Memorial Institute
Buffalo, NY 14263

James K. Wamsley, Ph.D.
Departments of Psychiatry, Anatomy and Pharmacology
University of Utah School of Medicine
Salt Lake City, Utah 84132

Henry I. Yamamura, Ph.D.
Departments of Pharmacology, Biochemistry, Psychiatry
and the Arizona Research Laboratories
University of Arizona
Health Sciences Center
Tucson, AZ 85724

Hiroshi Yoshida, M.D.
Department of Pharmacology I
Osaka University School of Medicine
4-3-57 Nakanoshima, Kitaku
Osaka 530, Japan

INDEX

Printed in the United States
by Baker & Taylor Publisher Services